The New Barbarism?

A PORTRAIT OF EUROPE

EDITORS MARY R. PRICE AND DONALD LINDSAY

1	300–1300	*From Barbarism to Chivalry* by Mary R. Price and Margaret Howell
2	1300–1600	*Authority and Challenge* by Donald Lindsay and Mary R. Price
3	1600–1789	
4	1789–1914	*Machines and Liberty* by Martin Roberts
5	1900–1973	*The New Barbarism?* by Martin Roberts

A Portrait of Europe 1900–1973

The New Barbarism?

Martin Roberts

Oxford University Press · 1975

Oxford University Press, Ely House, London W. 1

GLASGOW NEW YORK TORONTO MELBOURNE WELLINGTON
CAPE TOWN IBADAN NAIROBI DAR ES SALAAM LUSAKA ADDIS ABABA
DELHI BOMBAY CALCUTTA MADRAS KARACHI LAHORE DACCA
KUALA LUMPUR SINGAPORE HONG KONG TOKYO

Text set in 'Monophoto' Plantin by London Filmsetters Limited
and printed in Great Britain at the University Press, Oxford
by Vivian Ridler, Printer to the University

Editors' Preface

Recently there have been marked changes in the methods of teaching history and in our conception of what it is possible for amateur historians to experience and enjoy. The changes are designed to encourage them to make use of sources very early, to penetrate as deeply as they can into historical topics, to enjoy discovering for themselves what life was like in the past, and to develop their own individual interests. All this has had a very stimulating, in many cases a re-vitalizing, effect upon our presentation of the subject, and not least upon our attitude to the kind of books we need. There must now be few places where students are provided with only a single book for a year's work. Instead they are introduced to a multiplicity of publications dealing with separate topics, movements, and personalities, and the study of history is much enriched for them.

In view of this trend it may be asked if there is any place today for a series of background books such as these. Are they not quite out-moded and useless, if not positively harmful? We do not think so, for we are convinced that, if historical knowledge at any level is to be of lasting value and interest, such knowledge must not be piece-meal, but in the end set in a firm framework. Thus, in addition to books dealing with separate topics, young historians need books which will help to create this framework. It is not sufficient to relate topics solely to the history of our own country; historians will want books about the larger units with which their own country is par-ticularly and obviously linked, about Europe in the first place, and ultimately about the world. The need for as much knowledge about the history of Europe as possible in order to lead to better mutual understanding is today more important than ever before.

In these new Portrait books we have tried to avoid the superficiality of a brief chronological recital of events, and instead have chosen to highlight significant movements and people. Above all we have, wherever possible, introduced in the text and in the illustrations the sources of history, believing that this is one of the best ways to kindle the minds and imagination of the readers.

M.R.P.
D.D.L.

Author's Preface

To Diana, Sarah and Tom

Many friends have helped me with this volume. I am specially indebted to Christopher Black of the Department of History at the University of Glasgow whose careful criticism of almost all the original typescript brought about its substantial improvement. Edgar Jenkins of the Centre for Science Education at the University of Leeds did the same invaluable job on Chapter 10. The difficult task of converting my handwriting into typescript, often at short notice, was carried out with efficiency and grace by Ann Fritchley and Jean Sanders. To all of them my thanks.

The title needs some explaining. Working on this book has significantly increased my anxieties for the future. No one can write a book like this, covering as it does two World Wars with the attendant horrors of concentration camps and mass bombing, and which ends at a time when violence and atrocities come near to all of us, without realizing that barbarism is still abroad in the world. For nearly two centuries, few men had and have doubted the colossal potential of urban, industrial, scientific and technological society to benefit its members, yet equally few in the second half of the twentieth century can doubt its colossal potential to harm. Personally, I can see no good reason to suppose that the last twenty-five years of this century will be any more successful than the previous seventy-five in moving forward to a genuinely better civilization rather than backwards.

'Those who cannot remember the past are condemned to repeat it' wrote Santayana. If this book helps in any way to lessen the danger of our repeating recent past mistakes, it will have been worth writing.

Contents

		Acknowledgements	8
		List of Maps	9
1.		Europe at the beginning of the Twentieth Century	11
2.		World War I	23
3.		The Versailles Settlement	50
4.		The European Economy 1900–1939	65
5.		Russia: From Revolution to World War II	78
6.		Liberal Democratic Europe	104
7.		Dictators	120
8.		Hitler's Triumphs	147
9.		The Destruction of Nazism	172
10.		Science and Technology	188
11.		Urbanization	202
12.		Religion and Psychology	217
13.		The Arts in he Modern World	230
14.		Women and Society	250
15.		The End of European Empires	260
16.		The Cold War	284
17.		Economic and Social Developments 1945–1973	308
18.		Western Europe and the EEC	325
		Epilogue Whither Europe?	349
		Bibliography	351
		Index	356

Acknowledgements

Black and white photographs are reproduced by kind permission of the following:

Archives Documentation Français, 46, 161, 167 top, 181, 333; Associated Newspapers Group Ltd., 165 bottom, 317; Associated Press, 171; Association pour la Diffusion des Arts Graphiques et Plastiques, 118; Bank of England, 68; F. Behrendt, 301, 306; Bibliothèque Nationale, 117, 331; Bertolt Brecht Archive, 244; Brecht-Einzig Ltd., 238 bottom; Bundesarchiv Koblenz, 135, 138 top, bottom; Camera Press, 198–9, 220, 280, 306 top, 307, 322; Central Press Photos, 110–111, 336; Daily Express, 330, 335; Fotografia Ferruzzi, 215; Fox Photos, 287; Harlow Development Corporation, 208; Imperial War Museum, 47, 48, 165 top right, 165 top left, 178, 179, 191 centre, 285 top; Keystone Press, 75, 80 top, 89 top, 95 left, 96–7, 160 top, 182, 200–201 bottom, 252, 269, 272, 285 bottom, 289 top, 289 bottom, 298, 300, 315 top, bottom left, bottom right, 316, 327, 329, 338, 339, 342; Sir David Low, by arrangement with the trustees and the London Evening Standard, 137; London Transport Executive, 207; Federico Arborio Mella, 203; Marburg, 238 top; Mercedes-Benz, 200 top; Henry Moore, 236; Museum of Modern Art, New York, 232, 233, 234; National Film Archive, 245, 246 bottom; National Library of Ireland, 18, 108–9; Novosti Press Agency, 39, 83, 89 bottom, 168 top, bottom, 201 top, 256, 262, 305, 312; Oeffentliche Kunstammlung, Basel, 231; Paul Popper Ltd., 112, 129, 195, 276; *Punch*, 31, 155, 163; Radio Times Hulton Picture Library, 13, 33, 44, 51, 60, 64, 74 top, bottom, 80 centre, bottom, 81, 87, 94, 95 right, 100, 109, 113, 122, 198, 222, 227; Science Museum, 191 top; Ronald Searle, 266; Snark International, 160 centre, 202 top, 251; S.N.I.A.S., 191 bottom; Sovexportfilm, 246 top; Staatsbibliothek Berlin, 30; Süddeutscher Verlag, 37 bottom left, bottom right, 40, 42, 61, 134, 136, 140, 143, 149, 151, 162, 176, 184, 187; J. R. Tabberner, 214; Ullstein Bilderdienst, 12, 34, 37 top, 128, 141, 167; U.N.E.S.C.O., 216; U.S. Navy Department, 170 top, bottom; Victoria and Albert Museum, crown copyright, 242; H. Roger Viollet, 119, 200 bottom left, 321; A. Vollard, 230.

List of maps

1.	Europe at the beginning of the 20th century	10
2.	The Western Front 1914–18	28
3.	The Eastern Front 1914–18	35
4.	Europe in 1924	54
5.	The Russian Civil War	91
6.	The Spanish Civil War 1936–39	145
7.	German expansion 1933–39	149
8.	Fall of France 1940	160
9.	Nazi power at its peak 1942	173
10.	The North African campaigns 1942–43	173
11.	Stalingrad 1942–43	174
12.	The Allied invasion of France 1944	178
13.	London after World War II	208
14.	The urbanization of the Eastern Netherlands	210
15.	Expanding Paris	214
16.	Venice and its environs	215
17.	Southern Africa in the 1960s	276
18.	Cold-War Europe	297
19.	The economic groupings of postwar Europe	347

Triple Entente	Associates	Britain & France clash with Germany in Morocco ★
Triple Alliance	Associates	
Area of advanced industry	▶¦◀ ▶¦◀	Austria-Hungary and Russia clash in the Balkans

Chapter 1
Europe at the beginning of the Twentieth Century

Map 1 *Europe at the beginning of the 20th century*

Europe's Place in the World

Europe is a very small continent—'in reality a little promontory of the continent of Asia' is how the French poet, Paul Valéry, once described it. In area it is but one twenty-eighth of the land surface of the globe. However, at the start of the twentieth century, it was much the most powerful of the world's continents. Its power made itself felt in a number of ways. First, many parts of the world, mainly in Africa and in Asia, were directly ruled by Europeans. The British possessed the greatest empire in the history of mankind. The French empire was also large and Germany, Holland, Italy, Portugal, Belgium and Spain all had colonies of various shapes and sizes. Secondly, other areas, while not directly ruled by Europeans, were economically at their mercy. For example, the major ports of the vast country of China, through which most Chinese exports and imports had to pass, were by 1900 under European control (British, French, German and Portuguese) and Persia was divided by Britain and Russia in 190' into three zones, one of British influence, the second of Russian influence and the third neutral. Thirdly, Europe was the cultural centre of the world. Parisian artists might show interest in Negro carvings and British army officers might take up the Indian game of polo, but in the years before World War I the flow of ideas and method, whether in politics, philosophy, economics, religion, dress, sport or mass communications, tended to be one way, from Europe to the rest of the world.

Two countries alone preserved a real independence, the U.S.A. and Japan. The U.S.A. still contained some of its original inhabitants (the Red Indians) but in 1900 they numbered less than 1% of the population of 76 million. Roughly a tenth of the population were negroes who had originally been brought as slaves from Africa to work in the cotton plantations of the southern states. Most Americans however were white, emigrants from Europe. By 1900 they had created the most powerful and most industrially advanced society in the world whose economic, political and cultural institutions were usually similar to, and often copied from, those of Europe. Nonetheless most Americans believed that their new world was much superior to the old Europe which they had left behind. They were determined to keep the European nations out of South and Central America

11

and, in the last years of the nineteenth century, were busy expanding their influence throughout the Americas and into the Pacific.

The Japanese, who were the only non-white race successfully to resist European and American domination before 1914, did so by becoming more European than the Europeans. Determined not to be treated like the Chinese, the Japanese governments from 1867 energetically copied European methods with the aim of transforming Japan into a modern industrial state. Such was their success that in 1904–5 they fought and defeated a major European power, Russia, and within a century expanded their economic power so fast that they were second only to the U.S.A., leaving the nations of Europe well behind. In the early twentieth century, however, the success of the U.S.A. and Japan seemed only to emphasize the domination of the world by Europeans and their methods.

It is not easy to explain this extraordinary dominance. Partly it was a matter of population. The continent of Europe may only be one twenty-eighth of the world's surface but it is fertile and by 1902 was supporting 420 million people, approximately a quarter of the world's population. This population had been increasing rapidly for more than a century. Land therefore grew scarce while life in the fast-growing industrial towns was often harsh. Millions of Europeans sought a new life overseas and their vigorous emigration was an important cause of the European takeover of so much of the world. More important still was the economic strength of the nations of Europe. By 1900 virtually all of them, including even backward

The deck of the Japanese cruiser 'Hi Yei' at the time of the 1904–5 war. Note the sailors' uniform which, like the ship itself, was modelled on the most up-to-date European examples

Russia, had experienced an 'industrial revolution'. Western Europe, especially the area running diagonally south and east from the lowlands of Scotland through England, the Low Countries and Northern France, the Rhineland and Northern Italy, had been industrialized for nearly a century. Meanwhile the rest of the world except the U.S.A. and Japan had stood still. Consequently in 1900 66% of international trade, 54% of the pig-iron and 40% of the manufactured goods of the world were European. The inventions and technological skills essential to modern industry were either a European or an American monopoly. The major advances of nineteenth century science and technology—the steam engine, the electrical dynamo, dyestuff chemistry—were first made in Europe. This technological superiority was as marked in military matters as elsewhere and played an important part in the conquest of Africa and much of Asia. In the Opium War of 1839–1842 British ironclads had contemptuously smashed to pieces the wooden sailing-junks of the Chinese. In the 1880's and 1890's a few hundred white adventurers of the British South African Company armed with machine-guns and rifles had seized the area now known as Rhodesia from thousands of Matabele tribesmen armed only with spears and swords.

When Europeans of the early twentieth century themselves looked for an explanation of their worldwide successes, many found the answer in simple racial superiority. The achievements of nineteenth century Europe were so colossal compared with those of the rest of the world, European empires were created so easily and with so little bloodshed (European blood, that is), and non-Europeans were so

The Master Race : this picture from the Daily Graphic *of October 3rd, 1903, has as its caption "The ABC of signalling. Teaching men of the King's African Rifles the morse alphabet in Somaliland"*

impressed by and so quick to imitate European behaviour that it is not really surprising that Europeans assumed that they were superior. When Captain Binger, a Frenchman, explored West Africa in 1888 he had no doubt that the European 'should come there as master, constituting the upper class of society, and should not have to bow his head before native chiefs, to whom he is infinitely superior'. Moreover since Europeans were generally pale-skinned while the peoples they dominated were for the most part black or brown, it was easy to conclude that European superiority was based on colour. 'Scarcely anyone will have the hardihood to deny', wrote H. S. Chamberlain in his *Foundations of the Nineteenth Century*, widely-read and praised in Germany and England before World War I, 'that the inhabitants of Northern Europe have become the makers of the world's history.' He later quoted with approval the views of an American psychiatrist that 'even for their own good, the blacks must be treated for what they are, an absolutely inferior, lower type of man . . .'.

So it seemed really in the interests of the lesser races of the world that they be ruled by Europeans. They could not know what was good for them. 'We cannot foresee the time', wrote Sir John Strachey, a British government official in India from 1843 to 1880, 'in which the ending of our rule would not be the signal for universal anarchy and ruin . . . The only hope for India is the long continuance of the benevolent but strong rule of Englishmen.'

Yet in 1914 this advanced, arrogant and domineering continent fought, on its own soil for the most part, the most murderous war in human history, the consequences of which were catastrophic. Such a war could occur because, however much Europeans might share a sense of superiority to the non-white world, they had little else in common with each other. 1500 years previously much of the continent had been united under the strong rule of ancient Rome. After the collapse of the Roman Empire in the fifth century A.D. the only times it had looked at all united were briefly in 800 under the rule of Charlemagne and again a thousand years later when Napoleon ruled by force of arms. Otherwise differences of language, race and religion complicated by personal and family ambitions had caused war after war which kept Europe divided and grew no less murderous and bitter as century followed on century.

The Various European States

In 1900, Europe was divided into 22 states (excluding the tiny city-states of Andorra, Monaco and San Marino) between which there was no love lost. These states were of two distinct kinds. On the one hand there were countries like France and Italy whose inhabitants spoke a common language and, sharing a common history and traditions, thought of themselves as members of a single nation. Such states can be described as nation-states. On the other were states like the Habsburg (Austrian) Empire or the Ottoman (Turkish) Empire which for all their long histories were made up of different peoples with different languages and traditions and had in common only the same (often hated) ruler. Such states can be described as multi-national empires. The exact meaning of words is important here. A state is an area ruled by a single government. France is a state, so was the Habsburg Empire. A nation is a people having a common language and history and a sense of common identity. Thus France is a nation, the Habsburg Empire was not. An empire may be either various scattered areas ruled by one government (e.g. the British Empire) or simply an area ruled by an emperor (e.g. the German Empire or the Russian Empire). In fact both Germany and Russia whose inhabitants shared a common language and history are often described as both empires and nation-states. There is one other related term the meaning of which is vital to an understanding of modern European history—nationalism. It means the desire to make one's nation strong and respected.

In 1900, nationalism was an extremely powerful and disrupting force in Europe. It took two forms. In areas ruled by foreign powers—like the South Slav areas of the Habsburg Empire, or the Greek and Bulgar areas of the Balkans ruled by the Turks, or Poland ruled by Germans, Austrians and Russians, or Ireland ruled by Englishmen—nationalists demanded independence, the right of their nation to exist in its own right and run its own affairs without interference by anyone else. In countries which already had their independence—like France, Germany or Italy—nationalists demanded that their nation should be strong, respected and never outsmarted by another nation. Nationalism encouraged revolutions within multi-national empires and wars between nation-states.

Let us look now at the various countries of Europe in more detail. In the south-west of the continent were situated the old-established monarchies of Spain and Portugal. Both had proud histories. In the sixteenth century both had possessed extensive overseas empires and had been numbered among the great powers of Europe. The bulk of these empires had been lost by the early nineteenth century but while Portugal had managed to hang on to and even extend her remaining possessions (mainly in Africa) in the last quarter of the nineteenth century, Spain lost almost all hers (Cuba, the Philippines, Puerto Rico and Guam) to the U.S.A. after a disastrous war in 1898. Both in 1900 were theoretically constitutional monarchies but in practice the monarchs were usually weak and the constitutions often ignored. In Portugal, a republic replaced the monarchy in 1910. In Spain political power lay more in the hands of a few wealthy politicians in Madrid and of local political bosses in the provinces than of either the king or parliament. Corruption, especially the 'fixing' of election results, was widespread. The Catholic Church was a powerful political force, usually resisting change. Most of the peninsula was rural and economically backward. There was some industry, especially in the area round Barcelona, but relations between the workers of this area and the Spanish government were bad. In 1909 there was a major industrial and political crisis. After a general strike vicious street fighting raged in Barcelona for a week which became known as the 'Semana Tragica' or Tragic Week. Portugal played a small part as an ally of Britain in World War I. Otherwise, with the exception of the Spanish Civil War of 1936–39 (see pages 144–6) both countries remained outside the mainstream of European affairs.

Firmly in the mainstream, indeed one of the major states of Europe with Britain, Germany, Austro-Hungary and Russia, was France. Since 1870, when the Emperor Napoleon III had abdicated, France had been a republic. The new constitution of 1871 had established what became known as the Third Republic (that is, the third in France since the great revolution of 1789). Instead of a king or emperor, it had a president and a parliamentary assembly elected at regular intervals. All Frenchmen had the vote. Governments came and went rapidly—the average length of a government between 1880 and 1914 was only eighteen months, chiefly because there were so many political parties that no one party could maintain a majority in parliament. Consequently governments were usually made up of coalitions of various parties who were unable to bury their differences for any length of time. Nonetheless the Third Republic, which was finally destroyed by Hitler in 1940, turned out to be the most long-lasting form of government which France had experienced since 1789. It was stronger than it appeared to be. A powerful civil service kept the country running smoothly whatever coalition was in power and the party politicians disliked each other so much that they preferred to keep a form of government they understood rather than risk a new one which might give an advantage to their opponents. The French economy advanced steadily in the years before 1914. It included heavy industry—France exported more iron ore than any other European country—but a better balance was kept between industry and agriculture than in neighbouring Britain, Belgium or Germany. The population in 1910 numbered only 41.5 million, which was a matter of grave national concern. A hundred years earlier, France had been the most populated country in Europe after Russia. A large population had made possible the large army with which Napoleon had terrorized the continent. Now France was the least populated of the major states and so less powerful (Germany had

The German Empire which had come into existence in 1871 was a federation of mainly German-speaking states though in the east it included about 2,500,000 Poles. Since the military power of the Prussian state had created the Empire, it was dominated by Prussia. The ruling dynasty of Prussia, the Hohenzollerns, provided the German Emperors; Bismarck, the Prussian Prime Minister, became the first imperial Chancellor. The civil service and the railways tended to be Prussian-run. There was an imperial parliament, the Reichstag, elected by universal manhood suffrage, but its powers were much less than those of its British or French equivalents. Government ministers were appointed by the Emperor and could and did remain in office even if a majority of the Reichstag disapproved of their policies. In times of crisis, the Reichstag had little or no control over the Emperor, his ministers or his generals. By 1914 Germany was the most powerful nation in Europe. Her growing population numbered close to 60 millions. Her industry, already the most advanced in Europe, was expanding at a faster rate than any in Europe except the Russian, which was still extremely primitive. Her army, an enlarged and improved version of the triumphant Prussian army of 1871, was the most formidable in Europe and Admiral von Tirpitz was busy trying to create a navy to match it. Yet Germany was both restless and frustrated. In a period when nationalism was generally an active force, it was especially so among Germans. They had no doubt about their strength but they felt they had not achieved either in Europe or in the world the position their strength entitled them to. Their lack of colonies was a particular disappointment. Four uninteresting African territories and a few Pacific islands looked feeble in comparison with the British and French empires. France and Britain seemed part of a conspiracy formed to prevent Germany winning any more colonies outside Europe and, with Russia, to encircle her inside Europe to prevent her gaining even the continental dominance which, because of her strength, skill and efficiency, she deserved. With the Emperor William II, an indecisive theatrical figure at her head and with an army adored by the public and politically most influential, Germany was very dangerous.

The Scandinavian countries—Denmark, Norway and Sweden—were all constitutional monarchies with extensive male suffrage. Norway and Sweden were united until 1905 when Norway won her independence. Economically the most advanced country was Sweden with huge high quality deposits of iron ore in the north. All these countries avoided involvement in European power politics. So too did Switzerland, a republic made up of a number of self-governing cantons. Although the Swiss were divided into three distinct language groups—French, German and Italian—they possessed a strong sense of national unity strengthened by their Alpine geography and by their history. The mountains helped to preserve their neutrality. They remained at peace throughout the twentieth century and prospered.

Italy, another constitutional monarchy, had been united since 1860. From 1912 all men over the age of 30 had the right to vote in parliamentary elections. In the 50 years which followed unification most Italians showed little enthusiasm for the new state. Italian society was deeply divided. While the north was industrializing and comparatively prosperous, the south was rural, backward, over-populated and poverty-stricken. Most Italians were Catholic but, during unification, the Pope and the Italian government had fallen out, and the Catholic Church was reluctant to co-operate with successive governments until 1929. The governments themselves tended to be short-lived coalitions and politicians had a reputation for corruption. Despite these divisions, however, the Italians had a proud history going back to the ancient Roman Empire and, with a population of 36 millions in 1910, they considered themselves a major European power and tried to behave as one although the essential economic and military strength was lacking. The fashion was that major powers had overseas empires, so between 1880 and 1914 Italy pursued an aggressive colonial policy, the aim of which was to create a substantial Italian empire in Africa. In 1896 it met with disaster. An Italian army invading Abyssinia was ambushed at Adowa by Menelek, the Abyssinian leader, and driven out. It was the only time before 1914 that a European army had been defeated by an African one and was a dreadful humiliation for Italy. Some consolation was found in 1912 with the seizure of Tripolitania (modern Libya) from the decaying Turkish Empire. As a result of colonial rivalry with France, Italy had joined an alliance with Germany and Austria-Hungary (the Triple Alliance) in 1882. In 1914, this alliance still stood but was no longer obviously in Italy's interest. An influential nationalist movement in Italy demanded that all the Italian-speaking areas near Italy's borders should be handed over to Italy. The chief areas affected were the Tyrol and Dalmatia, parts of the Austrian Empire which was most reluctant to hand them over to anyone. The alliance of 1882 concealed, therefore, a real enmity. When war broke out in 1914, Italy ignored the Triple Alliance and stayed neutral. The year after, she joined in on the side of France and Britain, against Austria-Hungary and Germany.

So far the states we have been describing looked on the map very much as they do today. Indeed the frontiers of Western and Central Europe have altered only fractionally since 1900. Eastern Europe however is quite different, chiefly because the Austrian Empire has broken up and has been replaced by separate nation-states.

The Austrian Empire was often known as the Habsburg Empire since it was ruled by members of the Habsburg family. From 1867 to 1918 it was also known as the Dual Monarchy since a new constitution of 1867 divided it into two virtually independent halves—Austria governed from Vienna and Hungary governed from Budapest. The link between them was their common ruler, the Habsburg Emperor. The most important characteristic of the Dual Monarchy was the multitude of peoples within it. In the Austrian half, Germans were dominant but there were also many Slav groups—Czechs, Slovaks and Poles for example. In the Hungarian half, the Magyars were dominant but again there were large numbers of Slavs like the Serbs and Croats. Throughout the nineteenth century Magyar and Polish nationalism were active forces and Slav nationalism grew more vigorous. The more powerful nationalism became the less chance the Dual Monarchy had of surviving. Austria looked like a constitutional monarchy in that it possessed a parliament elected by universal manhood suffrage which on paper seemed to possess the power to limit the Emperor's authority. In practice, however, the work of parliament was so upset by the racial rivalries of the Czechs and Germans and the Emperor could so easily encourage these rivalries and rule by decree that his power remained considerable. In Hungary the right to vote depended on a property qualification which made sure that the Magyars kept control. Hungary was run in the Magyar interest at the expense of the other peoples. Economically, the Habsburg Empire was immensely varied, ranging from the advanced industries of Bohemia to the almost mediaeval agriculture of Transylvania. With a population of 51.4 millions in 1910, a large army and distinguished history, Austria-Hungary still ranked as a major power. The Habsburg Emperor, Franz Josef, who had been emperor since 1848, was however very conscious of the weaknesses of his country. His chief interest was to keep his dynasty and Empire in existence. Fearing Russian interference in the Balkans, Austria-Hungary had made an alliance with Germany (the Dual Alliance) in 1879. This alliance still stood in 1914.

For much of the nineteenth century the south-east of Europe, usually called the Balkans after the mountain ranges that run across it, had been part of the Turkish (Ottoman) Empire. As the years passed, the various Balkan peoples won their independence, and by 1914 Turkey in Europe had been reduced to a small triangle of land to the north and west of the great city of Constantinople. While the Turks were a danger, the Balkan peoples were usually able to co-operate against the common Turkish enemy, but once the danger was passed they found many reasons for quarrelling amongst themselves, the chief of which was where the frontiers between them should be. In such quarrels, Austria and Russia took opposing sides, so increasing the danger of a minor crisis turning into a major one. The Balkans was the most backward part of Europe. Industrialization had barely begun, agricultural methods were primitive and education almost non-existent.

The most easterly and by far the largest European state was the Russian Empire. In 1900 it included Finland and half of what is now modern Poland as well as extending as now right across Asia to the Pacific. Its population was nearly 140 millions. Its government was the most autocratic in Europe; in other words its Tsar (Emperor) knew few limits to his powers. From 1894 to 1917 the Tsar was Nicholas II, the last monarch of the Romanov dynasty which had ruled Russia for centuries. He was determined to keep up the autocratic methods of his predecessors. Though a serious revolution in 1905 forced him to agree to the setting up of a Duma or elected assembly he allowed it to have little influence on his rule. Russia however was changing rapidly. Between 1880 and 1910 her population increased by more than 50%. Moreover though she remained a predominantly agricultural and poverty-stricken country, her 'industrial revolution' was well under way by the time Nicholas came to the throne. In the 1890's the Russian economy expanded faster than any other in Europe and though it slowed slightly, it was still growing fast in the years before World War I. This industrial expansion did not however bring any general prosperity; rather it led to the rapid growth of industrial cities and created a new class of urban poor. The government of the Tsar had shown itself so firmly against major social reform for so long that most opponents of the government despaired of bringing change by peaceful means. By 1900 terrorism had become part and parcel of Russian life. In the years 1906 and 1907 about 4,000 state officials were killed or wounded in terrorist attacks. But to the problems that caused such terrorism and the revolution of 1905, the Tsar had no answer except repression. He did not really understand them. Instead, he allowed his government to pursue an active foreign policy which led first to a disastrous war with Japan in 1904–5 and then to an involvement in the Balkans which proved even more disastrous since it led Russia in 1914 into a European war which she was really in no position to fight, so serious were her internal problems.

Chapter 2
World War I

'Revelation', the last book of the Bible, tells in its sixteenth chapter of a great conflict against the forces of evil which raged at 'a place called in the Hebrew tongue Armageddon'. It was the Day of Judgement, the end of the world of ordinary mortals. 'And there were voices, and thunders, and lightnings; and there was a great earthquake, such as was not since men were upon the earth, so mighty an earthquake and so great. And the great city was divided into three parts, and the cities of the nations fell.' World War I was much the most disastrous war Europe had ever experienced and not only did it destroy millions of soldiers and civilians but also the confidence in progress, prosperity and the reasonableness of civilized men which was so characteristic of the nineteenth century. In a sense it was the end of a world and was soon referred to as Armageddon.

What made World War I particularly horrifying was the degree to which it was caused and continued by the stupidity and ignorance of those who were thought of as 'experts', particularly the military planners. Equally horrifying was the inability of politicians to fully impose their wills on their military experts both before and after the outbreak of war.

Causes
There were many reasons why a war should break out in 1914. One of the most important was that too few Europeans had any real hatred of war itself. The wars which the generation of 1914 remembered were the short decisive wars of 1859 to 1871 which had been fought by professional soldiers with no great loss of life or property and which had made the nations of Italy and Germany. 'War', read a German cigarette advertisement not long before 1914, 'is an element in God's natural order of things.' There was bound to be a war before long but it was not something to worry about since it too would be short, sharp and glorious for the victor. Such an attitude to war encouraged the fierce nationalism which had taken deep root in Europe by 1914. When disputes arose between nations, politicians backed by public opinion were more and more ready to threaten war and to regard peaceful compromises as dishonourable.

An elaborate system of alliances which had been built up since 1871 helped to make the war, when it came, a big one. In 1879

Germany had signed an alliance (the Dual Alliance) with Austria-Hungary which became the Triple Alliance of 1882 with the addition of Italy. France, determined to break out of the diplomatic isolation which Bismarck had imposed upon her and gain some security against the ever-increasing danger from Germany, had in 1893 made an alliance with Russia, who was also fearful of Germany and suspicious of Austria-Hungary's intentions in the Balkans. For most of the nineteenth century Britain steered clear of continental alliances but in 1904 she had come to an understanding (entente) with France and in 1907 with Russia. These ententes began as agreements concerned mainly with colonial matters but as Britain became more and more worried about the expansion of the German economy and navy, they became more anti-German. So, in 1914, the Triple Alliance (Germany, Austria-Hungary and Italy) was balanced by the Triple Entente (France, Russia and Britain).

In the decade preceding 1914, relations between these two hostile camps steadily worsened. Two areas gave particular trouble. The first was Morocco in North Africa. Morocco was an independent state ruled by a Sultan, but Algeria, immediately to the east, was a French colony and the French were building up their influence in Morocco too. Simultaneously, German businessmen were trying to establish themselves there and the German government, keen to increase German overseas influence, was ready to back them. Two serious crises, one in 1904–5 and the second in 1911, nearly led to war between France and Germany. In both cases, however, France was firmly backed by Britain, and Germany, having made warlike noises, eventually decided to retreat. The result of the Moroccan crises was to strengthen French influence in Morocco and to convince Germany that there was a Franco-British conspiracy to prevent her winning colonies, which could only be overcome by war.

The second troublesome area was the Balkans. Here the explosive question was the position of the South Slavs. The creation of a South Slav nation round Serbia would mean the end of the Austrian Empire, so Austria regarded Serbia and Russia (Serbia's champion) with great suspicion. After her defeat by Japan in 1905, Russia became more involved in the Balkans. There was a major crisis in 1908 when Austria-Hungary took over Bosnia and Herzegovina (a province bordering on Serbia and mainly occupied by South Slavs) in circumstances which convinced the Russians that they had been double-crossed. Two Balkan Wars which only involved the small Balkan states then followed, one in 1912, the other in 1913. One of their results was that Serbia emerged stronger and more confident than ever. 'The first round is won', said Pasic, the Serbian Prime Minister. 'We must now prepare for the second, against Austria.' While the Austrian government felt that it must soon go to war to crush the threat from Serbia, the Russians were more convinced than ever that they must strongly support Serbia should Austria attempt to bully her.

War was brought closer by an arms race which grew more intense as the twentieth century progressed. The most spectacular part of this race was the naval rivalry between Germany and Britain. Admiral Tirpitz, the creator of the German fleet, believed that 'without seapower Germany's position in the world is like a mollusc without a shell', and built his fleet to challenge the British fleet (the strongest in the world) in European waters. His aim seems to have been to frighten Britain into co-operating with Germany, but Britain reacted very differently. Since her army was tiny, her navy was absolutely essential to her security. In 1906 she began building a fleet of 'Dreadnought' battleships which were so much stronger and faster than any previous ship that their existence, the British hoped, would persuade Germany to give up her challenge. The Dreadnoughts had the opposite effect. The Germans saw it as a test of their technological skill and determination and before long were producing dreadnoughts of their own. The naval race had merely accelerated. Where armies were concerned, the Germans had the best in Europe but, from 1913, the French had begun major military reforms which would be fully effective by 1916. Consequently the Germans felt ahead in the race but in danger of being caught from behind. 'War is inevitable', said General von Moltke to Kaiser William, 'and the sooner the better'. The Austrian generals felt the same way. The Serbian menace grew with the passing of time. It must be nipped in the bud as soon as possible.

The final cause of war was the political influence of generals, especially in Germany, and the failure of governments to control them and their military machines in the summer of 1914. What happened was this. On 28 June 1914 the Archduke Ferdinand, heir to the Austrian throne, paid a visit with his wife to the Bosnian capital of Sarajevo. There he was assassinated by a Bosnian student, Princip, with the aid, the Austrian government insisted but never proved, of the Serbian government. The Serbian newspapers greeted the news with delight, and though the Serbian government was prepared to accept almost all Austria's demands, the Austrian government decided to take this opportunity to fight the war she needed to weaken Serbia. Sure of Germany's backing she declared war on Serbia on 28 July. In such a situation, Russia could not leave Serbia in the lurch. Now the military men began to take over from the politicians. They believed, having carefully studied the victorious Prussian campaigns of 1864 to 1871, that a modern war was won by that nation which mobilized its armies most rapidly and by skilful use of railway timetables concentrated them in overwhelming force at the critical strategic points. The dilemma facing Tsar Nicholas of Russia on July 29 and 30 was whether to order a partial mobilization against Austria alone or a general mobilization to secure his western frontier against a possible German attack. He finally decided on a general mobilization and on July 31 the Russian army began to move. At once Germany prepared for war and now the German military planners forced the hand of their political leaders. They had only one plan for this situation—to fight Russia defensively on the border of Germany while their main armies smashed France in a rapid campaign of a few weeks. Only when France was defeated would the offensive be taken against Russia. Moreover, the plan to smash France, the Schlieffen Plan, assumed that much of the army would take the easiest route to Paris, even though this meant marching through Belgium whose neutrality had been guaranteed by all the major powers of Europe in 1839.

The terrible consequences of so rigid a plan became clear on 1 August. German mobilization had begun the day before. Then the Kaiser learnt that Britain would remain neutral so long as Germany refrained from attacking France. The Kaiser was delighted. He had no desire to fight Britain and little to fight France. 'This calls for champagne,' he said, 'we must halt the march to the west.' To do so at this stage, however, meant halting 11,000 troop trains and throwing away all the strategic advantages which the German military planners with their carefully timetabled railway trains were after. 'It is impossible,' said General Moltke, 'the whole army will be thrown into confusion'. As in Russia, so in Germany, the military men triumphed.

From then on there was no more hesitation in Germany. That same day, August 1, Germany declared war on Russia. Two days later, German planes bombed the German city of Nuremberg and claiming the planes were French, declared war on France. On August 4, true to their railway timetables, German armies invaded Belgium. This action finally brought Britain into the war. Previously, to the horror of the French, who had assumed with some justification that Britain would come to her aid as soon as she was attacked, the British government had hesitated. It was very aware that British public opinion was deeply divided about the necessity of war. However, the cynical invasion of Belgium with its complete disregard of international treaties united public opinion in favour of war and, at midnight on August 4, Britain too was at war with Germany. Sir Edward Grey, the British Foreign Secretary, found words at that time which proved only too appropriate. 'The lamps', he said, 'are going out all over Europe; we shall not see them lit again in our lifetime.'

Much printer's ink has been spilt in discussing who was most to blame for the outbreak of this huge war. Plainly Austria must take much of the blame since she began it by declaring war on Serbia. Germany too had much to answer for. Her noisy nationalism, her schemes for spreading her power in Central Europe and her unpredictable leadership combined with her great strength drove the Entente powers together for security, and in the crisis of 1914 the Kaiser and his ministers were fatally indecisive. Nonetheless, it would be too simple and unfair to pile all the blame on these nations alone. Every major power had managed in the previous twenty years to contribute something to the explosive atmosphere of suspicions and hatred of 1914. 'There was a strange fever in the air', wrote Winston Churchill, First Lord of the Admiralty in 1914. 'Unsatisfied by material prosperity the nations turned restlessly to strife.' All over Europe, crowds cheered the troops as they marched to the railway stations. After all, it was a war that would be over by Christmas, and no-one doubted that his side would win.

The war was not over by Christmas. In fact it lasted four years and three months. It was an immense and complicated affair fought on many different fronts and changing its nature as time passed. Both sides chopped and changed during its course. In August 1914 Japan joined the Allies, in November Turkey the Central Powers. In 1915 Italy joined the Allies and Bulgaria the Central Powers. Roumania joined the Allies in 1916, as did Greece in 1917. The U.S.A. declared war on Germany in April 1917 while Russia finally made peace in March 1918. While the main theatres of war were in Europe, there were others—hence the title World War I. When war was declared, German warships were scattered all over the world. Some of these did some spectacular raiding before they were hunted down by the British navy. There was also a little fighting in Africa. Most of the German colonies were in Allied hands by 1915, though in German East Africa Colonel von Lettow-Vorbeck fought a guerrilla campaign of sustained brilliance which cost the British £72 million and more than 62,000 men and he was still undefeated in 1918. The severest fighting outside Europe was between the Allies and Turkey in the Near East. After some setbacks, of which the most serious was the British defeat at Kut-al-Amara in 1916, the Allies gained the upper hand and forced the surrender of Turkey a month before the final defeat of Germany. It was in this desert war against the Turks that T. E. Lawrence carried out the legendary exploits that made Lawrence of Arabia a household name.

1914

The crucial question in the first weeks of the war was whether the Germans could make the Schlieffen Plan work. If they could, so as to

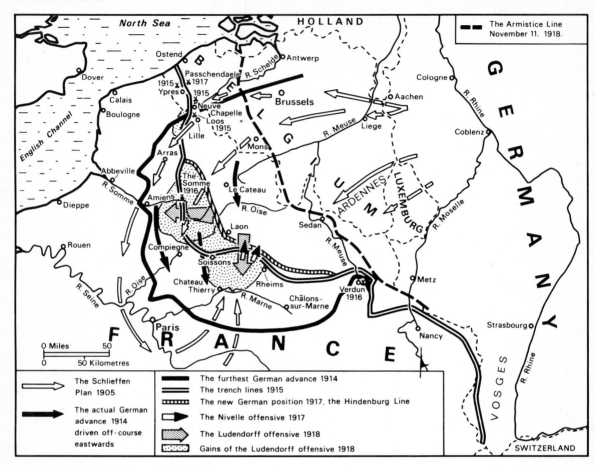

The Schlieffen Plan 1905 — The actual German advance 1914 driven off-course eastwards — The furthest German advance 1914 — The trench lines 1915 — The new German position 1917, the Hindenburg Line — The Nivelle offensive 1917 — The Ludendorff offensive 1918 — Gains of the Ludendorff offensive 1918

Map 2 *The Western Front 1914–18*

capture Paris and force a French surrender in a few weeks, then they would almost certainly win the war. If they did not, and had to fight a war on two fronts for any length of time, then they were likely to lose it. Schlieffen's Plan was bold and basically simple. The key was the right wing of the German armies which was to sweep in overwhelming force through Belgium down through northern France to the west then the south of Paris.

The German armies, commanded by the younger Moltke, flooded into Belgium at the beginning of August and advanced speedily. The Allied armies were immediately in confusion. The French, having begun a major offensive against Alsace and Lorraine, realized after five days that the massive German attack was coming through Belgium and had to reorganize. The tiny British army (the British Expeditionary Force) led by Sir John French first went into action at Mons in Belgium, but hugely outnumbered by the advancing Germans had to retreat inside the French border first to Le Cateau and then to Compiègne, less than 50 miles from Paris. By the end of

August German troops were advancing deep into French territory and complete victory seemed close at hand. But it was not. In a number of important ways the Plan had gone wrong and the further the Germans advanced, the weaker became their position. Moltke had modified Schlieffen's original plan before the war began by strengthening the left at the expense of the right, and unexpectedly fierce resistance from the Belgian army and from the B.E.F. had driven the potentially deadly movement of the right off course. Von Kluck, commander of the army on the far right, was coming down from the north to the east rather than to the west of Paris. Moreover if the Schlieffen Plan was to be carried through to success, it had to be executed by a commander of ice-cool nerves and deep self-confidence. The younger Moltke was not such a man. He was 66 and in poor health. He was nervous and had no confidence in his own leadership in such a crisis. He tinkered further with the Plan and, on August 25, committed what may well have been the decisive tactical error of the whole war. Disturbed by reports of a rapid Russian advance into East Prussia, he detached two army corps from his vital right wing and sent them to reinforce the Eastern Front. On September 2 Galliéni, commander of the Paris garrison, realized from the reports of his reconnoitring airmen that Kluck's army was passing to the east of the city and was vulnerable to a flank attack. The French commander-in-chief, Joffre, resolved on a major counter-attack. After a stormy interview on September 5 with Sir John French, Joffre won his co-operation, so Joffre recalled, by striking the table with his fist and crying, 'Monsieur le Maréchal, the honour of England is at stake!' The counter-attack began at once across the Marne river. A wide gap opened between the German armies on the right. Into this gap the B.E.F. marched and with their supply lines threatened, the Germans were forced to retreat. The battle of the Marne ended, Paris was safe, the Schlieffen Plan had failed.

The rival armies then raced towards the English Channel each trying to outflank the other and digging trenches as they went to hold the territory which they already occupied. Neither got past the other and, by mid-November, a line of trenches was established from the Vosges mountains to the Belgian coast. The Germans might not have won Paris and might have to fight on two fronts, but at the end of 1914 they had conquered almost all Belgium and about a tenth of France, including some of her richest coal and iron producing areas.

Meanwhile, in the east the speed of the Russian mobilization took everyone by surprise and we have already seen how the invasion of East Prussia weakened the German attack on France at a vital moment. In all other respects the Russian invasion of Germany

was a major disaster. The two Russian armies which lumbered westwards were ill-prepared and ill-led. They sent each other un-coded radio messages so the Germans knew their plans. Their commanders, Rennenkampf and Samsonov, proved incapable of co-ordinating their movements and squandered their huge advantage in numbers. The German commander who first faced them was soon sacked for dithering and was replaced by the most formidable military partnership of World War I, Hindenburg and Ludendorff. They struck at once against the divided Russian armies. The more southerly force of Samsonov was encircled in the woods and marshes near Tannenberg and smashed in three days fighting from 26 to 29 August. 100,000 prisoners were taken and Samsonov committed suicide. A week later it was the turn of Rennenkampf's armies near the Masurian Lakes. Another 100,000 Russians were lost and Rennenkampf retreated across the Russian border.

Further south, against the Austrians, the Russians did better. Their 5th army invaded Galicia, captured Lemberg and cut off the great fortress of Przemysl. The Austrians lost 300,000 men among whom were some of their finest regiments.

In the north, Hindenburg and Ludendorff went onto the attack. An October offensive took them to the walls of Warsaw where only bayonet clashes beneath the city walls kept the German army at bay. Another November offensive won the city of Lodz for the Germans and demonstrated that in everything except courage the German army was immensely superior to the Russians.

For the Austrians, however, 1914 continued to be a humiliating year. Their Galician defeats by Russia were followed by a failure to

Stylish, pampered, and incompetent, officers of the Tsarist army picnic on the Eastern Front in 1914

crush the tiny Serbian army. A leisurely and over-confident invasion met with stiff resistance. Not until December did Belgrade, the Serbian capital, fall into Austrian hands. It remained there for one day only. The next day with their old King Peter marching in the infantry with a rifle and forty rounds of ammunition, the Serbs rallied. The Austrians were taken by surprise, Belgrade was recovered and, on December 15, the Serbian government could proudly announce 'on the whole territory of Serbia, there remains not one free enemy soldier'.

1915

The most obvious and remembered feature of the Western Front was the trenches. These miles of ditches, dug so deep that a soldier could walk along them without being exposed to enemy fire, were strengthened by two newish weapons, the machine-gun and barbed wire. This combination placed the advantage in a battle firmly on the side of the defenders who, hidden in their trenches, could quickly

Serbia's successes at the expense of 'big bully' Austria delight Britain's Punch *magazine*

PUNCH, OR THE LONDON CHARIVARI.—December 23, 1914.

FULFILMENT.

Austria. "I SAID ALL ALONG THIS WAS GOING TO BE A PUNITIVE EXPEDITION."

and easily pick off the attackers with machine-gun and rifle fire as they struggled through the barbed wire. If the wind was blowing in the right direction, waves of poisonous gas (chlorine) would be launched towards the enemy. Neither side was prepared for warfare of this type. The British had ordered their officers to make sure that their swords were sharpened and issued only two machine-guns per battalion. (At the end of the war each had 50.) The French first went into action wearing their traditional red trousers. Nor had the generals any strategy to deal with the trench defences. They were wedded to the doctrine of the offensive which had worked so well for the Prussian armies forty years before. 'Whatever the circumstances', wrote the French general Foch, 'it is the intention to advance with all forces to the attack.' So massed infantry attacks against a narrow portion of the trenches followed one after another and the slaughter of soldiers on a scale hitherto unknown in warfare began and continued, month after month, year after year.

The Allies made a series of offensives in 1915—at Neuve Chapelle, Aubers, Ypres and Loos which made clear the essence of the new warfare, hundreds of thousands of casualties in order to gain a few yards of mud. A German officer described a British advance in his regiment's diary. 'Ten columns of extended line could clearly be distinguished, each one estimated at more than a thousand men and offering such a target as had never been seen before or even thought possible. Never had the machine-gunners such straightforward work to do nor done it so effectively. They traversed to and fro unceasingly. As the entire field of fire was covered by enemy infantry, the effect was devastating.' The poet and writer, Robert Graves, took part in some of this fighting as a twenty-year-old officer in the Royal Welch Fusiliers. In his autobiography *Goodbye to All That*, he described part of one such offensive. 'A few minutes later, Captain Samson with 'C' company and the remainder of 'B' reached our front line. Finding the gas-cylinders still whistling and the trench full of dying men, he decided to go over too—he could not have it said that the Royal Welch had led down the Middlesex ... One of the 'C' officers told me later what happened. It had been agreed to advance by platoon rushes with supporting fire. When his platoon had gone about twenty yards, he signalled them to lie down and open covering fire. The din was tremendous. He saw the platoon on his left flopping down too so he whistled the advance again. Nobody seemed to hear. He jumped up from his shell-hole, waved and signalled 'Forward'. Nobody stirred. He shouted: "You bloody cowards, are you leaving me to go on alone?" His platoon-sergeant, groaning with a broken shoulder, gasped: "Not cowards, sir. Willing enough. But they are all f---ing dead." The Pope's Nose machine-gun, traversing, had caught them as they rose to the whistle."' On the Eastern Front, things were even worse for the Allies. Throughout the summer the Germans

On the narrow beaches and overlooked by the steep slopes of what became known as Anzac Cove, Australian and New Zealand troops struggle to get their artillery ashore; Gallipoli, 1915

hammered at the Russians in Poland, inflicting fearful casualties. By August, 750,000 prisoners had been taken.

In 1915 the Allies tried, with mainly British forces, to open up a new theatre of war. The place they chose was Gallipoli, a peninsular of Turkey in Europe beside the Dardanelles Straits. The strategy was an intelligent one and owed much to Winston Churchill. He and some other members of the British government (the 'easterners') felt that there was deadlock on the Western Front, that Germany could most swiftly be defeated by 'knocking away her props' (Austria-Hungary and Turkey) through an attack from the south-east. If Constantinople could be won, Turkey would be knocked out of the war. Austria would be open to an attack through the Balkans and an all-the-year round supply route to a Russia desperate for arms and ammunition could be opened up through the Mediterranean and Black Seas. Churchill managed to persuade a hesitant British government to back the plan and in the spring of 1915 a mainly British, Australian and New Zealand force appeared in the Eastern Mediterranean. It got off to a bad start and never recovered. It had been badly organized with poor security, and was poorly led. The navy was unable to clear the Dardanelles minefields and when the army landed on the rocky shores of the Gallipoli peninsula, it found the

Turks, ably led by Mustapha Kemal and the German Liman von Sanders, entrenched and well-armed on the steep slopes above them. Despite heroism and heavy casualties (particularly among the Australian and New Zealand troops—the ANZACS), the Allies hardly got off the beaches. There they hung on grimly despite the heat of summer and the ravages of disease until the final evacuation in the winter of 1915–16. It was a major defeat. Casualties numbered 200,000. The Allied position in the Balkans weakened and the warm water supply-route to Russia remained shut for the rest of the war. Churchill had to resign and the British government, abandoning the easterners' strategy, decided once more to try to break through on the Western Front.

The final Allied disaster of 1915 was the collapse of Serbia. In October a combined German, Austrian and Bulgarian army attacked the Serbs, who, despite their appeals, received no effective aid from Britain and France. The contest was quite unequal, but rather than surrender, the Serbs, led by General Putnik and the 71-year old King Peter, marched out of Serbia across the mountains of Albania towards the Adriatic coast in the hope that Allied ships would evacuate them to safety. It was more a nation than an army on the march and

Deep into snow-bound Serbia advances an Austrian unit, closely watched by its commanding officer in the foreground, late 1915

Map 3 The Eastern Front
1914–18

The Eastern Front
early 1916

Land temporarily gained
by Brusilov 1916

DENMARK

Baltic Sea

R. Dvina

Moscow

R. Elbe

R. Oder

Königsberg

EAST PRUSSIA

Tannenberg
1914

Masurian
Lakes 1914

Berlin

R. Vistula

Brest-
Litovsk

GERMANY

Lodz

Warsaw

Kovel

R. Dnieper

Lutsk

AUSTRO-HUNGARY

Przemysl

R
U
S
S
I
A

Vienna

R. Danube

CARPATHIAN MOUNTAINS

Caporetto
1917

Vittoria Veneto
1918

R. Piave

ITALY

ROUMANIA

Bucharest

Black
Sea

Belgrade

R. Danube

Adriatic Sea

SERBIA

ALBANIA

BULGARIA

Constantinople

Salonica

1915

Gallipoli

GREECE

Corfu

Athens

Aegean
Sea

TURKEY

The Strategic Plan
of Gallipoli 1915

The Brusilov
offensive 1916

The Salonica
offensive 1918

0 Miles   200

0   200 Kilometres

there was appalling suffering in the winter conditions. In one contingent alone 20,000 men, women and children died of cold, disease or starvation. Harassed along the coast by the Austrians, the survivors were eventually evacuated to Corfu where, their spirit unbroken, they prepared for the day when they would fight again to free their country.

1916

Astride the river Meuse 150 miles to the east of Paris stands the historic town of Verdun. For centuries a fortress guarding the north-east border of France, it now held an important position in the trench-line. It was the centre of a large salient (or bulge in the front) and was therefore open to German attack from three directions, rather than one. Early in 1916 Falkenhayn, who had succeeded Moltke as commander of the German forces in the west, decided to attack the Verdun salient. It was the way, he believed, 'of bleeding France white'. So great was the historic reputation of Verdun and so important to French morale that a German attack would lure the best French regiments to the area. He would then batter them to pieces with his artillery firing from all three directions. German casualties would be comparatively small.

Lulled into a sense of false security by the strength of their fortifications and by months of inactivity, the French garrison in Verdun was small in number and ill-prepared. On 21 February 1916 the outer defences were hit by the heaviest bombardment in history. Similar bombardments fell in the two following days. The outer French defences crumbled away on February 24 and the day after, to his amazement, a twenty-four year old German lieutenant with a handful of infantry entered Douaumont, the strongest fort of the northern defences, barely five miles from the centre of Verdun and found it undefended. Somewhat late in the day, Joffre realized how serious the situation was. Pétain, his finest defensive general, was placed in command of the town and ordered to hold it, 'whatever the cost'. Pétain quelled the panic which was spreading after the fall of Douaumont and with 190,000 reinforcements rushed in along the Bar-le-Duc road—the one road left open to the south which came to be called the 'La Voie Sacrée (the Sacred Way)—he hung on. Falkenhayn forgot his original aim of preserving his own troops and threw them in ever greater numbers against the stubborn French defences. Through May and June Verdun was a hell on earth. 'One eats, one drinks beside the dead, one laughs and sings in the company of corpses', wrote a French army doctor. June was the critical month for Verdun. By the 23rd the Germans were moving along the ridges overlooking the city itself. They got no further. On June 25 the British launched a huge offensive in the Somme valley, and on the eastern front the Russian general Brusilov was

In this man-made hell on earth . . .

. . . comradeship was the one redeeming feature both during and between the fighting. These are German soldiers

sweeping all before him. To meet these threats, the German High Command had to weaken the German armies round Verdun. The French counter-attacked and were able to clear the whole salient. The struggle for Verdun cost the French 377,000 casualties of whom 162,000 were dead or missing. The Germans lost almost as many, 337,000.

Haig, who had replaced Sir John French as commander of the British forces, agreed to bring forward the date of the long-planned Somme offensive to take the pressure off Verdun. The Somme valley was not a good choice for an attack. The British had to advance uphill against trenches and dugouts so placed that they commanded a wide view of the approaching enemy. Moreover, so slack had been British security that the Germans had a good idea when the attack was coming. After a five day bombardment which did only minor damage to the enemy position, 100,000 troops advanced against the German line on 1 July. 'They came on', a German reported, 'at a steady easy pace as if expecting to find nothing alive in our front trenches.' The German machine-gunmen however were very much alive. An average British battalion had about 30 officers and 800 men. On 1 July the 2nd Middlesex lost 22 officers and 601 men, the 8th Yorks and Lancs 21 and 576. Total casualties for that day were 60,000 of whom 20,000 were dead. Probably not more than 100 machine-gunners did the damage. Yet the offensive continued, through the heat of the summer to the gales and mud of autumn. At most, five or six miles of shell-shattered land was gained in indescribable fighting conditions. 'We have just come out of a place so terrible', wrote an Australian officer at the end of September, 'that a raving lunatic could never imagine the horror of the last thirteen days.' The Somme offensive petered out at the end of November. British casualties numbered 420,000, French 200,000, German 450,000. Between them, the battles of Verdun and the Somme had killed or wounded nearly two million soldiers. When they ended, the Western Front had hardly moved and neither side was nearer victory.

In complete contrast to the stalemate on the Western Front were events in the east in 1916. There one of the outstanding and most imaginative generals of the war, the Russian Brusilov, persuaded the Tsar to allow him to attack, despite the defeats of the previous year. The Brusilov offensive began on a broad front on 4 June, the main attack falling on the Austrians between Lutsk and Kovel. The Austrians were overwhelmed. By 10 August, when the attack ended, 375,000 prisoners had been taken and the Habsburg Empire severely shaken. Brusilov always maintained that if he had been properly supported by the Russian armies in the north, he could have driven Austria out of the war and forced Germany to ask for peace. He may

General Brusilov

well have been right. As it was, his offensive made a major contribution to the course of the war by the indirect aid it gave to the French at Verdun. As a leading military historian put it, 'Even if Brusilov had not won the war, he probably stopped the Allies losing it.'

Naval Warfare 1914–1918

Before the war, the naval experts had believed that the war at sea would centre round the great dreadnought fleets of battleships and battle-cruisers, huge, heavily-armoured fast-moving ships which could sail for thousands of miles without refuelling. The British and German governments had spent millions of pounds on these fleets and the British in particular looked forward to a twentieth century Trafalgar which would annihilate the enemy in a single day. Like the army planners, the naval planners were wrong. Three newish inventions—the submarine, the torpedo and the mine—proved to be so dangerous to dreadnoughts that they stayed shut up in their heavily guarded bases for most of the war. The really effective weapon was the submarine, especially the German U-boat. At the beginning of the war there were only 30 of them and at the end of it barely 100. They were enough, however, not only to hamper gravely the British Grand Fleet but to bring Britain close to surrender, such destruction did they cause to the merchant shipping bringing food supplies to the British Isles. In fact the most important question the naval chiefs had to answer was less how to bring about and win the decisive battle of the dreadnoughts but who could mount the most effective blockade of the enemy.

The naval war began with some spectacular German successes. In November 1914 von Spee's East Asia Squadron of modern cruisers, making its way home to Germany, met a British squadron of older cruisers off the Chile coast near Coronel. The British force, outpaced and outgunned, was destroyed. The last time a British naval force had been defeated was 130 years previously during the American War of Independence and a shocked public opinion demanded revenge. A strong force, headed by two modern battle-cruisers, was sent into the South Atlantic. Von Spee was unlucky enough to come into sight of this force while it was coaling in the Falkland Islands. It was now the turn of the East Asia Squadron to be outpaced, outgunned and destroyed (December 1914). Other German raiders were hunted down and, from the beginning of 1915, the British navy controlled the surface of the oceans of the world.

In European waters, the main fleets watched and waited. British strategy was to prevent the German High Seas fleet from breaking out into the Atlantic. The Dover strait in the South was mined

and the Grand Fleet from its Orkney base at Scapa Flow guarded the northern exit. The German High Seas Fleet tried to work out a means of engaging and defeating the Grand Fleet. Since the British force was larger, it had somehow to be lured into the North Sea in sections and defeated bit by bit. The battle of the Dogger Bank (1915) occurred when German battle-cruisers were trying this luring tactic. Since they only narrowly avoided destruction, the Kaiser ordered his admirals to stick to submarine warfare. A year later, in May 1916, Scheer, Admiral of the German High Seas fleet, tried once more to bring the Grand Fleet to action. Hipper with the German battle-cruisers was to lure Beatty's battle-cruisers from their Rosyth base into the jaws of Scheer's battleships before Jellicoe with the main Grand Fleet could come to his aid. The British Admiralty realized that something was afoot and the whole Grand Fleet was at sea before Scheer could set his trap. The result was the

The British Grand Fleet sails from Scapa Flow, its main base in the Orkneys

40

Battle of Jutland on 31 May, the only major action between the two dreadnought fleets during World War I. It was a confused affair. The Germans were outnumbered and, on two occasions at least, on the verge of destruction. However, Jellicoe, the British commander, was acutely aware that he was the one man who could 'lose the war in a day'. If he miscalculated and the Grand Fleet was lost, Britain was defenceless. If Britain had to make peace, the Allied cause was doomed. He acted very cautiously and Scheer was able to get his fleet safely home. The result of the battle has aroused keen controversy. When the rival navies came to tot up their losses, it became clear that the British had lost more than the Germans (14 ships, including 3 battle-cruisers, of 112,000 tons and 6,000 men, against 11 ships, including one battle-cruiser, of 61,000 tons and 2,500 men). The Germans, therefore, claimed a victory. On the other hand, their naval chiefs decided that the High Seas Fleet could never again risk such an action. They relied more and more on submarines and the High Seas Fleet mouldered in port. Its morale slumped. There was one mutiny in 1917 which Scheer was able to quell. The second, in 1918, was the beginning of the end for Germany. In the long run, therefore, the British could claim Jutland as a victory too.

From 1915 to the end of the war both sides concentrated on improving their methods of blockade. The British set up a Ministry of Blockade. Aware that important supplies for Germany were being carried by neutral ships and landed in neutral ports in Holland and Scandinavia, they insisted on the right to stop, search and, if necessary, seize neutral ships. Despite strong protests from the U.S.A. a blacklist was made of neutral firms trading with Germany and a network of information centres established all over the world which provided the Admiralty with increasingly accurate information as to which ships were worth intercepting. This blockade slowly strangled Germany. The winter of 1916–17 in Germany became known as the turnip winter. Food was so scarce that much of the population survived on turnips. In 1917, meat consumption fell to a quarter of its pre-war level and there was significant malnutrition. The death-rate of young children rose by 50% and civilian morale began to collapse.

The German blockade of Britain was almost entirely the work of U-boats. Since submarines by their nature could only sink, they could hardly search, confiscate or seize, their activities caused the deaths of thousands of civilians, sometimes of neutral countries, (for example, the sinking of the Cunard liner Lusitania which was carrying more than 100 American citizens as well as arms and ammunition) and outraged world opinion in a way the British blockade did not. So it was not until 1917 that the German govern-

ment ordered completely unrestricted U-boat warfare (i.e. the sinking at sight of any merchant ship approaching British ports). Even so, they caused havoc to British supplies. In 1915 they sank 750,000 tons of Allied shipping and in 1916 1,237,000 tons. With the monthly figures for sinkings moving steadily upwards, Britain, which was far more dependent on food supplies from overseas than any other country, faced defeat by starvation. The Admiralty did not know what to do. Towards the end of 1916 it issued this helpless memorandum, 'No conclusive answer has as yet been found to this form of warfare; perhaps no conclusive answer ever will be found.'

The most deadly of the warships—a German U-boat ready for action

Faced with stalemate on the Western Front and confident that the U-boat was the ultimate weapon, the German government announced unrestricted U-boat warfare on 1 February 1917. 'We will frighten the British flag off the face of the waters and starve the British people until they, who have refused peace, will kneel and plead for it', declared the Kaiser. It looked as if he would be right. In April 1917 881,000 tons were sunk. Such a loss if continued would mean the defeat of Britain in a few months. There was one method of dealing with U-boats which stood a chance of success—the convoy system where merchantmen sailed in groups (convoys) escorted by destroyers. The Americans (at war with Germany since 6 April, mainly as a result of unrestricted U-boat warfare) were in favour, so was the British Prime Minister, Lloyd George, but only after a series of bitter meetings was he able to get the reluctant British Admiralty to try the system out. It worked. Though losses continued at quite a high level, the crisis figures of spring 1917 were not repeated. Moreover in 1917 Britain at last developed an effective mine—by exactly copying a German design—and was able to make the Channel and the North Sea more and more dangerous for U-boats. In 1918 the U-boat blockade of Britain was less effective than it had been and the danger of defeat by starvation a thing of the past. The British civilian population never suffered as the Germans did.

1917

The U.S.A. declared war on Germany on 6 April 1917. By the autumn of 1917, due to the continuing internal turmoil which followed the March revolution of 1917, the Russian army was no longer an effective fighting force. The character of the war was therefore totally transformed.

When the war began, most Americans were happy to remain neutral and felt that there was little to choose between the two sides. As time passed, however, they became increasingly anti-German. The conquest of Belgium seemed atrocious, worse still was the sinking of the Lusitania in May 1915 with 128 Americans on board. 'The torpedo which sank the Lusitania', commented the *Nation* magazine, 'also sank Germany in the opinion of mankind.' Moreover, American businessmen invested extensively in the Allied war-effort, 3,000 million dollars worth by 1916, while the British blockade severely limited their opportunities of trading with Germany. The major American complaint against the Germans however was U-boat warfare and, when the Germans began unrestricted U-boat warfare in 1917, they knew that it was likely to bring the U.S.A. into the war against them. Their calculation was that even if the U.S.A. declared war, they would manage to win the war in Europe before the power of America could play any significant part. It was an enormous gamble.

Unrestricted U-boat warfare persuaded Woodrow Wilson, the American President, to break off diplomatic relations but he was still reluctant to declare war. Two remarkably stupid German acts finally overcame his reluctance. Another British liner, the Laconia, was torpedoed in February soon after it left New York and an American mother and daughter died a miserably slow death by drowning. Soon afterwards, Zimmermann the German Foreign Secretary proposed by coded telegram to give Mexico the states of Texas, Arizona and New Mexico in return for an anti-American alliance. British intelligence agents cracked the code and made sure the message got into American hands. 'Zimmermann shot an arrow in the air', remarked an American wit, 'and brought down neutrality like a dead duck.' War was declared within a month.

The U.S.A. was economically the most powerful nation in the world. Her declaration of war was therefore a tremendous psychological boost to the Allies whose position was otherwise extremely bleak. But its immediate effect on the course of the war was limited. American industry had been feeding the Allied war effort since 1914 and the U.S.A.'s entry into the war had little effect on Allied war production. Nor were the Allied armies quickly reinforced. At the end of October 1917 there were less than 100,000 Americans on the Western Front. In 1918, however, things were different. The Germans knew that, unless they won a quick victory, defeat was certain.

Once the Brusilov offensive ended, the exhausted Russian armies had nothing left. Neither they, nor the Russian civilian population nor the economy could stand the strain of war any longer. In March 1917 a revolution forced the Tsar to abdicate. A Provisional Government, of which Kerensky was the dominant figure, tried vainly to keep the Russian armies fighting. Whole divisions walked away from the front and after a summer of turmoil a second, Communist, revolution, led by Lenin, got rid of Kerensky. Lenin was ready to make peace at almost any price and eventually, in March 1918, the Treaty of Brest-Litovsk brought the war between Russia and Germany to an end. The Germans demanded a high price from their defeated enemy, seizing hundreds of thousands of square miles of Russian territory. The timing of peace with Russia was of great significance to the course of the war. At last Germany was free of a war on two fronts. She could switch her Eastern Front forces, comparatively fresh after only sporadic fighting in 1917, to the Western Front. Could they overwhelm the exhausted Allied forces before American reinforcements arrived in large enough numbers? It was a race against time.

Meanwhile, for the Allies, 1917 was if anything grimmer than 1916. War weariness affected both France and Britain. A French officer summed up a widely-shared feeling when he wrote 'the year is opening in a grim atmosphere. Promises and hopes are followed by too many disappointments.' But the generals kept on making promises. By this time Joffre had been replaced as French commander by Nivelle, a man not lacking in self-confidence. 'We have the formula', he declared. 'We shall break through the German front when we wish.' In April he launched his offensive against the Craonne plateau between Laon and Rheims. He could not have chosen a worse place. It was the Somme all over again. A few strategically worthless miles were gained for 200,000 casualties. The French army cracked. Much of it, including units with superb fighting records, mutinied. At one point in June there were only two dependable divisions between the Germans and Paris. Amazingly the Germans remained ignorant of the mutinies, though they lasted six weeks. Pétain, the saviour of Verdun and a favourite of the troops, replaced Nivelle. He made many visits to the discontented regiments and by a combination of firmness, reforms and mercy, managed to restore order. 412 mutineers were eventually sentenced to death but only 55 actually executed. The brunt of the fighting on the Western Front fell increasingly on the British Army.

Lloyd George. This photo was taken in 1914 outside the Houses of Parliament when he was a minister in Asquith's Liberal government

Haig, the British commander, also kept making promises. In June he was telling the British cabinet that 'if the fighting were kept up at its present intensity, Germany would soon be at the end of her available manpower.' What evidence he had for this belief is not clear. He began another offensive in the Ypres area in July which he continued although prolonged rain turned the battlefield into a swamp. In August, he was as hopeful as ever. 'The time is fast approaching' he told the British government, 'when Germany will be unable to maintain her armies.' The third Ypres campaign ended at Passchendaele in November. For about six miles of 'a porridge of mud', the British lost 265,000 men, the Germans 206,000. Wounded men drowned in the quagmire. 'It was no longer life at all', wrote General Ludendorff, 'it was mere unspeakable suffering.'

Fortunately in this desperate year, both Britain and France had exceptional leaders, David Lloyd George and Georges Clemenceau. Lloyd George was a Welshman of humble birth whose quick wits, eloquence and energy had made first a successful lawyer and then a leading Liberal politician. In the early part of the war he was Minister of Munitions and then Secretary for War and his energy and decisiveness appeared in marked contrast to the over-passive approach of Asquith, the Prime Minister. In 1916 he noted in his diary, 'we are going to lose this war.' Asquith clung like a limpet to power but a crafty plot engineered by Lloyd George, Bonar Law (the leader of the Conservatives) and Max Aitken (the owner of the *Daily Express*) got rid of him. Lloyd George became Prime Minister and immediately reorganized the government. The day-to-day running of the war became the responsibility of a small war cabinet of five. New ministries were established and experts from all walks of life called in to speed up the war effort. The country was given a new will to victory. We have seen how Lloyd George dealt with the Admiralty over the convoy question. He was unable however to deal so firmly with the army leadership. In fact his greatest weakness was his relationship with Haig. They both distrusted each other but such was Haig's reputation and so strong his political connections that Lloyd George felt unable to sack him. Nonetheless he guided Britain through some of the darkest years in her history to final victory. He must be rated among our greatest war leaders.

Clemenceau was 76 when he became Prime Minister of France in the crisis following the failure of the Nivelle offensive and the mutinies. His fiery character had won him the nickname 'Tiger' and ensured for him a stormy political career. In 1892 he had seemed politically ruined by the famous Panama scandal, which had shown many French deputies, including Clemenceau, to be involved in crooked financial dealings. However, he made a comeback and hit the headlines in 1906 when as Minister of the Interior he took ruthless action against striking workers. He was Prime Minister from 1906 to 1909 and, since the outbreak of war, had been a noisy and effective critic of the government. Two things were never in doubt about Clemenceau, his patriotism and his obstinacy. He was just the man France needed when defeatism was everywhere. A series of purges sorted out the traitors and faint-hearted at home. Regular visits to the front-line encouraged the troops. When asked what were his policies, his answer was simple. 'Home policy? I wage war! Foreign policy? I wage war! All the time I wage war!' Leadership of such quality was essential if the Allies were to survive the climax of the war.

1918

In November 1917 the German High Command met at Mons in Belgium to review the war situation. Ludendorff, now commander on the Western Front and virtually dictator of Germany, summed up the meeting as follows. 'Our general situation requires that we should strike at the earliest moment, if possible at the end of February or the beginning of March, before the Americans can throw strong forces into the scale.' On 21 March 1918, after an immense artillery barrage of 6,000 guns, Ludendorff's first offensive hit the British lines. Aided by the fog, the Germans broke through and gained 1200 square miles of land before the Allies could stabilize their line. In April, Ludendorff struck the hard-pressed British again, this time just south of Ypres. The line bulged dangerously and Haig, fully aware of the desperate consequences of another breakthrough issued his most famous Order of the Day on April 12. 'With our backs to the wall and believing in the justice of our cause, each one of us must fight to the end.' The German attack was held.

At this stage, the French General Foch was made commander-in-chief of all the Allied armies. A man of great intelligence and calm, he out-thought Ludendorff through the summer of 1918, usually predicting the directions of his attacks and holding them by the swift movement of his reserve troops. In May Ludendorff concentrated on the French sector. The line bulged—momentarily the Germans were across the Marne—but it did not break. Americans troops were now pouring into France while irreplaceable German divisions were destroyed. The famous German discipline now showed signs of breaking down. Military police were used increasingly to keep the troops moving forward. A single Allied tank—the one weapon in which the Allies had real superiority and used increasingly skilfully after its introduction in 1916—might find itself able to capture a whole company. On 8 August the Allies counter-attacked. It was, said Ludendorff, 'the Black Day of the German army.' Though the Germans retired in good order, the Allied counter-

This sketch of Clemenceau visiting soldiers at the front captures well his determination and vigour, despite his age

Tanks—a new weapon which gave a considerable advantage to the Allies—counter-attack with the infantry in close support (1918). These ones are British

attack gained momentum. The German High Command knew that the war was lost. It hoped however to keep fighting long enough to bargain for an honourable peace. Events far away from the Western Front put an end to these hopes.

In northern Greece, near the port of Salonika, a small Allied force had been stationed since 1915. It had achieved very little but in the summer of 1918 had gained a new French commander, Franchet d'Espérey. The British officers under his command aptly nicknamed him 'Desperate Frankie'. 'I expect from you savage vigour', he told his men. Under his leadership French, British, Serbian and Greek forces attacked northwards. A combined infantry, cavalry and air attack forced Bulgaria to make peace on 30 September. While the French and Serbians attacked the Austrians in Serbia, the British hammered the Turks who, defeated also in Syria and Mesopotamia, asked for peace on 30 October. The Habsburg Empire was now tottering. The Italian army had forced the Austria-Hungarians to keep many divisions on the Italian Front since 1915 and, though it had been badly defeated in 1917 at Caporetto, it rallied in 1918 to overwhelm the Austrian defences along the Piave river. By the end of October, they had captured Vittorio Veneto and half a million prisoners. The chance of the various national groups within the Habsburg Empire had now come. With Allied encouragement, they declared their independence. The Habsburg Empire ceased to exist and on 3 November the government in Vienna ordered what troops still obeyed it to cease fighting.

Revolution spread to Germany. The civilian population had suffered much since 1916 as a result of the Allied blockade. There had been a wave of strikes in 1917. The successful Communist revolution in Russia encouraged extreme socialism and anti-government feeling among some groups of workers. The news of military setbacks brought critics of the government and of the German High Command into the open. By the end of September Ludendorff realized that defeat was certain and resigned, leaving a civilian government the impossible task of securing an honourable peace. On 29 October the High Seas Fleet mutinied and disorder quickly spread to other ports. There was a revolution in Munich. On 9 November revolution spread to Berlin and the Kaiser abdicated. Prince Max of Baden, the former Chancellor, resigned in favour of Ebert, a Socialist who asked for an armistice. At 11 a.m. on 11 November 1918 the guns at last went quiet along the Western Front.

Conclusions

Four years of war showed that the European balance of power achieved in the years before 1914 was very even. Germany and Austria-Hungary could sustain a war on two fronts against France, Russia and Britain, and come close to winning it. The reasons why they eventually lost it were many and various but the most important were these. First, the Allied naval blockade slowly throttled Germany. Secondly the effects of this blockade forced the violent reaction of unrestricted U-boat warfare which brought the U.S.A. into the war. The American war effort was eventually though not immediately decisive. Thirdly, the German High Command made too many major miscalculations. The First World War is full of huge and horrible blunders but the Germans blundered most. One was von Moltke's weakening of his right during the Battle of the Marne. Another was the constant underestimation of the fighting power of the British army. Before 1914, the Germans feared the British navy but they assumed that the British army would play no significant part in a European War. In 1917 and 1918, however, it was the British not the French army that bore the brunt of the fighting on the Western Front. Moreover, the German High Command tended to exaggerate its own strength. It misjudged the effectiveness of U-boat warfare in 1917 and 1918, and, crucially, it misjudged the timing and direction of the final attacks in the West. It made a calculated gamble which failed.

The costs of this war—caused and prolonged by so much ignorance and stupidity—were these. At least eight million soldiers were killed. A larger number were wounded, many of whom were mutilated for life. Nearly five million civilians lost their lives directly as a result of the war. Another six million died in the influenza epidemic of 1918–19. The resistance of many of these victims must have been weakened by the hardship of the war-years. The destruction of property was on a similar scale. Thirteen million tons of shipping lay on the ocean-bed. In Northern France, 10,000 square miles of land lay waste. 1200 churches, 1000 schools, 377 public buildings, 1000 industrial plants and 246,000 other buildings were destroyed. On the Eastern Front, there was similar destruction over a much larger area. There were other costs which cannot be put into figures but were just as serious. The European economy was so dislocated that pre-war standards of co-operation and prosperity were difficult to achieve for many years (see Chapter 4). Furthermore the suffering of the war-years were so great that few who had lived through them were prepared to forgive and forget. Hatred, bitterness, and the desire for revenge were to blight the next twenty years of Europe's history and help bring about the catastrophe of World War II.

So the sacrifice of so many men, mainly young men, had for the most part been in vain.

> When you see millions of the mouthless dead
> Across your dreams in pale battalions go,
> Say not soft things as other men have said.
>
> Charles Sorley (killed in the Battle of Loos 1915, age 20)

Chapter 3
The Versailles Settlement

The most important part of the long and complicated discussions which worked out what shape postwar Europe would take occurred at Versailles, just outside Paris, in 1919. The leader of the British delegation was Lloyd George and a member of his negotiating team was a comparatively young economist, J. M. Keynes. Keynes was exceptionally intelligent, one of the greatest economists of all time. He took an active part in the negotiations but the longer he stayed, the more appalled he became. When the terms of the peace treaty became known, he resigned and sent the following letter to Lloyd George. 'I ought to let you know that on Saturday I am slipping away from this scene of nightmare. I can do no more good here. I've gone on hoping even through these last dreadful weeks that you'd find some way of making the Treaty a just and expedient document. But now it's apparently too late. The battle is lost. I leave the twins (two senior British officials who approved of the Treaty terms) to gloat over the devastation of Europe'. Soon after he returned to England, he wrote a fierce and brilliant criticism of the Versailles Settlement called *The Economic Consequences of the Peace* in which he predicted that it could only bring further disaster to a Europe which had already suffered far too much. From the first, therefore, the Versailles Settlement aroused much controversy. Before considering the strength of Keynes' arguments, we must first study the details of the Settlement itself.

The Settlement was based on fourteen points which President Wilson of the U.S.A. had first listed in January 1918 as the possible basis of a peace treaty. The German government was not interested in these Fourteen Points until it knew the war was lost. Then, since they seemed the best terms Germany was likely to get, it agreed to accept them as a basis for settlement and the armistice of 11 November followed. The eventual settlement included most of the Fourteen Points but, both in spirit and in letter, it was tougher on the defeated enemy.

The Makers

Thirty-two nations were represented at Versailles. Germany was not allowed to negotiate. Only three men really mattered, Wilson for the U.S.A., Clemenceau for France and Lloyd George for Britain. They had very different aims. Wilson was a man of high ideals. The

The Big Three, 1919—from right to left Wilson, Clemenceau and Lloyd George, with Orlando of Italy

Allied victory, he believed, had provided an opportunity which mankind could not afford to let slip away. The war had been 'a war to end wars' and 'the world must be made safe for democracy'. As the leader of the New World of the Americas, he felt that he could and should give a new lease of life to the Old World of Europe. When he first arrived in Europe, he had received a tremendous popular welcome which convinced him that he was right, and in the negotiations he proved very stubborn. Meanwhile in the U.S.A. itself support for his policies was ebbing away and he became an increasingly lonely and forlorn figure.

In complete contrast was Clemenceau, deeply cynical about human nature in general and German nature in particular. His only concern was the security of France and France would only be secure if Germany was weak. For him and for the French electorate whom he represented and who far more than the British or the Americans had suffered from the war, the idealistic Wilsonian policy of 'no annexations, no contributions and no punitive damages' was wrong. Clemenceau was the chairman of the negotiations and he too could be very obstinate. Lloyd George was in a difficult position. He agreed with Wilson that a harsh peace such as France wished for was unlikely to bring a lasting peace to Europe but he had just fought and won an election during which it became clear that, like the French, the British electorate wanted the defeated enemy to be crushed. 'We will get out of Germany all you can squeeze out of a lemon

51

and a bit more. I will squeeze her until you can hear the pips squeal', was one popular comment. 'Hang the Kaiser' was another. Though he was on the side of moderation, Lloyd George felt that he could not appear too moderate. During the negotiations 370 Conservative M.P.'s, on whom his government depended for support, sent him a telegram warning him against being too soft. In the House of Commons debate which followed, he summed up his attitude like this. 'We want a stern peace because the occasion demands it. But its severity must be designed not to gratify vengeance, but to vindicate justice ... (our) duty is not to soil the triumph of right by indulging in the angry passions of the moment but to consecrate the sacrifice of millions to the permanent redemption of the human race from the scourge and agony of war.' He devoted his considerable negotiating skills to preventing the final settlement from turning out too harsh but had only partial success.

Six separate treaties signed between 1919 and 1923 made up the final settlement. The Treaty of Versailles (1919) made peace with Germany, of St.-Germain (1920) with Austria, of Trianon (1920) with Hungary, of Neuilly (1919) with Bulgaria, and of Sévres (1920) and of Lausanne (1923) with Turkey. Two treaties were needed with Turkey because the terms of 1920 were upset by a war between Turkey and Greece from 1920 to 1923.

The Terms

The Europe of 1923 looked quite different from the Europe of 1914, especially in the east and centre. Beginning in the south-east, Turkey in Europe was limited to Constantinople and its approaches. Bulgaria lost her outlet to the Aegean to Greece and some small pockets of land to the new nation of Yugoslavia.

The Habsburg Empire disappeared. Its south-western provinces of Bosnia, Herzegovina, Croatia and Slovenia were united to Serbia and Montenegro to create a state inhabited mainly by South Slavs— Yugoslavia. A large section which had formerly been part of Hungary was given to Roumania, and, Hungary, now much smaller in size, became independent. In the northern part of the old empire, the Czechs and Slovaks were united in the new state of Czechoslovakia. Galicia was given to the revived nation of Poland and Istria, the South Tyrol and Trentino to Italy. All that was left of Austria was a small German-speaking area of about six million inhabitants lying mainly to the east of Vienna, the old Habsburg capital.

The Poles gained land from Russia and Germany as well as from Austria. In 1919 a civil war still raged in Russia and Lenin's Communist government was regarded with great suspicion by most European leaders. Russia was not therefore represented at Versailles and a war had to be fought between Poland and Russia before their common border was finally established by the Treaty of Riga in 1921. Three former Russian provinces bordering the Baltic Sea— Lithuania, Latvia and Estonia—also declared their independence as did Finland which had been ruled by Russia since 1815. The Poles took Vilna from Lithuania in 1920 while Lithuania, for her part, took Memelland (previously German) from its League of Nations administration in 1923.

Germany lost bits of land on all her frontiers. Posen, the Polish Corridor and part of Upper Silesia went to Poland and the great port of Danzig became a free city. Part of Upper Silesia went to Czechoslovakia. In the north, Northern Schleswig went to Denmark and, in the west, Eupen and Malmédy to Belgium and Alsace and Lorraine to France. The Saar coalfields were also handed over the the French while the Saar itself was to be run by the League of Nations. (It was returned to Germany after a plebiscite in 1935.) Finally, a wide strip of territory on both sides of the Rhine was forbidden to German troops. This area was known as the Demilitarized Zone.

President Wilson firmly believed that no lasting peace could come to Europe unless the principle of self-determination was made a reality in Central and Eastern Europe. By this he meant that every people with a sense of common nationality based on a common language or history should have the right to govern themselves, to determine their own futures. The new map of Europe attempted to give some reality to this ideal of self-determination. The Poles, the Czechs and Slovaks, the South Slavs (in Yugoslavia), the Magyars (in Hungary), the Latvians, Lithuanians, Estonians and Finns governed themselves in 1923 when in 1914 they had been governed by foreigners. The mainly Italian-speaking Istria and Trentino was transferred from Austria to Italy, the mainly Danish Schleswig from Germany to Denmark and the mainly Roumanian Eastern Hungary from Hungary to Roumania. However, the pattern of racial settlement in Eastern Europe combined with the need to please the victors at the expense of the defeated caused rough justice to be done and many discontented groups were left under the rule of other races whom they despised and feared. For example, millions of Magyars as well as Roumanians were transferred to Roumanian rule. Only two-thirds of Poland's population of 32 millions were Poles. Of the 14 million inhabitants of Czechoslovakia less than 10 million were Czechs or Slovaks. The German-speaking South Tyrol was handed over to Italy. Moreover there was one race to whom the principle of self-determination did not apply, the Germans. Millions of Germans passed under Polish and Czech rule and Austria, now a German-speaking state was expressly forbidden to unite with Germany. Such rough justice could easily be explained in 1919. After all Germany and Austria had lost the war. In the long-run

Legend:
- Territory lost by Germany
- Territory lost by Austria and Hungary
- Territory lost by Russia
- Territory lost by Bulgaria
- To Greece 1920 Recovered by Turkey 1923
- National Boundaries at 1924
- Frontiers of New or Revived States

NORWAY
SWEDEN
FINLAND
Lake Ladoga
Bergen
Oslo
Åland Islands
Viborg
Leningrad
Stavanger
Stockholm
Åbo
Helsinki
Göteborg
ESTONIA
Tallinn
Pskov
North Sea
DENMARK
Malmö
Copenhagen
Bornholm (Dan.)
Baltic Sea
Riga
LATVIA
Libau
Dvinsk
Smolensk
N. SLESVIG
Flensborg
Memel
Königsberg
LITHUANIA
Kaunus
Minsk
Hamburg
Danzig (Free City)
Vilna (seized from Lith.)
The Hague
HOLLAND
Bremen
Stettin
EAST PRUSSIA (GERMANY)
Bialystock
London
Essen
RUHR
Berlin
Thorn
Warsaw
Kief
BELGIUM
Brussels
Düsseldorf
Cologne
Leipzig
Dresden
Posen
POLAND
Lodz
Lutsk
Malmédy
GERMAN REPUBLIC
Breslau
LUXEMBURG
UPPER SILESIA
Przemysl
Lvov
Paris
Verdun
SAAR
LORRAINE
Cracow
GALICIA
Kamenets Podolsk
ALSACE
Ulm
CZECHOSLOVAKIA
Czernowitz
Belfort
Basel
Munich
Pilsen
Prague
Brno
BESSARABIA
Geneva
Berne
Linz
Vienna
Bratislava
Debreczen
BUKOVINA
Lyons
Locarno
SOUTH TYROL
Innsbrück
Graz
AUSTRIA
Budapest
Grosswardein
Milan
ISTRIA
SLOVENIA
HUNGARY
ROUMANIA
Ploesti
Venice
Trieste
Zagreb
VOYVODINA
TRANSYLVANIA
Galatz
Turin
Fiume (It.1924)
SLAVONIA
Genoa
CROATIA
JUGOSLAVIA
Belgrade
Bucharest
Black Sea
Monaco
Zara (It.)
BOSNIA
Sarajevo
SERBIA
DOBRUJA
Marseilles
Toulon
Florence
San Marino
HERZE-GOVINA
Nish
Sofia
Varna
CORSICA (FR.)
Lagosta (It.)
Durazzo
Monastir
BULGARIA
Burgas
Rome
Bari
ALBANIA
MACEDONIA
Adrianople
Constantinople
Naples
GREECE
Salonika
Dedeagach
Chanak
SARDINIA (IT.)
Taranto
Athens
TURKEY
Smyrna
Mediterranean Sea
Palermo
SICILY
Messina
Corinth
Bizerta
Dodecanese (It.)
Tunis
Malta (Br.)
CRETE
ITALY
FRANCE
U.S.S.R.

0   100   200 Miles
0   300 Kilometres

Map 4
Europe in 1924

however its consequences were serious. The new boundaries of Central and Eastern Europe proved unstable and the discontent of the German minorities in Poland and Czechoslovakia was seized upon by Hitler to justify his aggressive foreign policy between 1936 and 1939 and played a significant part in bringing about World War II.

Germany lost all her colonies and Turkey her provinces in the Middle East. These were distributed to the victorious allies in the form of mandates supervised by the League of Nations (see Chapter 15). The German African colonies were divided between Britain, France, Belgium and South Africa. In the Middle East, Britain gained Palestine, Iraq and Transjordan and France Syria and the Lebanon from Turkey. German possessions in the Far East and Pacific north of the equator went to Japan, south of the equator to Britain, Australia and New Zealand.

As well as suffering these considerable territorial losses, Germany was also forced to agree 'to make compensation for all damage done to the civilian population of the Allied and Associated Powers and their property'. These compensation payments or reparations as they came to be called had not been mentioned in the original Fourteen Points but had been included in the armistice terms on the insistence of France and Britain. The peacemakers had great difficulty in working out how large these reparations should actually be. The French first suggested £15,000,000,000 but the sum eventually fixed in 1921 was £6,600,000,000. Germany also had to surrender all her merchant ships over 1600 tons and some smaller ships too; give free coal for ten years to France, Belgium and Italy and horses, sheep and cattle to France and Belgium.

The Allies intended that German armed forces should be cut down to a size which was adequate only for defending her borders against attack. At the end of the war the German High Seas Fleet had surrendered to the British Navy. (When it arrived at Scapa Flow, it was scuttled by its captains.) By the Treaty of Versailles, Germany was allowed six battleships, some smaller craft but no submarines. Nor could she have an airforce. The army was limited to 100,000 men. Conscription, tanks and armoured cars were all forbidden. The western bank of the Rhine was occupied by Allied forces. This occupation was originally intended to last until 1935 but ended in 1930.

Finally, Article 231 of the Versailles Treaty read as follows: 'the Allied and Associated governments affirm and Germany accepts the responsibility of Germany and her allies for causing all the loss and damage to which the Allied and Associated Governments and their nationals (citizens) have been subjected as a consequence of the war imposed upon them by the aggression of Germany and her allies.'

The Allies included this article to justify their demand for reparations. The Germans, however, read it to mean that they alone were responsible for causing the war and greatly resented it.

An Evaluation

We can now consider Keynes's criticism of the Versailles settlement which was published in England in December 1919. He made three main points. The first had nothing to do with economics. The peace terms were merciless when they should have been merciful. They had been inspired too much by the desire for revenge. A merciless peace could not be a lasting peace for Europe. His second point concerned reparations. They were, he argued, quite beyond the means of a defeated Germany whose defeat had been brought about to a large extent by economic strangulation. His third point was that the peacemakers had concentrated too much on political and too little on economic problems. The European economy before 1914 had been inter-dependent—i.e. the nations of Europe had prospered or stagnated together. While the break-up of the Habsburg empire would dislocate the economy of Eastern Europe the demand for reparations would slow down the economic recovery of Germany which in turn would slow down the recovery of Central and Western Europe. The peacemakers had managed to make rapid and durable economic recovery very difficult.

There can be little doubt that Keynes was right on all three counts though his arguments were sharply questioned in 1919 and later. The defeated enemy, and with the passing of time many of the victors, regarded the terms as humiliating and merciless. If you humiliate someone, he will rarely co-operate with you again. The Germans—who took no official part in the negotiations and had had the peace terms dictated to them—did their utmost to avoid fulfilling them and part of the appeal of Hitler to the German people was his determination to defy the Versailles Treaty. When he began to challenge the Treaty openly, Britain and France, lacking the conviction that their cause was just, had neither the unity nor the will to enforce it. Reparations did turn out to be impractical. German attempts to pay them were never more than half-hearted. When the French did try to force repayment in 1922–23 by occupying the Ruhr industrial region, the result was an inflationary crisis in Germany so severe that not only was the value of the German mark destroyed but the French economy also began to suffer and the occupation had to be brought to an end. With Britain and the U.S.A. increasingly unhappy about the effect of reparations, an international committee, chaired by the American General Dawes, drew up a new plan for repayments in 1924. The Dawes Plan avoided fixing either a definite final sum or a final date for repayments to be completed. Instead it ordered that payments should be in annual instalments graded according to Germany's ability to pay. The U.S.A. provided a loan towards Germany's economic recovery and payments were to begin again after two years. Further discussions in 1929 led to the Young Plan. Reparations were to be reduced to a total of £2,000 million to be repaid by 1988. 1929 however was the year of the Wall Street Crash and before long Europe was in the grip of the Great Depression. In 1932 when a conference at Lausanne finally ended the demand for reparations, Germany had failed to pay off more than a fraction of the original reparations sum. Moreover she had received more than three times as much as this from the Allies in loans, the chief aim of which had been to make her economically strong enough to afford to pay reparations! Keynes's third argument also turned out to be accurate. The Versailles Settlement increased rather than lessened the economic dislocation of Europe and both the 1920's and 1930's were, at least partly because of it, deeply troubled both economically and politically.

The Versailles Settlement can also be criticized for reasons other than those suggested by Keynes. While it humiliated the defeated, it failed to please the victors. The U.S.A., the most powerful country in the world, refused, despite President Wilson, to have anything to do with it. Wilson led the Democratic Party. While he was at Versailles the other major American party, the Republicans, had won control of Congress. American public opinion had been shocked by the losses of the American army in 1918, irritated by the squabbles of the Versailles peacemakers and worried by Wilson's apparent determination to keep the U.S.A. involved in European affairs by the creation of a League of Nations, and it now turned isolationist. Wilson was out of touch with the changing mood of America. He handled the Republican leaders tactlessly with the result that Congress, without whose approval no treaty was valid, refused to approve the Versailles terms. Wilson, however, was a fighter. In September 1919 he began a great speaking tour of his country to convince the people of the importance of the settlement, particularly the setting-up of a League of Nations. The strain was too great. After three weeks of campaigning, he had a stroke which paralysed one side and affected his powers both of speech and reasoning. Even as an invalid however he refused to compromise with his opponents. 'Better a thousand times to go down fighting', he declared, 'than to dip your colours in dishonourable compromise.' He went down fighting. Despite his campaign, the Versailles Treaty failed to win the necessary two-thirds Congress majority and the U.S.A. took up an isolationist position. In the presidential election of 1920 the Republican Harding won. Soon after he made his position clear in relation to the Versailles Treaty and the League of Nations. 'We seek no part,' he said 'in directing the destinies of the world.'

This American refusal to stand by the Treaty was a terrible blow to France. Clemenceau had agreed to the final terms—which to him seemed dangerously soft on Germany—because he had Wilson's guarantee of American military support to make the Treaty effective. Now the guarantee was useless. Not surprisingly, the French felt betrayed and insecure. They became even more grimly determined to keep Germany weak. The American refusal to sign also put Britain in a difficult position. She too had promised military support to France but with no U.S.A. alongside and with France taking a tougher line than ever against Germany, she was uneasy and as time passed ever more reluctant to give France active backing.

Other allies were unhappy with the Settlement. The Italian delegates left the negotiations half way through in a huff. They felt that they were not being treated with enough respect nor were they gaining enough of the old Habsburg Empire. The Japanese were hardly more content. They had to hand over to China the port of Tsingtao which they had won from the Germans in 1914 after months of hard fighting.

Finally, the Settlement ended the balance of power in Europe which had helped to keep the continent comparatively peaceful between 1815 and 1914. Whereas before 1914 Eastern Europe was ruled by three great empires—the German, Austrian and Russian—in 1923 a large Germany and a huge Russia were separated by many small and militarily feeble nation-states. In 1919 these states had looked safe enough because Germany was exhausted and Russia distracted by civil war. Twenty years later, however, Hitler's Germany and Stalin's Russia were the most powerful states in Europe and the nations in between them lay at their mercy.

To be fair to the peacemakers, one must remember that they often had little choice but to do as they did. In Eastern Europe, for example, the new nations proclaimed themselves as the Habsburg Empire collapsed. The peacemakers could adjust their frontiers but they could not make major changes without threatening another war which they were not prepared to do.

Moreover there was one part of the settlement which, if it had worked, would have made up for all the shortcomings of the rest of it —the establishment of the League of Nations. The League was the brainchild of Wilson. Its aim was to defend the territories of each of its members from attack by another nation and to provide the means whereby international disputes could be settled peacefully. If the League of Nations had been strong, determined and effective then the Versailles Settlement might have lasted longer and provoked less trouble. Unfortunately it never was.

When all is said and done, Versailles was a bad peace. 'Suspicions, resentments, misunderstandings and fear' Lloyd George noted later, 'had poisoned the mind of mankind.' The poison worked its way deep into the veins of postwar Europe and was to prove fatal to millions of the next generation.

The League and its work

The fourteenth of the Fourteen Points read like this: 'A general association of nations must be formed under specific covenants for the purpose of affording mutual guarantees of political independence and territorial integrity to great and small states alike.' The set of rules, or Covenant as it was called, of the League was approved at the Versailles Conference in 1919 and a Briton, Sir Eric Drummond, was appointed its first Secretary-General with headquarters in Geneva. The League was made up of a General Assembly and a Council. All member-states sent a representative, usually their Foreign Minister, to the Assembly which met once a year. The Council, which dealt with the most important matters requiring prompt treatment, usually met three times a year and consisted of some Permanent Members and some Non-Permanent ones. The major nations provided the Permanent Members—in 1920 Britain, France, Italy and Japan—and these were balanced by four Non-Permanent Members elected at intervals from all the member states. Later the number of members of the Council rose with Non-Permanent members outnumbering Permanent ones. This made it hard for the major powers to get things their own way and their influence was further limited by the rule that decisions in both the Council and the Assembly had to be unanimous. So any member nation however small could use its veto and prevent the League taking action. This rule often prevented the League from taking decisive action in international crises and was an important cause of its eventual failure.

Another part of the League was the Permanent Court of Justice made up of fifteen judges and based at the Hague in Holland. Its main job was to pass judgement on those international legal disputes which were submitted to it for judgement. The League also set up a number of special commissions. Some of these dealt with short-term political problems like the administration of the Saar or Danzig, others like the Mandates and Minorities Commissions dealt with more long-term problems. Some were concerned with particular social and humanitarian problems like refugees and drugs, others like the International Labour Organization (I.L.O.) with industrial and economic problems. Most of the constructive work of the League of Nations was achieved by these special commissions.

The Versailles Settlement and the League of Nations

For an institution intended to prevent the nations of the world from going to war, the League suffered from serious weaknesses from its very start. The most serious was the fact that some of the most important nations in the world were not members. As we have seen, the U.S.A. was not. Communist Russia, extremely suspicious of the rest of the world, did not join until 1934 (and was expelled in 1939 for attacking Finland). Germany was not allowed to join until 1926 but was taken out again seven years later by Hitler. Another important weakness was the lack of any military forces to make its decisions effective.

Nonetheless, the League began with some successes. It established Danzig as a Free City despite the rivalries of Germans and Poles. In 1921 it ended a dispute between Sweden and Finland over the Aaland Islands, returning them to Finland. In 1922 it proved timely financial aid to Austria, saving her from economic collapse. Simultaneously it did useful work in non-political matters, lessening the refugee problem, returning former prisoners to their homelands and checking the spread of epidemic diseases in Eastern Europe. Before long, however, the League's lack of authority began to show. Disputes between Poland and Lithuania over Vilna in 1920 and between Italy and Greece over Corfu in 1923 were side-stepped by the League. In 1925 however it forced the Greeks to end an invasion of Bulgaria and pay £45,000 in compensation.

At the summit of his fame and popularity, in Europe at least. President Wilson in Paris, June 1919

In the 1930's the League began to be openly defied by major states, beginning in 1931, when Japan, a Permanent Member of the League's Council, invaded Chinese Manchuria after a number of border clashes. China appealed to the League which held an on-the-spot inquiry and condemned the Japanese action. Japan invaded more of China and left the League in 1933. It was open defiance. Unless the leading nations of the League—Britain, France and Italy—were prepared to take firm action against Japan, the League's authority would be severely weakened. None of them had the will. The Japanese invasion of Manchuria was a watershed in the League's history. Not only was it weak but it had been seen to be so.

Further blows soon followed. Claiming that Britain and France were reluctant to disarm, Hitler took Germany out of the League and began a rapid rearmament programme defying both the Versailles Treaty and the League. None of the members was prepared to act against him. On the contrary, Britain, without whose positive leadership the League was bound to fail, proceeding to sign a naval agreement with Germany (1935), in effect recognizing Germany's right to ignore the Versailles Treaty and underlining the impotence of the League.

The final blow to the League was struck by the Italian dictator,

At the League of Nations, the Abyssinian Emperor, Haile Selassie, appeals for aid against Mussolini's aggression—in vain

Mussolini. In October 1935 an Italian army invaded Abyssinia. It was open aggression arising from a border dispute between Abyssinia and Italian Somaliland which had already been considered by the League Council. For a moment, it looked as if the League would act decisively since both Britain and France seemed ready to lead collective action against Italy. Three days after the invasion Italy was condemned as an aggressor and the League ordered 'economic sanctions' against her, placing restrictions on Italian trade and on financial aid to Italian banks and companies. Before long, however, Britain and France got cold feet. Their main fear was Hitler and they wished to avoid an open conflict with Italy. They prevented sanctions being applied to Italy's vital oil supplies and Hoare and Laval, the Foreign Secretaries of Britain and France respectively, worked out a scheme behind the scenes—the Hoare-Laval Pact—which would have given two-thirds of Abyssinia to Mussolini. News of the Pact leaked out to the Paris newspapers and so strong was disgust in Britain that Hoare had to resign. But no more positive action was taken against Italy. Abyssinia was conquered. The Emperor Haile Selassie was driven into exile and in June 1936 appeared in person to make a final appeal to the League Assembly. 'It is a question', he said, 'of collective security; of the value of promises made to small states ... In a word it is international morality which is at stake. God and History will remember your judgement.' The League however would do nothing more. A month later it ended sanctions against Italy. Mussolini had been allowed to triumph. He later admitted that oil sanctions would have forced him to retreat.

Thereafter the League was powerless and usually ignored. Japan and the major European powers were preparing again for war and had no use for it. In 1937 China appealed once more and without success against further Japanese attacks. The last act of the League was to expel Russia in 1939 for attacking Finland. It was a final despairing gesture since the Russian attack was one of the opening moves of the second World War, the prevention of which had been the chief reason for the founding of the League.

The League of Nations failed mainly because the major nations of the world were not prepared to make it work. Either they did not belong, like the U.S.A., or they became actively hostile to its ideals, like Japan, Germany and Italy, or they were never more than half-hearted about it, like Britain and France. In 1919 President Wilson made a prophecy which turned out to be dismally true. 'I can predict with absolute certainty', he said, 'that within another generation there will be another world war, if the nations of the world do not concert the methods by which to prevent it.'

The Versailles Settlement and the League of Nations

From Versailles to Locarno

The League dealt or tried to deal with many international disputes. However there were many events affecting European nations which did not become the League's concern.

The period 1919–23 was a period of deciding peace terms and of putting them into practice. Two areas gave immediate trouble which in each case ended in war. The Turks, now led by Kemal Ataturk, refused to agree to the Greek occupation of Eastern Thrace and Smyrna as laid down by the Treaty of Sèvres. Their victories in the war of 1921–22 forced the Allies to agree, by the Treaty of Lausanne (1923), to the return of these areas. The Poles also refused to accept the boundaries allowed them at Versailles. In April 1919 they attacked Russia with the aim of seizing the Ukraine. Though the Communist government was still fighting a civil war, Trotsky led an army against the Poles which nearly captured Warsaw before the Poles regained the upper hand. Eventually, in 1921, the Treaty of Riga was signed. Poland gained more territory than in 1919 but not the Ukraine.

The relations of the U.S.S.R. with the rest of Europe were very strained between the wars. The 1917 Revolution had brought to power the first Communist government the world had ever known. It called itself a workers' government. It had abolished private property. It was dedicated to the spread of working-class revolution the world over and to the destruction of capitalism. In 1919, the Communist International was set up in Moscow to co-ordinate the activity of Communist groups all over the world. Not surprisingly, other European states and the U.S.A. which were thoroughly capitalist in their economic organization, viewed the U.S.S.R. with concern. For its part, the Russian government was convinced that the capitalist West was plotting its downfall. This conviction stemmed to some extent from the fact that British, American, Czechoslovak and Japanese forces aided the anti-Communist forces during the civil war.

On both sides, the suspicions were exaggerated. The assistance given to the anti-Communist forces by the West was lukewarm and ineffective. On the other hand, despite the chaos in Germany and in the Habsburg Empire immediately after the war, Communism made very limited progress in Europe outside Russia. In fact the only non-Russian Communist government in Europe between the wars was Bela Kun's in Hungary which only lasted a few months in 1919–20. But the suspicions existed and friendly relations were impossible in the years immediately after the war.

In 1921, however, things began to improve. The West grew less anxious that Communism might spread like wildfire. Britain signed a trade agreement and the following year, when a war-weary Russia was ravaged by a terrible famine, the U.S.A. sent 750,000 tons of food and propped up the creaking Russian economy. Despite the anti-capitalist propaganda of the Commintern, Chicherin, the Russian Minister for Foreign Affairs, began to win the confidence of European diplomats. In 1922 he scored a great success. German and Russian representatives met at Rapallo in Northern Italy and agreed to exchange ambassadors, to trade with each other, and to abandon any claims each might make against the other because of the recent war (Germany therefore no longer needed to fear that Russia might make reparations claims). They also agreed in secret that German soldiers, in defiance of the Versailles Treaty, should train in Russia. In 1923, moreover, the British Labour party formed a government for the first time, and, though the Communists still refused to pay the debts owed to Britain from Tsarist days, gave diplomatic recognition to the U.S.S.R. Italy did the same. This thaw however turned out to be temporary. Lenin died in 1924 and as Stalin's power grew inside Russia Chicherin's influence declined. Before long Russia moved into isolation once more.

Another feature of the 1920's was France's search for greater security: she succeeded in building up a network of alliances, with Belgium (1920), Poland (1921), Czechoslovakia, Roumania and Yugoslavia (1924–27). She also took tough but not, in the long run, very successful military measures against Germany to ensure that the Versailles Treaty and reparations agreements were respected. French troops occupied Frankfurt and Darmstadt in 1920, the Saar during a coal strike in 1923 and the Ruhr from 1923 to 1925 (see page 116).

The years 1919 to 1925 were therefore years of crises and war. In contrast, the following four years, 1925 to 1929, seemed to be ones of co-operation and of hope. An important cause of the new optimism was a conference held at Locarno in Switzerland in 1925. Britain and France were represented, so too was Germany as well as Belgium, Italy, Poland and Czechoslovakia. The western frontier of Germany was accepted and guaranteed not only by Germany, France and Belgium (whose common border it was) but by Britain and Italy too. France promised to defend Poland and Czechoslovakia and Germany was at last allowed to join the League of Nations. What was so encouraging about Locarno was Germany's apparent readiness to accept much of the Versailles Settlement and to co-operate whole-heartedly in keeping Europe at peace. Mme. Taberin, a French journalist, described her feeling when the Locarno Pact was signed. 'I was literally drunk with joy. It seemed too good to be true that Germany, our enemy of yesterday, had actually signed the pact with its eight clauses of reconciliation! From now on, no more fears

for the future! No more war! ... I was not alone in my enthusiasm. Everyone in Locarno was jubilant.' But on her way back to Paris she called in on her uncle, a retired diplomat. He could not understand his niece's excitement. 'Can't you see', he asked her, 'that in spite of all those fools who are congratulating themselves on Locarno, nothing has been really altered? If our safety depends on that institution (the League of Nations) then we are indeed badly off.'

Unfortunately it was the pessimistic uncle not the optimistic niece who was the more far-sighted. The Locarno spirit was fragile and could not survive in the bleak climate of Europe after 1929. After the Wall Street Crash of that year, the worst economic depression in modern history spread from the U.S.A. to Europe. It helped dictators to flourish who believed not in reconciliation, collective security and peace but in revenge, national expansion and war.

In London soon after the Locarno conference and still pleased with their efforts, Baldwin, British Prime Minister, is on the right, Briand, French Prime Minister, is in the centre holding his top hat, Stresemann, the German Chancellor, is immediately above him slightly to the right, while Churchill, then Chancellor of the Exchequer, is at the back leaning on the railings

Chapter 4
The European Economy 1900-1939

Conditions in 1900

Nature has given freely to the continent of Europe. Though less than a thirtieth of the world's land surface, it contains one third of the acreage naturally suitable for cultivation. In 1900 it possessed rich mineral supplies, particularly of coal, then the most useful source of energy, and of iron, the most useful metal. It was also densely populated with inhabitants possessing both the energy and the education necessary for the exploitation of those natural advantages. Another advantage was Europe's geographical position. Much the cheapest form of transport was by sea. Europe has a long coastline and many excellent harbours. The Atlantic provides a link with the Americas and with West Africa, the Mediterranean and Suez Canal (completed in 1869) with the rest of Africa, Asia and Australasia. In 1900 the combined fleets of European nations were six times greater than those of the rest of the world (including the U.S.A.) put together. Consequently Europe was the centre of world trade and, after North America, the most economically advanced and wealthy of the world's continents.

Most of this wealth came from industry and many parts of the continent had been industrialized for more than a century. A wide strip of land, beginning in Scotland and Northern England and running south and east through London, Belgium, Holland and Northern France, into the Rhineland and then on into Northern Italy, contained a high proportion of this industry. There were other advanced industrial areas elsewhere, e.g. Sweden and Bohemia (then part of the Habsburg Empire, now western Czechoslovakia), and previously backward areas like parts of Russia were beginning to industrialize rapidly. Certain industries like iron and steel, ship-building and textiles, had been familiar for more than a century having been pioneered by Britain at the end of the 18th century, but newcomers like chemicals and electricity (with Germany the leader) and automobiles (with France and Germany the leaders) already rivalled or were soon to rival the older industries in size and importance.

In 1900, so many of the European population were working in industry that in order to feed them a considerable amount of food had to be imported from other continents. This trend continued in the 20th century and by 1939 the only foodstuffs in which Europe was self-sufficient were olive oil and wine. In the last thirty years of the 19th century improved methods of farming in places like North America, the Argentine, New Zealand and Australia coupled with improved methods of transporting food in refrigerated steamships meant that not only could foreign producers meet all Europe's needs but they could also undercut the prices of European farmers. Different European nations reacted to this challenge to their farmers in different ways. France and Germany put tariffs on foreign food, so protecting their farmers while preventing their citizens from enjoying cheaper food. Denmark turned to more specialized agriculture, producing bacon and diary products at competitive prices. Britain however preferred cheap food and refused to protect her farmers by tariffs. Unable to compete with overseas prices, many of them went out of business. Consequently there was a considerable variation in the proportion of the population employed in farming in 1900. In Britain it was only 8%, in Germany 35% and in France 45%, whereas in Eastern Europe, with industry less advanced, the proportion was much higher. In the Balkans and in Russia it was still in the region of 90%.

Farming methods also varied greatly. In Denmark and East Prussia farms were specialized, well-equipped and producing for export. In South-East Europe peasants, i.e. villagers having a small plot of land from the produce of which they maintained themselves and their family, used methods unchanged for centuries and were seldom far from starvation. Peasant farming was practised in many parts of Europe. In France 2½ million peasants farmed on average 2.5 acres, in Germany much the same number tended to own larger holdings. In Britain, however, the peasant class had long since disappeared. Farmers with an average holding of 66 acres usually employed agricultural labourers to whom they paid a wage. In southern and eastern Europe large estates owned by the nobility were also worked by landless wage-earners. Generally speaking, peasant farming was conservative and inefficient. The most productive farming in Europe was usually the most specialized.

The nations of Europe can be classified into three main groups according to the balance between their industry and agriculture in 1900. The first group, in which can be placed Britain, Belgium and Germany, were heavily industrialized, so much so that the economic fortunes of each was closely tied to the success of its manufacturing industries. The second group, which included France, Holland and Sweden, still had agriculture as an important part of their economy and, though they possessed extensive heavy industries, were less dependent upon them than were the first group. The third group, which was still the largest in 1900, included all those countries where agriculture was still dominant. The less industrialized the country, the poorer it tended to be.

The strongest economy in the world at the beginning of the twentieth century was not European at all but American. The U.S.A. was not only the strongest but it was forging ahead leaving its European competitors further and further behind. By 1913 the U.S.A. was producing 32% of the world's manufactured goods, nearly as much as Britain, Germany and France put together (36%). In certain key industries the advance of the U.S.A. was even more marked. Between 1880 and 1910 Europe's share of the world's steel production fell from 63% to 48%. In the same period that of the U.S.A. rose from 29% to 43%.

1900–1914

Nonetheless, the period 1900 to 1914 was a golden age for the European economy and one which later generations looked back to with nostalgia. World trade was expanding rapidly and manufacturing production within Europe expanded similarly. Every European nation enjoyed greater prosperity, though some more obviously than others. The pace-setter was Germany which had overtaken Britain to become the most powerful economic state in Europe. Between 1880 and 1910 her steel production multiplied twenty times. In 1900 she produced between 80% and 90% of the world's dyes. Her electrical and chemical industries were unrivalled.

In this golden age, the European economy was unusually un-affected by national rivalries, in comparison with earlier and later times. Though there were tariff barriers between European nations and between Europe and the U.S.A., these were much less disrupting to the world trade than those to be erected between 1919 and 1939. The economic system rested on certain beliefs and methods which encouraged economic co-operation. The European and American economies were capitalist, dominated by privately-owned businesses, whose chief responsibilities were to their individual shareholders, many of whom might well be foreigners. The French invested heavily in Russia, the Germans in South America, the British the world over. European businesses set up branches in every part of the globe. Huge volumes of goods—both in the form of raw materials and of manufactured products—poured across national frontiers inside Europe and between Europe and the rest of the world. Whether national governments liked it or not, national economies prospered or were depressed together. In other words, they were interdependent. Governments knew it and tended to act accordingly. Moreover, the system was further unified by the special role played by the City of London and of sterling, the British

currency. The City had unrivalled skill and experience in international finance and London was the financial centre of the world. As a result of the strength and stability of the British economy for most of the nineteenth century, sterling was trusted the world over to such an extent that it acted as an international trading currency. Never before or since has the international business community been able to function so easily. Keynes, looking back in 1920, neatly caught the character of the period in a few sentences. 'The inhabitant of London', he wrote, 'could order by telephone, sipping his morning tea in bed, the various products of the whole earth, in such quantity as he might see fit and reasonably expect delivery on his doorstep; he could at the same time and by the same means venture his wealth in the natural resources and new enterprises of any quarter of the world and share, without ... even trouble, in their prospective fruits and advantages ... Most important of all, he regarded this state of affairs as normal, certain and permanent.' It was not permanent at all. World War I destroyed it.

The cornerstone of British finance in its golden age, the Court of Directors of the Bank of England, 1903

World War I

The economic effects of World War I were enormous. During the war, about 65,000,000 troops were mobilized and 200 billion dollars spent on the war-effort. Moreover, once it became clear that neither side would win a quick victory, both sides settled down to waging economic as well as military warfare. The Entente powers aimed to strangle the German economy by naval blockade, Germany to starve Britain into surrender by her U-boats. Obviously interdependence was a thing of the past since the greater economic self-reliance each of the warring nations could achieve, the greater her chance of winning. Governments therefore took control of their national economies to an extent which would never have been tolerated in peacetime. Germany acted most promptly, spurred on by one of her leading industrialists, Walter Rathenau. Three days after England had declared war, Rathenau called on the Chief of the War Department. Both agreed that the war was likely to go on a long time and, as Rathenau put it, 'the supply of absolutely essential raw materials (for German industry) could probably last only a limited number of months.' A Raw Materials Department was soon established under Rathenau's direction with the power to take control of vital sections of German industry. Much of its energy was directed to increasing the production of raw materials obtainable in Germany and finding substitute (ersatz) materials for those cut off by the enemy blockade. Some sensational discoveries were made. For instance, the scientist Haber perfected the process of extracting nitrogen from air in order to provide a substitute for Chilean nitrates which were vital in the manufacture of explosives. Cellulose was invented as a substitute for cotton. As time passed, Rathenau's 'war economy' affected more and more of German life. By the National Service Law of 1916, all males between the ages of 17 and 60 came under the control of the Minister of War to be used as workers or soldiers, as the government directed. A special government department controlled prices and rationed food.

Britain acted less swifty, 'business as usual' being the slogan for 1914. Not until 1916, after a scandalous shortage of shells had been made public and compulsory military service introduced, did the government move to take control of the economy. With the setting up of five new departments, the British economy was almost as directly organized as the German by the time the war ended. Since Britain's greatest danger was defeat through starvation, agriculture was given every assistance. 'After long years of neglect its vital importance . . . has at last been recognized', an official report noted in 1917. British women made a vital contribution to the economic war effort. In the munitions industry, for example, 60% of the workforce was female.

If the advanced economies of Germany, Britain and France were able to bear the strain of total war by careful planning and sustained effort, less advanced economies were not. A Russian government report showed that in 1915 the iron production there fell by 21% and conscription and the commandeering of most available transport for the army had brought such chaos to industry that many iron-smelting plants had to close down completely. Russia's economic collapse was a major cause of the 1917 revolutions and of her readiness to seek peace while the economic difficulties of the Habsburg Empire sharpened the racial conflicts which eventually destroyed the Empire in 1918.

1918–1929

When the war ended, most politicians, economic experts and businessmen looked back fondly to the prosperous, certain, pre-war days. The sooner they could return to those good old days and revive the smooth-running prewar economic system the better. The trouble was that Europe and the world were very different in 1919. The pre-war economic system had gone for ever. The men who mattered failed to realize this or to discover new methods for the solving of the new and awful economic problems of the 1920's and 1930's. This failure was an important cause of the miserable economic performance of Europe in the next twenty years. Europe in 1919 was economically different from Europe in 1914 in a number of ways. First, the war had been destructive on an unprecedented scale. The area bordering the trenches of the Western Front from the Channel to the Vosges were devastated, hundreds of square miles of Central and Eastern Europe were, if anything, worse. Sir William Goode, leading a British Relief Mission in Central Europe in 1919, sent a vivid report of the chaos he saw. The transport situation was particularly serious. 'In countries where I found wagons,' he wrote, 'I found ... a shortage of locomotives; where there were locomotives, there was a shortage of wagons; where coal lay at the pithead ... there were no wagons and where wagons waited, men were not available to work the coal.' Such chaos meant hunger or death for millions. 'In many parts of Poland', Goode continued, 'children were dying for want of milk and adults were unable to obtain bread or fats. In eastern districts ... the population was living upon roots, grass, acorns and heather.' Even two years after the war ended, Europe's industrial production was only three-quarters of what it had been in 1913.

Secondly, the war had left such suspicion and hatred amongst the European nations that they were reluctant to revive that co-operative interdependence which was so basic a part of the pre-war economic system. The war had encouraged economic self-sufficiency which few nations were in the mood to abandon. In 1920, the report of an International Finance Conference called by the League of Nations to analyse the international economic difficulties since the war noted that 'every country finds impediments to its international trade in the new economic barriers which have been imposed during or after the war'. Nothing was done, however, to improve the situation. The U.S.A.'s economic strength had expanded greatly during the war and she had lent millions of dollars to her allies so that they could maintain their war-effort. However, instead of playing an active part in reviving international trade, the U.S.A. moved towards isolation. She insisted on the regular repayment of the war loans—as Calvin Coolidge put it, 'they hired the money, didn't they?'—which European nations in the 1920's could ill afford and protected her industries, which were not obviously in need of protection by tariffs, so encouraging others to take similar measures. Russia, now Communist, moved into even greater isolation. The Communist government regarded every capitalist country as its enemy and refused to pay the debts of the Tsarist government. In Eastern Europe the new nations, anxious to establish their national identity, hastily raised tariffs to protect their most important industries. The main consequence was to destroy the economic unity and much of the prosperity of the Danube valley. In Western Europe, France and Belgium were as determined to extract every penny of reparations from Germany as Germany was to pay as little as possible. Reparations were in fact typical of the economic short-sightedness which plagued Europe in this period. France and Belgium believed that they could prosper while Germany stayed poor. The truth was that no lasting prosperity would return to any part of the continent until the major manufacturing nations prospered together.

The third important difference from pre-war days was the weakened economic position of Britain and of sterling. During the war, Britain had to change from being a lender of money to being a borrower, especially from the U.S.A., and the stability and strength of sterling as an international currency declined. The extent of this change was not fully appreciated either in New York or London. The U.S.A. was not ready to shoulder nor Britain to give up the job of being the world's international banker. Consequently in times of crisis the City could no longer act so decisively, currencies were much less stable and the world economic system more fragile than in pre-war days.

In the two years after the war, Western Europe enjoyed boom conditions but they did not last. A depression set in which badly affected traditional heavy industries like coal-mining and ship-building. Britain suffered particularly. In Eastern Europe the end of the war was followed by such disorder that an economic boom was out of the question. Before long, the currency situation got out of hand. Goode reported from Poland in 1919: 'The following notes were in circulation—Russian roubles notes of various issues, German mark notes, Austrian kronen notes, Ukraine notes, Polish mark notes issued by the Germans . . . and Polish mark notes issued by the new Polish government. Exchange was immensely complicated, not only by the immense variety of notes current but also by the prejudices of the peasant population who would not accept rouble notes if they were damaged even in the slightest degree.' Many currencies were destroyed by runaway inflation. Prices in Austria multiplied by 14,000, in Hungary by 23,000, in Poland by 2,500,000 and in Russia by 400,000,000! The worst inflation however was in Germany following the French occupation of the Ruhr in 1923. In 1921, an American dollar could be exchanged for 75 German marks, at the beginning of 1923 for 7000, in July for 160,000, in August for one million, in November for 4 billion. German banknotes were hardly worth the paper they were printed on which meant, as the Minister of Economics put it in August, 'the thorough destruction of business, of employer and worker. Markets contract daily, orders hardly come in any more.' Only the creation of a temporary currency (the Rentenmark) and of a new currency (the Reichsmark) backed by an international loan ended the German inflationary crisis. In Eastern Europe, the League of Nations helped to bring stability by taking control of the finances of Austria and Hungary until 1926.

In 1925 the worst seemed over. Production at last began to move above 1913 levels and Europe, which had previously been losing ground to the rest of the world, began to hold its own. Between 1921 and 1924 world exports had risen by 25%, European by only 18%. Between 1924 and 1929 the world figure was 30%, the European 29%. The U.S.A. was more ready to lend money to Europe and short-term loans flowed across the Atlantic, especially to Germany. Britain responded to the new more confident mood by returning in 1925 to the gold standard (the rule that the Bank of England should be able to exchange gold for its banknotes at a fixed rate and a much respected characteristic of the pre-war financial system) which had been abandoned in the face of post-war difficulties in 1919. Moreover, the exchange rate of sterling was fixed at the good old pre-war rate of 4.87 dollars to the pound. Other countries followed Britain's example. 'I believe,' said Mr. Churchill, who as Chancellor of the Exchequer was responsible for the step, 'that it will facilitate the revival of international and inter-Imperial trade.' He was wrong. The rate against the dollar was too high. It immediately weakened Britain's already depressed economy and helped to bring about the General Strike of 1926. It also revived an international financial responsibility for Britain which she was not strong enough to bear, so increasing rather than lessening the fragility of the world economy.

The Wall Street Crash and the Great Slump

How fragile it really was became clear between 1929 and 1931. A stock market crash in the U.S.A. began the Great Slump which made the 1930's the period of the greatest economic depression since the industrial revolution began. The catastrophe—for it was nothing less when measured in terms of human suffering and of blighted hopes—began in the autumn of 1929 with what is usually known as the Wall Street Crash (the collapse of the New York stock market). By September 1929 over-confident American speculators had been able to borrow more than 7 billion dollars to finance their deals and pushed share prices far above realistic levels. In late October panic-selling began. On Thursday 24 October, nearly 13 million shares changed hands; on Tuesday, 29 October, more than 16 million. Prices plummeted. The result of this crash was not merely the ruin of individual speculators. More than 5,000 banks and finance houses also went out of business. The amount of money available for investment was cut back drastically, most companies had to reduce production and many went bankrupt. Unemployment rose and with it the purchasing power of the American public collapsed. At the depth of the slump, twelve million Americans were out of work and industrial production was barely half the 1929 level.

Trouble soon spread to Europe. Much of the 1925–29 prosperity had been built on American loans and between 1929 and 1932 many of the lenders demanded their money back. In 1931 Europe's banks began to fail, the first to go being Austria's largest and most respectable, the Kredit-Anstalt. Before long Germany, who had borrowed heavily from the U.S.A., was struggling and, by August 1931, Britain too. Never before had governments had to deal with such a crisis nor were they ready for it. Generally they reacted similarly, cutting government spending and the amount of money in circulation in the hope of stabilizing the situation. They also played the game of 'beggar my neighbour' as far as other countries were concerned, particularly after the failure of the World Economic Conference in 1933. Home industries were protected by new and higher tariffs. Such policies deepened the Depression and increased unemployment. In 1932–33, world unemployment, of which about half was European had risen to about 30 million. Between 1929 and 1935, world trade fell by 65%.

The following table gives comparative figures of industrial production for the U.S.A. and the major European states:

	U.S.A.	France	Germany	Italy	U.K.
1929	100	100	100	100	100
1932	52.7	72.2	53.3	66.9	83.5
1936	93.6	76.3	106.3	87.5	115.8

Of these European states, Germany suffered the most, her industrial production being virtually halved and unemployment rising to 6 million. Britain suffered least, with industrial production falling less than 20%. Britain, however, had prospered less in the late 1920's so the fall was a serious one. In 1932–33 more than 3 million were unemployed. For France, the slump came later less dramatically than to her more heavily industrialized neighbours, but it was real enough and her recovery slow. The country least harmed of all was Russia, isolated, self-sufficient and totally involved in her Five-Year Plans (see page 99). Eastern Europe was badly affected by the world-wide drop in the price of foodstuffs and other raw materials but the ill-effects though just as real were less noticeable than in the more industrial West.

The pace of recovery was uneven. The most rapid recovery was that of Nazi Germany. Between 1933 and 1937 Hitler's policies of public works (see page 141) and rearmament virtually eliminated unemployment and pushed industrial production over the 1929 figure. Britain too made a substantial recovery, though without any obvious help from her government. Most of the recovery was in the newer industries of the south, the older heavy industries of the north remaining depressed. Whatever recovery individual nations might make, however, world trade stagnated. In 1939, world exports were still at less than half the 1929 level and European exports were much the same.

The consequences of the Great Depression both for individuals and for civilized political behaviour were terrible. In old industrial areas like the English ship-building town of Jarrow, men found them-selves thrown out of work in their forties with virtually no hope of ever finding full employment again and as this contemporary des-cription shows, their families suffered dreadfully. 'A.K. hates enforced idleness and is continually trying for work. He had three children, a boy aged 16, getting 8 shillings a week sick pay, a boy of 14 chronically rheumatic and a child of eight. Mother is ailing and bloodless and gets 2/6d. a week extra sick allowance. The eldest boy went to work as an errand boy at 6 shillings a week. He became ill simply because he was not getting enough food to stick the long hours and carrying the weights required ... What is really wrong with all of them is the effect of a long period of semi-starvation.'

In Paris in the winter of 1932–33 a journalist reported how 'on street benches and at tube entrances, groups of exhausted and starving young men would be trying not to die. I don't know how many never came round. I can only say what I saw. In the Rue Madame one day, I saw a child drop a sweet which someone trod on, then the men behind bent down, picked it up, wiped it and ate it.' Such misery could not but have political results. In 1932 an American journalist pointed out how in Germany, young people were des-

One of the 'hunger-marches' on London during the Depression Years. This one was in 1932, from Scotland

Above *Human talent squandered—one of Britain's millions of unemployed in the 1930's.* Right *French hunger-marchers in 1933. 'Bread and Work' is the demand of many of the banners*

pairingly turning to extreme political parties: 'With 60% of each new university graduating class out of work, with over half of all Germans between 16 and 30 unemployed, with a dole system that favoured the elder jobless at the expense of ... youth, young Germany was an easy victim of the political demagogue.' This was the situation in which the Nazis were able to make their successful bid for power (see pages 133–134).

The Great Slump was a major event in modern history. Both its severity and length made clear that there was something radically wrong with the international economic system and that the old methods of businessmen and governments were no longer adequate. If Nazi-type preparations for war were the only solution to modern economic problems, then the outlook for Europe was grim. New ideas and new policies were desperately needed.

★ ★ ★

A Note on J. M. Keynes
John Maynard Keynes was the outstanding economist of the first half of the twentieth century. He was born in Cambridge in 1883 where his father was a lecturer in economics and logic. Both at Eton and at Cambridge he displayed an intelligence which was widely regarded as exceptional and having gained a First in Mathematics, became increasingly interested in economics. For a time he taught economics at Cambridge but during World War I moved to the Treasury where he quickly became a key member of the planning team supervising the war economy. There he won a reputation for skill and for being no respecter of persons. On one famous occasion Lloyd George, then Prime Minister, asked for comments on a statement he had made about the situation in France. 'With the utmost respect' said Keynes, 'I must, if asked my opinion, tell you that I regard your account as rubbish.' He was with Lloyd George at the Versailles negotiations, resigned and wrote *The Economic Consequences of the Peace* (see Chapter 3) upsetting the British government but making his international reputation. In the 1920's he wrote and taught a great deal and campaigned to prevent Britain returning to the gold standard. When Churchill took the decision to return, he produced *The Economic Consequences of Mr. Chuchill* which predicted accurately its disastrous outcome. During the 1930's he was on a number of economic advisory committees. The government of the time was seldom prepared to take his advice since it was usually for greater government spending, particularly on public works to reduce unemployment, the opposite to what the government believed in doing. A heart attack in 1937 made him a semi-invalid but he was active during World War II, being Britain's chief financial negotiator with the U.S.A. first over lend-lease, then at the Bretton

Woods Conference which set up the International Monetary Fund and the World Bank, and finally over the U.S. loan to replace lend-lease. He died in 1946.

His most important book was *The General Theory of Employment, Interest and Money*, first published in 1936. The year before he had written to the playwright G. B. Shaw, 'I believe myself to be writing a book on economic theory which will largely revolutionize . . . the way the world thinks about economic problems.' It did.

In *The General Theory*, Keynes investigated the analyses by earlier economists of the causes and nature of unemployment. He found them inadequate and put forward his own explanation, which—and this was what was revolutionary about it—demanded a quite different approach by governments if unemployment was to be cured. *The General Theory* placed the economic events of the previous two decades into a completely new light. Past economists had argued that full employment was the normal state of affairs. Large-scale unemployment was a temporary situation which would be put right by the self-adjusting mechanism of the economic system. During depressions which bring temporary unemployment wise governments should sit tight and balance their budgets. Keynes argued on the contrary that full employment was not necessarily normal. It was perfectly possible for the economic system to adjust itself to a level where large-scale unemployment continued year after year as in Britain in the 1920's and 1930's. The level of unemployment, he argued, was largely determined by the levels of investment and of consumption, both of which could be changed by government policy. In times of economic crisis, the last thing a government should do is sit tight and balance the budget. It should actively interfere and head off the threatening depression by appropriate action, often by unbalancing the budget and overspending.

Virtually all the governments of capitalist industrial states are Keynesian now. They interfere continuously and profoundly in economic matters, changing interest rates, personal taxes, investment grants and so on to maintain high employment and encourage growth. Some governments were moving in this direction before World War II without reading Keynes—notably Roosevelt with his 'New Deal' in the U.S.A.—but *The General Theory* gave to postwar governments the confidence to move much more decisively in the right direction. There are many reasons why the 1950's and 1960's were years of increasing prosperity and high employment. *The General Theory* of Keynes is one of them.

Chapter 5
Russia: From Revolution to World War II

Recent Russian history is particularly hard to write. In November 1917 the Bolshevik (Communist) Party seized power in Russia and today Communist Russia (the U.S.S.R.) is one of the two super-powers of the world, of enormous military strength. Communist society is not only different from the capitalist society which existed and exists to this day in Western Europe and the U.S.A. It is hostile to it. Communists believe that capitalist society is basically evil and will eventually be destroyed by revolution. The duty of a good Communist is to hurry this revolution along. Consequently, for most of the 20th century, Western Europeans and Americans have generally regarded Russia as a most dangerous enemy while most Russians have been convinced that there has been a capitalist conspiracy against her since 1917. The result is that Russian history written by anti-Communists tends to minimize the Communist achievements while that written by Communist sympathizers tends to exaggerate it. There are two other serious difficulties. Karl Marx, the founder of Communism, taught that there was a pattern in history, the central feature of which was the struggle between the classes. In modern capitalist society, the middle-classes (whom he called the bourgeoisie) had both economic and political power which they used to exploit and grind into the dust the industrial workers (or proletariat). By Marx's set of values, bourgeois power was evil, the working-class revolution against it good, so that any action or event which benefited the working classes should be praised and any which benefited the bourgeoisie criticized. Communist historians have accepted both Marx's pattern and his set of values. Most British and European historians, however, do not fully accept either his pattern or his ideas of class conflict. Events of the past appear to them more haphazard and the achievements of groups or individuals less open to praise or blame. Consequently a Russian history of 20th century Europe would read very differently from this one. Another difficulty has been the huge exaggerations and falsifications of the 'official' Russian history while Stalin was alive, one characteristic of which was to make both Lenin and Stalin out to be persons of superhuman qualities who never made mistakes. History became propaganda and the disentangling of the truth—which was often nightmarish—from the propaganda myths was (and often still is) very difficult.

78

The Effects of World War I

War brought revolution to Russia. The desperate struggle against the Central Powers caused strains which neither the economy nor the civilian population nor the army for all its courage and exceptional powers of endurance could bear any longer. Serious economic problems appeared soon after war was declared. The mobilization of fifteen million men brought vital factories to a standstill. In 1915, 573 factories stopped work and in contrast to the other warring European nations whose economic production was rising rapidly, Russia's began to fall. The situation was made worse by chaotic transport conditions. The Trans-Siberian railway was single track with so limited a supply of rolling-stock that though the Allies had got plenty of supplies to the Pacific port of Vladivostok by 1917, it would have taken another year to transport them across Russia to the front. At the northern port of Archangel, things were equally bad. There, vital supply crates were lost because so many other crates were piled on top of them that they became embedded in the ground. Food rotted in railway stations as the railway system disintegrated into chaos. A former minister later remembered how 'there were so many trucks blocking the lines that we had to tip some of them down the embankment to move the ones that arrived later'. Simultaneously there was rapid inflation with prices speeding ahead of wages. Between 1914 and 1916 while wages doubled, the price of food and many other goods increased four- or five-fold. Cities began to starve as early as 1915 and major strikes began further to disrupt the economy.

At the front the Russian army was led and organized with the same terrible inefficiency and paid for it in blood. 'We go about in ragged uniforms,' a soldier reported, 'and without boots I have to go practically barefoot, just in my socks'. More seriously, their weapons were out of date and they seldom had enough ammunition. While they could for the most part outfight the Austrians, they were really no match for the Germans. In 1914 and 1915 their casualties were 3,400,000, and 1,500,000 were taken prisoner. The Brusilov offensive of 1916 was the last successful campaign of the Imperial Russian army. Once it was back on the defensive, its morale began to crack and the news of the suffering and discontent at home and the numerous strikes and demands for a new government weakened its readiness to continue fighting.

Ruler of this despairing country was Tsar Nicholas II of whom it has been said that he could not have run a post-office let alone Russia in the greatest crisis of her recent history. Nicholas was not stupid but he lacked self-confidence in his abilities as a ruler. He was no judge of the right men for political responsibility and relied too much on his strong-willed but foolish wife Alexandra. She had a

The Romanov tragedy

The Tsar and his family before the revolution

After his abdication and under house-arrest at Tobolsk, Nicholas was photographed with his son and four daughters, enjoying what they can of the low Siberian sun

The room in Ekaterinburg where, almost certainly, Nicholas, with all his family, was murdered in 1918

simple answer to Russia's problems—unquestioning obedience to the Tsar's orders. 'How they (the Russian people) need to feel an iron will and hand,' she wrote to her husband in 1915. 'You are the lord and master in Russia and God Almighty placed you there and they shall bow down before your wisdom and firmness.' She had no time for the Duma, the elected assembly set up after the 1905 Revolution to advise the Tsar. 'I have heard', she wrote again in 1915, 'that that horrid Rodzianko (President of the Duma) and others . . . beg that the Duma be at once called together. Oh please don't do it, it's not their business, they want to discuss things not concerning them and bring more discontent.' In 1915 Nicholas had gone to the front as commander-in-chief to play an active part in the war. It was a bad decision. For one thing, Nicholas was no general and his presence hindered rather than helped the war effort. For another, the political influence of the Tsarina grew even stronger in his absence. From this time onwards government policy can only be described as lunatic and, in the circumstances, suicidal.

Rasputin with two lady admirers, 1914

Since 1911 an extraordinary individual, Rasputin, had been an important figure at the Russian court. He was an illiterate wandering Siberian peasant who described himself as a priest though he had no connexion with any Russian church or sect. He possessed, however, genuine hypnotic skills with the help of which he could control better than the most distinguished doctors the haemophiliac condition of Prince Alexei, the heir to the Russian throne. This condition was such that any cut or bruising was liable to endanger his life. The Tsarina was convinced that his powers were miraculous, that he was a man of God and his opinions on all matters sacred. He became indispensable to her and a most powerful political figure. He was, however, a rogue, a thief and a drunkard, with sexual habits which would have been outrageous even if he had been an ordinary man with no claims to being a man of God. His political abilities were non-existent yet his influence led to the appointment of incompetent ministers and removed the royal family even further from reality. As Prince Yusupov, a young aristocrat, put it: 'The Emperor and Empress, cut off from the world, isolated from their subjects and surrounded by Rasputin's clique, decided matters of world shaking importance. One can only feel dread for Russia's fate.' On 16 December 1916 Yusupov and two accomplices murdered Rasputin. Wines and cake laced with arsenic were not enough to kill him nor were shots from Yusupov's revolver. Eventually they had to club him to death and push his body through a hole in the ice on the River Neva. His murder came too late to make much difference. The royal family had by now exhausted the loyalty of too many of the Russian people.

The Revolutions of 1917

The sequence of events in Russia in 1917 can be confusing if one does not allow for the fact that Russia used the Julian calendar and was almost a fortnight behind the rest of Europe which used the Gregorian version. The dates used here are the Russian unless in brackets.

In January 1917, the leaders of the Duma warned the Tsar that unless major reforms were promised immediately, there might be serious trouble. Nicholas took no notice. On 23 February by the Russian calendar (8 March in the rest of Europe) major strikes began in Petrograd (St Petersburg). 'I order that the disorders in the capital shall be ended tomorrow', Nicholas cabled from the Army H.Q. at Mogilov. The attempts of the Petrograd authorities to enforce this order were half-hearted. Soldiers began to fraternize with the strikers. Some shot their officers. While the Duma set up a Provisional Government, strikers and sympathetic soldiers formed the Petrograd Soviet or Council which demanded a Constituent Assembly for Russia 'elected on the basis of universal, equal, secret and direct suffrage'. On 15 March Nicholas II abdicated in favour of his brother, the Grand Duke Michael, who, however, refused to take his place. So ended the Romanov dynasty, which had ruled Russia for more than 300 years, and with it the Russian monarchy. It died with barely a whimper.

The February/March Revolution happened so suddenly that no political group had a clear plan of what to do next or by what kind of government the Tsar should be replaced. Eventually the Provisional Government led first by the indecisive Prince Lvov, a liberal, and later by the more positive Kerensky, a socialist, was widely recognized as the new government both inside and outside Russia. Alongside it and soon hostile to it, however, were the Soviets. These local democratically elected assemblies of workers and soldiers were closer to ordinary people than the Provisional Government. Particularly important was the Petrograd Soviet. It, not the Provisional Government, controlled the troops of the capital.

Lenin and the October Revolution

On 16 April Lenin arrived at the Finland Station of Petrograd, back after a ten year exile. His return transformed Russian politics almost at once and his successes in 1917 and the years that followed have transformed the world. He was born Vladimir Ulyanov in 1870. ('Lenin' was the pseudonym he first used when he became involved in secret revolutionary work.) He was the younger son of a comparatively prosperous inspector of schools and had an excellent education. He was strikingly intelligent. 'Very gifted, neat and industrious', read his final school report. 'Ulyanov was first in all

Lenin, easily recognizable because of his baldness, seated with a group of Communists in St Petersburg. This photo was taken in 1897

subjects and, upon completing his studies, received the gold medal as the most deserving pupil in ability, progress and conduct. Neither in the school nor outside it has a single instance been observed when he has given cause for dissatisfaction by word or deed to the school authorities and teachers.' In 1886 his father died and, the year after, far worse tragedy followed. His elder brother Alexander, to whom he was very close, became involved in a plot to assassinate the Tsar, was betrayed by one of the plotters, arrested and executed. Lenin seldom spoke of his brother's death but it must have affected him deeply. The rest of his life was dedicated to the destruction of Tsarism. From school he went to Kazan University which before long expelled him for organizing a student demonstration. He then became involved in revolutionary politics and had many brushes with the police. He spent some years in a Siberian exile and from 1900 to 1917 was driven to spend most of his time in Western Europe, first in London, then in Switzerland. An important influence in his life was Plekhanov who introduced him to Marxism.

83

The most important ideas of Karl Marx which Lenin shared are these. History is the story of conflict between different classes. Before the French Revolution of 1789, the aristocracy or upper classes aided by and aiding their kings and emperors, held most economic and political power. Through the French Revolution and changes associated with it, the bourgeoisie (or middle classes) over- threw aristocratic society and won for themselves both economic and political power. The bourgeoisie created the modern industrial state and controlled the 'means of production' within it—the factories, the banks, the railways and so on. It possessed the wealth (or capital) of this society and organized 'capitalist society' to increase its wealth. In order to set up a business or expand an existing one, an individual would need to borrow money. This money would be loaned in the form of shares in the new company bought by men with money to spare, on condition it was paid back with interest out of the profits that the new business eventually made in proportion to the amount of shares held. The bigger the profits a business made, the better the shareholders would do. There was, however, another immense class without which modern industrial society could not function, the proletariat or industrial working class. It owned no property, it did not even enjoy partial control of the means of production. The money it had came in the form of wages. But in any factory, the higher the wages paid to the workers, the lower the profits made by the company and the smaller the interest payments to the share- holders. Consequently the economic interests of the property-owning, shareholding capitalist bourgeoisie were in sharp conflict with those of the wage-earning proletariat. The bourgeoisie therefore used its economic and political power to keep wages as low as possible and to stifle attempts by the working class through trades unions or other means to improve their political and economic position. Eventually, Marx predicted, the huge army of exploited and downtrodden workers would rise in violent revolution and destroy capitalist society. After a short period of 'proletarian dictatorship' a genuine communist society would develop. All means of production would pass from private to public ownership to be used for the good of all not just the good of some. Such common ownership would allow a society of brotherhood and co-operation to grow, of genuine freedom where the weak were cared for, not trampled on, by the strong and where the principle 'from each according to his ability, to each according to his need' would become a reality.

Though they both believed that the revolution was bound to come, neither Marx nor Lenin were prepared to wait for history to take its course. 'Philosophers', wrote Marx in 1848, 'have only interpreted the world in various ways. The point, however, is to change it.' And Marx, most impractically and unsuccessfully, and Lenin, most practically and successfully, set about changing the world from Capitalism to Communism.

Russia, however, was the last place in Europe where Marx expected a successful Communist revolution to take place. Compared with Britain, France, Belgium and Germany, it was industrially backward, its bourgeoisie had yet to overthrow the aristocracy, and its proletariat (the urban industrial working-class) was small. It was still a country of peasants and Marx regarded peasants as anti-revolutionary since they usually owned and were devoted to their own plot of land. Perhaps the most important addition made by Lenin to Marx's teaching was his belief that the poverty-stricken peasantry of Russia had so much in common with the proletariat that it could be regarded as a powerful revolutionary force.

Together Lenin and Plekhanov took over the leadership of the exiled Social Democratic party and planned how a Marxist revolution might be successfully brought about in Russia. They produced an underground newspaper *Iskra* (the Spark) which they had printed in Germany and distributed secretly inside Russia. But the influence of this Social Democratic Party on events in Russia was never very great before 1917. An important cause of weakness was a bitter split in its ranks in 1903. One group, a minority (Mensheviks) led by Martov wished to broaden the membership of the party to include people who sympathized with its ideas but were not active revolutionaries. The majority (Bolsheviks) led by Lenin disagreed, arguing that the party could never achieve success unless it remained a tightly-knit band of professional revolutionaries. The disagreement was not a friendly one and its wounds were never healed. Partly as a result of it, Bolsheviks played no significant part in the Revolution of 1905 nor in the March Revolution of 1917. Early in 1917, their numbers in Russia were estimated at only 30,000.

Thus at the beginning of 1917 Lenin was a nobody—the exiled leader of a tiny extremist Russian political party of no obvious significance in either European or Russian affairs. Even he had little hope of achieving much in his lifetime. In January 1917 he gave a talk to some students in Zürich. 'We older men', he said, 'will not live to see the international Socialist revolution. But you youngsters, you will see it.' Once the news of the March revolution reached him, he was desperate to get back to Petrograd. Railway transport was provided by the Germans who argued, accurately as things turned out, that this troublemaker might cause so many difficulties inside Russia that the Provisional Government would give up its struggle to continue the war against them.

When Lenin arrived in Petrograd, the small Bolshevik party was considering to what extent it should co-operate with the Provisional

Government and other political parties. Lenin's view was simple and uncompromising. There should be no co-operation of any kind. The Bolshevik aim should be to 'transfer all power to the Soviets', build up Bolshevik strength in the Soviets and from there organize another revolution to establish the dictatorship of the proletariat. In April 1917 this plan seemed ridiculous. Many people, including some Bolsheviks, decided that Lenin had been out of Russia so long that he did not really understand the situation. Such, however, was his personality and powers of argument (someone who once had to argue with Lenin came away saying he felt as if he had been hit over the head with a flail) that he was able to convert the Bolshevik Party to this strategy.

Meanwhile the Provisional Government struggled to assert itself but it faced huge difficulties. In parts of Russia there was anarchy. In rural areas peasants were rioting and looting. Soldiers, delayed at a railway station for half an hour, beat up and killed the station-master. The war was going from bad to worse and the desperate economic situation showed no sign of improving. The Provisional Government now dominated by Kerensky tried to keep the war going and refused to carry out any major reforms until the Constituent Assembly met later in the year. These were fatal errors. The economic crisis deepened. The Bolsheviks increased their strength in the Soviets and undermined army discipline by continuous propaganda. When an attempt was made at the end of June to launch another offensive against the Germans, the entire 11th Army deserted. The Bolsheviks thought that the time had come for revolution but their July coup d'état failed and Lenin had to move to a temporary exile in Finland. In August, another coup d'état was tried, this time by a General Kornilov—'a man with the heart of a lion and the brains of a lamb' was how General Brusilov described him. Kornilov advanced on Petrograd with the aim of destroying the power of the Soviets and perhaps of overthrowing Kerensky but some of his troops refused to obey him, his train was switched off the main line by pro-Soviet workers and his attempt failed. However the Kornilov conspiracy seriously damaged Kerensky's government which appeared paralysed without any clear sense of direction. In contrast, the Bolshevik policy was short, simple and appealing. 'End the War: Land to the Peasants: All Power to the Soviets' was their slogan. Despite their failure in July, their popularity grew. 'Our Provisional Government attacks the Bolsheviks a great deal,' a Russian soldier commented in August, 'but we front-line soldiers don't find any fault in them. Earlier we were against the Bolsheviks, but now after the Provisional Government has promised to give freedom to the poor people but hasn't given it, we are little by little passing to the side of the Bolsheviks.' In October, food grew even scarcer, especially in Moscow

Above *Posing fiercely for the photographer, these Red Guards and their vehicle took part in the October Revolution*

and Petrograd. John Reed, an American in Petrograd, noted that 'Towards the end of October, there was a week without any bread at all . . . There was enough milk for half the babies in the city. For milk and bread and sugar and tobacco, one had to stand in a queue for hours in the chill rain'.

In October Lenin slipped back to Petrograd. At a secret meeting he convinced the Bolshevik leadership that the time was ripe for a second attempt to seize power. By now the Party numbered at least 300,000 and was particularly strong in the two vital cities of Moscow and Petrograd. Well-organized by the quiet but thorough Stalin, it was ready for action. Public opinion was swinging in its favour. Its influence in Soviets was strong. The important Petrograd Soviet, whose President was the dynamic Trotsky, Lenin's deputy, was under its control. Perhaps most important of all, it could count on armed support from soldiers, sailors and units of factory workers.

On 23 October (5 November), Trotsky's Revolutionary Committee quietly occupied the Smolny Institute and made it its H.Q. Formerly it had been a famous girls' school and occupied a strategic position in Petrograd. The following day Kerensky began counter-measures, sending troops to occupy the newspaper office of the Bolsheviks and closing down their printing press. The day of decision was 25 October (7 November). Early in the morning the Bolshevik 'Red Guards' took to the streets and began occupying the strong points of the city. Generally they met with little resistance. Only the huge Winter Palace held out. There a group of young officers and a women's battalion defended members of the government while Kerensky slipped out of the city to find help from loyal troops. On

26 October (8 November), the Winter Palace was bombarded by the cruiser 'Aurora' from the River Neva and as the Red Guards prepared to storm it, the government ministers ordered what few defenders were left to end the hopeless struggle. The conquest of Petrograd had been virtually bloodless. Six Bolsheviks had died in the attack on the Winter Palace but none of the defenders. Kerensky, who had only been able to find 700 troops to support him, tried but failed to win back the city on 29 October (11 November). He went into exile, never again to return to Russia. The struggle for Moscow was more bitter. It lasted five days but here too the Bolsheviks were successful. The twin capitals of Russia were in Communist hands. The October Revolution was over.

The Bolsheviks, though still a minority, had triumphed because in a situation of total chaos and widespread suffering, they alone of the Russian political parties had a clear policy which appealed to the most discontented and numerically the largest sections of the Russian people—workers, peasants and soldiers. In Lenin they possessed a leader of extraordinary determination, personality and sense of what was possible and what not; in Trotsky a man of action of great energy, intelligence and eloquence and in Stalin an unrivalled party organizer. Moreover, most Bolsheviks shared a crusading zeal to end past miseries and build a radically better future and a burning conviction that they alone were right. Since they were also disciplined and ready to be quite ruthless with anyone who disagreed with them, they were very hard to stop.

They needed to show all these characteristics in the terrible years that followed. They immediately faced many serious problems. Somehow peace had to be made with Germany. A civil war was brewing. The economy had collapsed and the Bolsheviks were still a minority party with only limited support outside the capitals. They first concentrated on tightening their hold on power. The day after the capture of the Winter Palace, a Congress of Soviets met in Petrograd. Lenin was given a hero's welcome and the formation of a Bolshevik government approved. Those few who disapproved got the sharp edge of Trotsky's tongue. 'You are miserable, isolated individuals', he declared. 'You are bankrupt. You have played out your role. Go where you belong—to the dustbin of History.' The new government, however, found itself in an embarrassing position when, at long last, the Constituent Assembly, planned since the March Revolution, met in January 1918. In this assembly, the most democratically elected in Russian history, the Bolsheviks won only 175 out of 707 seats. Their main rivals, the Social Revolutionaries, gained 370, an absolute majority. Lenin did not hesitate. When it refused to approve some of his measures, the Bolsheviks walked out and troops closed it down the next day. It never met again.

In scenes very similar to the early days of the 1789 Revolution in France, thousands of demonstrators assemble in the Field of Mars in Petrograd during the first revolution of 1917

Right *The cruiser 'Aurora' moored in the Neva. (This photo was taken in 1918)*

Then they had to make peace with Germany. The Russian army no longer existed as an effective fighting force. The Germans knew it and demanded the most humiliating peace terms. Trotsky, leading the Russian negotiators, began by refusing to consider them but Lenin overruled him. Lenin realized that without an immediate end to the war his Bolshevik government was likely to collapse as the Provisional Government had done. So in March, in Poland, the Treaty of Brest-Litovsk was signed. Russia had to surrender to Germany 26% of her population, 27% of her arable land, 26% of her railway system, 33% of her manufacturing industry, 73% of her iron industry and 75% of her coalfields. Peace was secured but at an enormous price.

Map 5 The Russian Civil

The Civil War

Hardly was the foreign war over when civil war began. The first fighting began in Central Russia around the Czechoslovak Legion. This was a force of ex-prisoners of war of the Habsburg Empire who had declared themselves ready to fight to free their homeland from Habsburg rule and were being moved along the Trans-Siberian railway to Vladivostok where they were to sail to join the Allied forces on the Western Front. The Czechs were a well-disciplined fighting unit and they were suspicious of the Bolsheviks especially since they had made peace with the Germans. An incident on their journey started war and before long they controlled a large area of the Volga region. Their success encouraged the many other enemies of the Bolsheviks to come into the open.

The anti-communists are usually known as the Whites, the Communists as the Reds. For much of 1918 and 1919, the Red position seemed hopeless. 75% of Russia was in White hands and they had to fight along a front which was sometimes 6,000 miles long. Their army numbered only 300,000, about half the number at the disposal of the White commanders, who, moreover, received French, British, American and Japanese aid. But as the fighting continued, the Reds began to gain the upper hand. Their position was stronger than it looked. They fought from a compact central area which included the two capitals of Russia. Their supply and strategic problems were therefore much less than the Whites'. They were also united, fighting for the survival of themselves and their cause. In contrast the Whites were united only by their hatred of the Reds. In the west, the Ukrainians, in the south General Denikin, in Siberia Admiral Kolchak, each had their own ambitions. They offered no clear political alternative to Bolshevism and their rivalries were a serious weakness. In November 1919 a White army reached the suburbs of Petrograd. It was the closest the Whites ever came to complete victory. Under Trotsky's command, the Red Army grew in size and improved rapidly as a fighting force. Kolchak was decisively defeated.

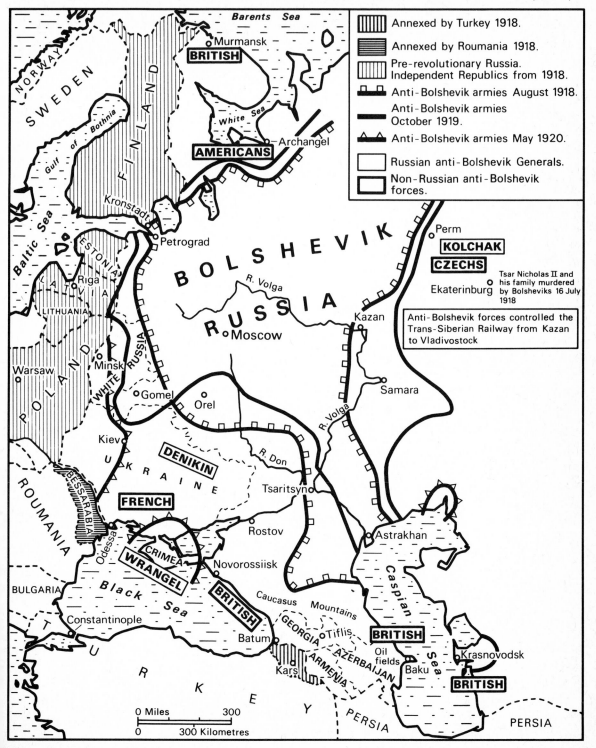

Legend:
- Annexed by Turkey 1918.
- Annexed by Roumania 1918.
- Pre-revolutionary Russia. Independent Republics from 1918.
- Anti-Bolshevik armies August 1918.
- Anti-Bolshevik armies October 1919.
- Anti-Bolshevik armies May 1920.
- Russian anti-Bolshevik Generals.
- Non-Russian anti-Bolshevik forces.

Barents Sea

NORWAY

SWEDEN

Murmansk

BRITISH

White Sea

Archangel

AMERICANS

Gulf of Bothnia

FINLAND

Baltic Sea

Kronstadt

Petrograd

ESTONIA

LATVIA

Riga

LITHUANIA

B O L S H E V I K

R. Volga

Perm

KOLCHAK

CZECHS

Ekaterinburg

Tsar Nicholas II and his family murdered by Bolsheviks 16 July 1918

R U S S I A

Kazan

Moscow

Anti-Bolshevik forces controlled the Trans-Siberian Railway from Kazan to Vladivostock

Warsaw

POLAND

Minsk

WHITE RUSSIA

Gomel

Orel

R. Volga

Samara

Kiev

U K R A I N E

R. Don

DENIKIN

FRENCH

BESSARABIA

Tsaritsyn

ROUMANIA

Odessa

Rostov

Astrakhan

WRANGEL

CRIMEA

Novorossiisk

BRITISH

Caspian Sea

BULGARIA

Black Sea

Caucasus Mountains

Constantinople

T U R K E Y

Batum

GEORGIA

Tiflis

BRITISH

Kars

ARMENIA

AZERBAIJAN

Oil fields

Baku

Krasnovodsk

BRITISH

PERSIA

PERSIA

0 Miles 300

0 300 Kilometres

The Czechs who had been fighting with him handed him over to the Reds and made their peace. Kolchak was shot. In the south Denikin was sacked by his former supporters and his successor Wrangel was defeated in the Crimea and the White forces there destroyed. Once the war with Poland which had begun in 1920 was brought to an end by the Treaty of Riga in 1921, White resistance ended. The Polish War was fought to decide where the frontier between Poland, revived as a separate nation by the Versailles Settlement, and Communist Russia, unrepresented at Versailles, should be. The Poles took the opportunity of the civil war to launch an attack on Russia in order to fix the frontier as far to the east as possible. At first they were victorious, advancing as far as Kiev, but a mighty Red counter-attack drove them back almost to Warsaw. Reorganized and supplied by the French, the Poles attacked once again and forced the Russians at Riga to agree to a peace which gave a considerable part of the Ukraine and White Russia to Poland.

No government could have succeeded in such dangerous times without ruthlessness but, between 1918 and 1921, the Bolsheviks turned savage. Like the French revolutionaries who in 1793 faced both foreign and civil war, they made terror a part of their policy. Instead of the guillotine, they used the firing-squad. The institution that spread the terror was the 'All-Russian Extraordinary Commission to fight Counter-Revolution' usually known from its Russian initials as CHEKA. The chief of Cheka was Dzerzhinsky, who, having spent fifteen years of his life in the hands of the Tsarist secret police, knew his business. The first victims were other political groups like the Anarchists and Social Revolutionaries who had played so important a part in the 1917 revolutions and still enjoyed popular support. In April 1918 600 anarchists were arrested of whom 30 were killed 'while resisting arrest'. In the summer of 1918 an attempted coup d'état in Moscow by Social Revolutionaries led to 13 of their leaders being shot. An assassination attempt on Lenin in August 1918 caused the Cheka to step up its activities. Not only did the Bolsheviks seem to be turning Russia into a one party state but within this one party the Cheka seemed to be developing an ever more secret and sinister influence. To many former sympathizers it looked as if the Bolsheviks 'drunk with power were starting bloody despotism all over again'.

Feelings of this kind helped to cause the most serious threat the young Bolshevik government had to face in this period, the Kronstadt rebellion of 1921. The naval base at Kronstadt had once been the favourite of the Bolshevik party. Trotsky had described the sailors as 'the pride and glory of the Revolution'. It had provided them with vital support in the October Revolution and the sailors and workers of the town were among the most politically conscious in Russia.

By 1921 they were very disillusioned with the way things were turning out. In the early spring the sailors, convinced that the Kronstadt Soviet had become, thanks to rigged elections, a mere mouthpiece of Bolshevik propaganda, demanded that new elections by secret ballot should be held throughout the country to make sure that the Soviets really did represent workers' feelings. They also demanded more freedom of the press and of trades-union activities. In fact they were demanding that the government should return to the ideas of the October Revolution and represent the interests of all workers, not simply those who agreed whole-heartedly with Communist ideas, and they were implying that the Bolsheviks were both unrepresentative and over-powerful. They began organizing their own elections to the Soviet and arrested local Bolshevik officials. The Bolshevik reply was swift and brutal. Acting on Trotsky's instructions, Tukhachevsky, the most successful Red general of the Civil War, led his troops against Kronstadt. After a day's fighting, the battleships were captured and the rebellion ended. The government proceeded to misrepresent the rising as a bourgeois or White conspiracy against the working-people of Russia—which was almost the opposite of the truth. Those sailors whose family links most easily allowed them to be framed as bourgeois or White sympathizers were publicly proclaimed as ringleaders and shot at once. The real leaders were imprisoned and shot in batches—without publicity—in the following months.

Bolshevik Economic Policies

The combination of foreign wars, civil war, and continuous political chaos since 1916 meant suffering for the ordinary people of Russia on a scale unparalleled elsewhere in Europe. Economic production had been falling before 1917 but after 1917 it dropped much more rapidly. In 1917 pig-iron production was 98% of the 1913 figure, in 1920 it was 18%. Coal production in 1917 was 72% of the 1913 figure, in 1920 it was 4%. Where 17,000 trains had been in service in 1917, only 4,000 were still in working order in 1920. Food production was falling too. The result was economic disaster, famine, disease, starvation and death. Something in the region of 20 million people lost their lives in Russia during these years.

The economic policy which Lenin first tried he called 'War Communism'. All major industries and banks were nationalized (passed from private to public ownership) and food supplies were strictly rationed. The peasants were ordered to bring any spare food they had to market and the hoarding of food—a customary practice in hard times—was strictly forbidden. 'War Communism' was not popular with the peasants and terror tactics were necessary to make them obey. Even so it did not work well. In 1921 cereal production was only 40% of 1916 levels, livestock production 25–35%.

In fact, the Bolsheviks were pushing the Russian people too hard, as Lenin realized in 1921. For him, the Kronstadt Rising was a terrible warning. It 'illuminated reality like a flash of lightning' was how he put it. He decided that the government must change course and despite considerable opposition from within the party, he got his way. The result was the New Economic Policy (N.E.P.) of 1921. The chief aim of the N.E.P. was to win back the support of the peasantry which War Communism had lost. Public control of trading and marketing was ended in rural areas and private trading restored. The peasantry were once again encouraged to produce food for their own profit, and the local merchant—a person much harassed by Bolsheviks between 1917 and 1921—was allowed to resume his business. Russia also began to trade with other European countries for the first time since 1917. The N.E.P. was a drastic retreat from Communist principles. Lenin knew it and did not attempt to hide the fact. 'It is necessary to a certain extent to help to restore small industry', he wrote, 'and the effect will be a revival of the petty bourgeoisie and of capitalism on the basis of a certain amount of free trade. That is beyond doubt.' One of Lenin's most outstanding qualities as a leader was his readiness—fanatical Communist though he was—to compromise some of his most dearly held principles in order that his government, the only Communist government in the world, could survive.

Left *Innocent victims of the Civil War; starving children in a camp near Samara, October 1921*
Above *Lenin in Moscow*

Education for Russian women railway workers, 1923

After further dark days—a terrible famine in Central Russia cost the lives of between two and three million people in 1922 and 1923 —recovery began. Aid came from abroad, mainly from the U.S.A., and the N.E.P. provided conditions in which the peasantry was ready to co-operate more closely with the government. In 1925 the potato harvest was 50% more than the 1913 level and industrial production also began to reach pre-war levels for the first time. Lenin did not live to see this recovery which he had done so much to make possible. He was never a strong man and, after a serious stroke in 1922, he died in 1924. He had held power in Russia for a mere seven years yet in that time he had achieved enough to rank among the most important men in modern history. His Communist beliefs were carefully thought out and offered hope to the millions of poor and under-privileged throughout the world. Moreover Lenin was much more than an inspired political thinker, he was a practical politician of remarkable skill. The use he made of the opportunities of 1917 was masterly. His sense of timing was acute and throughout the troubles that overwhelmed the Bolsheviks from 1918 to 1921 he kept his head. He was never a dictator. Communist Russia was run by committees of which he was the senior member. He dominated his associates by a combination of intelligence and personality. He was basically a modest man quite unaffected by success and he genuinely cared about the welfare of ordinary people. A fellow revolutionary described his way of life when he was effectively master of all Russia. 'In the Kremlin he still occupied a small apartment built for a palace servant. In the recent winter he, like everyone else, had no heating. When he went to the barber's he took his turn, thinking it unseemly for anyone to give way to him. An old housekeeper looked after his rooms and did his mending.' Nonetheless he never flinched from using ruthless methods against his political opponents when the Communist Revolution seemed in danger. By the time he died a powerful secret police force and rigid political censorship were part and parcel of the new Russia. By masterminding the first successful Communist Revolution and by piloting it through the treacherous waters of its first years, he changed the course of history.

Who should succed Lenin?

The major policy-making committee of Comunist Russia was the Political Bureau of the Central Committee of the Communist Party, usually known as the Politburo, and within the Politburo a vicious struggle took place to decide who should dominate it after Lenin's death. There were two outstanding members of the Politburo— Trotsky and Stalin. Realizing that a leadership crisis was likely to occur on his death and to involve these two rivals, Lenin made some notes about them for the guidance of party members if they had to choose between them. 'Comrade Trotsky', he wrote, 'is distinguished not only by his outstanding qualities. He is personally the most capable person on the Central Committee but he suffers from an excessive self-confidence and is too liable to concentrate on the purely administrative side of things.' Stalin worried him. 'Since he became general secretary', he continued, 'Comrade Stalin has concentrated in his hands immeasurable power and I am not sure that he will always know how to use that power with sufficient care.'

Trotsky was born Leib David Bronstein in 1879. (He took the name Trotsky from one of his Tsarist jailers.) His father, of Jewish origin, was a small landowner near Odessa. Like Lenin, Trotsky had a brilliant career at school and, after a year at university, turned to revolution. In 1905 he was chairman of the St. Petersburg Soviet and played an active part in the revolution of that year. Although a convinced Marxist, he refused for many years to join either the Bolsheviks or the Mensheviks after their split in 1903, for which some Bolsheviks never forgave him. At the beginning of 1917, he was in exile in America and did not get back to Russia until a month after Lenin. Then he joined the Bolsheviks and quickly became Lenin's right-hand man. Lenin was the planner, Trotsky the man of action. He was a superb speaker and in every crisis energetic and decisive. He was the chairman of the Petrograd Soviet in autumn 1917, the actual organizer of the takeover of Petrograd during the October

Colleagues in the heroic days of October 1917. Trotsky is on the extreme left, Stalin the extreme right

Right *In exile, Trotsky with his wife*

Revolution, the negotiator at Brest-Litovsk, the Commander-in-Chief of the Red Army who, hurtling round the theatres of war in an armoured train, won the Civil War and remorselessly destroyed the Kronstadt rebels. His weakness was his arrogance. He gave the impression of being too clever by half and tended to underestimate his colleagues. In particular he underestimated Stalin. They never got on. After their first meeting in 1913, Stalin referred contemptuously to Trotsky as that 'noisy champion with faked muscles'.

Stalin was very different: born Josef Dzhugashvili (Stalin or 'Man of Steel' was one of the many names he used when being hunted by the Tsarist police) in 1879, the same year as Trotsky, his father was a poor shoemaker and his mother a very religious woman determined to do well for her son. She got him to the local school where he did well and then to the training college for young priests in Tiflis. This college had a history of politically-centred difficulties with its students and he was soon in trouble, mainly for reading forbidden books, and expelled in 1899. He became a Marxist and deeply involved in revolutionary activities in his native province of Georgia. Between 1902 and 1917 he was arrested and imprisoned six times and came to be widely respected in the Bolshevik party. He returned to Petrograd from a Siberian exile in March 1917 and temporarily was among the obvious leaders. On the return of Lenin and Trotsky however he retired into the background, and became a tireless behind-the-scenes organizer, without, it seemed, a strong personality of his own. 'A grey blur' was how a colleague described him. So he continued during the Civil War but quietly and surely he was building up his own personal power. In 1922 he became General Secretary of the Communist Party and used this important administrative post to put his friends and supporters in strong positions within the party. Though he lacked the brilliance of Lenin and Trotsky, Stalin was no fool. Moreover he had a secretive nature and a peasant's cunning which made him a most dangerous political opponent. Unlike Lenin, he became intoxicated with power for its own sake and unlike Trotsky, he had little interest in the spread of Communism beyond the borders of Russia.

In the last few months of his life Lenin grew increasingly un-happy about Stalin. He suggested in writing that the Central Committee should find a way of removing him from the General Secretaryship and of replacing him by someone 'more patient, more loyal, more polite and considerate to comrades, less capricious etc.'. But it was too late. As soon as Lenin died, Stalin moved carefully and effectively against Trotsky. On one major political issue they openly disagreed. While Trotsky believed that an international Communist revolution was not far away and that the Communist government should do everything within its power to speed this revolution, Stalin believed that the government should concentrate first on making Russia fully Communist ('Socialism in one country' was his slogan) and only then should it try to export Communism. He ridiculed Trotsky's international ambitions and, winning most of the Central Committee to his view, isolated him first in the Politburo, then in the Communist Party. In 1925 he had him dismissed from his ministerial posts, in 1927 from the Communist Party, in 1929 he deported him from Russia and, vindictive to the last, had him murdered in his Mexican exile in 1940 by an assassin who smashed his skull with an ice-axe. By 1928 Stalin had not only eliminated Trotsky but all the other old Bolshevik leaders as serious rivals to him in both the Politburo and the Central Committee. Such power did his position as General Secretary give him over younger men rising within the Communist Party that he was effectively dictator of Russia and remained so until his death in 1953.

Stalin's Domestic Policy
Between 1928 and 1941 Stalin had four main aims. The first was to expand Russian industry as rapidly as possible. The second was to collectivize Russian agriculture (i.e. to end the traditional peasant system of individually owned small plots and replace it by much larger commonly-owned collective farms). The third was to make

Russia strong enough to defend herself against a foreign attack which, after 1933, seemed more and more likely to come from Nazi Germany. If these three aims could be achieved then 'Socialism in one country' would have been made effective. His fourth aim was to keep himself in power whatever the cost.

To bring about rapid industrialization he devised a method of Five-Year Plans. The first ran from 1928 to 1932, the second from 1933 to 1937. What the Plans did was fix targets which particular industries had to reach in the five-year period and all the resources of a one-party police state were mobilized to make sure that they did. Stakhanov, a coal-miner of the Donetz basin who improved his productivity by 1400%, was given massive publicity by the government information services and a Stakhanovite movement was started to encourage 'socialist competition' to boost production. Perhaps more effective was the instant dismissal and disgrace of those managers whose factories failed to meet their targets. The government decided to keep wages low and reinvest industrial profits to increase the size of industrial plant and of production. Between 1928 and 1938 average real wages in Russia fell, perhaps by as much as 20%. Nonetheless in 1937 80% of Russian factories had been built since 1918. The first Five Year Plan was most ambitious and was made out to be a great success by government propaganda. According to the official figures 93.7% of the targets set in 1928 had been reached in 1932. These figures are almost certainly a gross exaggeration and many targets were missed by miles. The achievement, however, was there and great new cities appeared in the eastern sparsely populated areas of Russia, e.g. Magnitostroi in the Urals and Kuznetsstroi in Western Siberia. The second plan was less ambitious and probably more successful. Major projects were completed like the Moscow–Donets railway, the Baltic–White Sea canal and the Moscow underground. In the ten-year period, Russian industry was transformed. Russian economists claim a staggering 20% annual growth rate, western economists prefer 12–14%. Whatever the correct figure it was a major economic achievement.

The policy of collectivizing farming followed a more chequered, bloodspattered course. The planners originally assumed that farming organized on N.E.P. lines would be able to meet the targets of the Five Year Plans. They were wrong. Moreover the peasantry became increasingly opposed to government interference in their lives—especially the kulaks, the comparatively prosperous farmers of the Ukraine. Stalin therefore decided to force collectivization throughout the country. There were two main types of collective farm, the 'kolkhoz' and the 'sovkhoz'. The 'kolkhoz' was and is jointly owned by its members and its produce is divided into two parts, one of which goes to the state, the other is shared out among the members accord-

One consequence of the Five-Year Plans—a blast furnace in Magnitogorsk

ing to the amount of work which each has contributed. The 'sovkhoz' was and is entirely owned by the state and hires the labour of the local peasantry. In 1929 there were 57,000 kolkhozy and 1500 sovkhozy, in 1932 the respective numbers had leapt to 211,000 and 4,337. This collectivization policy met with desperate resistance. The peasants reduced their output and slaughtered their animals. The 1931 harvest was a terrible failure and in 1932 and 1933 there was a severe famine in the Ukraine. Despite the resistance and the suffering, Stalin pressed on. Malcolm Muggeridge, then Russian correspondent of the *Manchester Guardian*, described a visit which he made in 1933 to a small market town in the Caucasus: '"How are things with you?" I asked one man. He looked round anxiously to see that no soldiers were about. "We have nothing, absolutely nothing. They have taken everything away," he said and hurried on . . . It was true. They had nothing . . . Everything had been taken away. The famine is an organized one. Some of the food which has been taken away—and the peasants know this quite well—is being exported to foreign countries.' In the winter of 1930, thousands of kulaks were shot for resisting collectivization, hundreds of thousands taken to forced labour camps and their farms compulsorily collectivized. The exact cost of collectivization is hard to calculate. A conservative estimate is two million dead and millions more made homeless. Yet the policy brought no immediate improvement in agricultural production. Only once (1938) between 1929 and 1940 did agricultural output improve on the 1928 figures, usually it was considerably lower.

There was a riddle often quoted in Russia during the 1930's. 'Why are Adam and Eve like Soviet citizens? Answer: Because they live in Paradise and have nothing to eat.' Cynical though this riddle is, it contains some truth, even in the Paradise part. Frightful though the sufferings were of millions of Russians and considerable the opposition to the Communist government, a very large section of the Russian people were ready to tolerate the hardships, accept low wages and long hours, near-starvation and constant government interference, because Communism, even Stalin's version, offered the hope of an earthly Paradise just round the corner. In Tsarist Russia, no ordinary Russian could really hope for a significantly better future, in Stalin's Russia, he could. It was a vital difference. For all the powers of a dictator of a one-party state and Stalin's terrifying ruthlessness, the Communist achievement of the inter-war years would have been impossible if government policies had not captured the imagination or at least had the firm backing of many of the working population. It should be remembered that though real wages might fall, the quality of Russian life was, in certain important ways, obviously improving. A massive investment in education made at least 4 years of schooling available to every Russian child by 1938 and 7 years the norm in the cities. Before World War I less than half the population was literate, by 1938 85%. There was also a similar improvement in health and welfare services throughout the country.

However, if one was a senior member of the Communist Party, the 1930's were a long drawn out nightmare. Stalin had always been an isolated, cunning, suspicious man but the crises which he faced and the huge powers which he had concentrated in his own hands appear to have unbalanced him. He saw conspiracies against him everywhere and lashed out murderously and indiscriminately. In November 1932 his gifted and sensitive wife Nadja committed suicide, partly for personal, partly for political reasons. There is no doubt that she was appalled by her husband's policies and the suffering which they were inflicting on the Russian people. Years later, their daughter Svetlana commented 'Because my mother was intelligent and intellectually honest, I believe her sensitivity and intuition made her realize that my father was not the New Man she had thought when she was young and she suffered the most terrible, devastating disillusionment.' Stalin refused to attend her funeral or to visit her grave. He hardly spoke of her for the next twenty years. By her action she had betrayed him and joined his enemies. There were also dedicated Bolsheviks who shared her disillusion. They too became enemies.

The first major purge of the Party began in 1933 but violence on a large scale came the following year. On 1 December 1934 Kirov, a senior, able and popular Party member and—or so it seemed—a close friend of Stalin, was assassinated in Leningrad. Stalin immediately presented the murder as a Trotskyite plot but the circumstances remain mysterious. Stalin may have had him killed because he seemed to be becoming too popular. Whatever the circumstances, Stalin then let loose his secret police. The assassin and thirteen accomplices were swiftly tried and shot. 103 other prisoners were also executed. Old Bolshevik leaders, heroes of 1917 and former close comrades of both Lenin and Stalin, were imprisoned. Another purge followed in 1936. At two great 'show' trials, senior members of the Party were accused and often confessed to plotting against the Party and against Stalin personally. In the same year the head of the secret police (the N.K.V.D.) was replaced and the purge was intensified. It was the army's turn next. Marshal Tukhashevsky, hero of the Civil War and hammer of the Kronstadt rebels, was arrested with many other generals in a lightning police swoop and shot. Between 1937 and 1939 the senior ranks of the Red Army were virtually destroyed—3 out of 5 Marshals, 14 out of 16 army commanders, 75 out of 80 members of the Supreme Military Soviet, 167 out of 280 corps and divisional commanders. The Terror had in fact run away with itself. No-one trusted anyone else. Finally when one in twenty Russians had been arrested, when a third of the Communist Party had lost its membership and 98 out of 139 of the Central Committee were in prison, when the prison-camp population had reached ten million, when Chinese were being arrested as Japanese spies, Jews as Hitler's agents and the N.K.V.D. began to purge itself, Stalin decided enough was enough. 'The purge was unavoidable', he told a Party Congress in 1939, 'and its results on the whole beneficial.' The main beneficiary was himself. Into the empty posts of party members destroyed by the purge, he promoted tens of thousands of younger men who had played no active part in the heroic days of 1917 and owed everything to Stalin. His dictatorship was more complete than ever.

In the 1930's Russia most dangerous enemies were Germany in the West and Japan in the East. Both detested Communism and in 1936 signed the anti-Comintern Pact, the aim of which was to weaken the power of Russia and of Communism throughout the world. In 1934 Stalin took Russia into the League of Nations and began to look for friends. So deep however were the suspicions aroused in Europe by Communism that neither Britain nor France nor Czechoslovakia nor Poland, although they too were menaced by Nazi Germany, were prepared to ally with him. So, in August 1939, the world was shocked to learn that he had signed a non-aggression pact with Hitler (the Nazi-Soviet Pact) which allowed the two dictators to carve up helpless Poland in the autumn of 1939. Stalin claimed in later years that he signed the Pact to buy time. Before long the Nazis would attack Russia who would need all the time she could possibly get to rebuild her armed forces—demoralized by the recent purges—to be ready for what must be a life and death struggle. In 1941 the attack came. The Communist achievement in Russia was put to its greatest test.

Chapter 6
Liberal Democratic Europe

Political Terms

First, some definitions. Between 1919 and 1939 there were three main types of government in Europe, Communist, Fascist and Parliamentary Democratic. A Communist government would follow closely the teachings of Marx and Lenin. It would strongly disapprove of private property and replace it as soon as possible by public or state ownership. It would regard itself as the government of the workers and would seek to destroy the privileges of the upper and middle classes. It would, in theory at least, be international in outlook, believing that workers throughout the world would eventually revolt successfully against their masters and unite in an international brotherhood. Since it believed that the teachings of Marx and Lenin were entirely correct, it saw no need to allow non-Marxists the right of political opposition or freedom of speech.

The term 'Fascist' was first used in Italy to refer to Mussolini's supporters who organized themselves in 'fasci di combattimenti' or fighting groups. A Fascist government would view Communism as a deadly enemy. Fascists were fiercely nationalistic and thought that Communism put their nation's existence in danger. Moreover, unlike Communists, they had no objection to private property and could usually count on the solid support of big business. Another characteristic of Fascism was contempt for parliamentary democracy which as a method of government was thought of as leading to the flabby, indecisive, stagnant rule of mediocre men. Instead Fascists demanded the strong rule of one party of tough single-minded men (themselves) under an inspiring leader. They believed in violence, excitement and national glory. The main belief which they shared with the Communists was the conviction that they were right and no opposition or freedom of speech should be allowed. The term Fascist was first used in Italy. The Italian Fascists and the German Nazis were the most successful Fascist parties.

Communism is usually described as an extremely *left-wing* political philosophy, Fascism an extremely *right-wing*. In between them lies parliamentary democracy. In a parliamentary democracy, a parliament or assembly holds supreme power and is made up of members who are chosen by regular elections in which every adult can vote. The government is formed by that political party or parties which have the most members of parliament after an election. Unlike Communist and Fascist states, parliamentary democracy cannot work without an effective political opposition. If no opposition exists, the electors have no real choice of government and have no effective check on the government in power. Freedom of speech and expression are equally vital. In most parliamentary democracies private property is generally regarded as an important safeguard of individual freedom so it is carefully preserved, although some public ownership—e.g. of railways, coal, gas and electricity—has also been introduced. Because parliamentary democracies between the wars placed a much higher value on individual freedom than either the Fascists or the Communists, they are often described as liberal democracies, a liberal being someone who regards individual freedom as an essential part of civilized society.

Apart from Bela Kun's brief rule in Hungary, there was only one Communist government in Europe between 1919 and 1939, the Russian. Nonetheless, Communism was a powerful political force in other parts of the continent, notably in France and in Germany. In 1919 there were no Fascist governments in Europe. In 1939, however, not only were Mussolini and Hitler firmly established as dictators of Italy and Germany respectively but they had formed an alliance which terrified the rest of Europe. Their successes encouraged the formation of Fascist parties all over the continent, though only one of these, Gombos' party in Hungary, proved strong enough to form a government (1931–1936).

There is one other term, the meaning of which we need to keep clear—dictator. A dictator is a ruler whose power is unrestricted either by an elected assembly or, like some modern presidents, by a fixed term of office. By 1939 there were dictators all over Europe of various types—Communist dictators like Stalin, Fascist ones like Mussolini and Hitler, military ones like Franco in Spain or Metaxas in Greece, and royal ones like King Boris III of Bulgaria or King Carol II of Roumania. A dictatorship is in most ways the opposite of a parliamentary democracy, the chief aim of which is to represent all the people and to prevent any one man or group of men from becoming too powerful.

Between the wars parliamentary democracy withered while dictatorship flourished. President Wilson had been determined 'to make the world safe for democracy' and in 1919 there was hardly a country in Europe which did not call itself a democracy or was without a parliament. Twenty years later this was true of only a handful of Western European countries—Britain, France, Belgium, Holland, Luxemburg, Switzerland and Scandinavia. The rest were dictatorships.

It is not easy to explain this swing to dictatorship. Partly it was due to the newness of parliamentary democracy to Central and Eastern Europe. Not until after 1880 had it begun to be accepted that all men whatever their education or social class had the right to vote and only in Western Europe had parliaments actually possessed both in theory and practice supreme political power before 1914. Moreover the economic and political problems of the interwar years were exceptionally difficult and quickly appeared to be too difficult for the new multi-party parliamentary democracies. Only in Western Europe where there was already long experience of parliamentary democracy and in industrially advanced Czechoslovakia did it survive. Elsewhere it seemed to breed only muddle and indecision. Beset by poverty and despair, most Europeans turned to the security and strength which one-party and one-man rule seemed to offer.

Great Britain 1919–1939

Britain had possessed a strong parliament longer than any other country in the world except Iceland. At the beginning of the 20th century virtually every male Briton over 21 had the vote, and the franchise was extended to all women over 30 in 1918 and to all women over 21 in 1928. At no time between the wars was British parliamentary democracy seriously threatened. The failure of the British working class to rise in revolution was a source of continual puzzlement to European Communists. After all, Britain had experienced the industrial revolution earlier than anywhere else and her oppressed workers should surely have been among the first to arise against their capitalist masters. In 1926 many Communists believed that the inevitable British revolution had come at last when a general strike brought the country to a standstill. However, there was little violence. Strikers and policemen were seen to play football beside the picket lines and the strike leaders never attempted to overthrow the government. Even at the height of the 1930's slump, Communism failed to win any sizeable support in Britain. The same was true of Fascism. An ex-cabinet minister, Sir Oswald Mosley, did found a Fascist party, the British Union of Fascists, in 1932 which between 1932 and 1936 captured the headlines by organizing uniformed rallies which often ended in violent clashes with anti-Fascist groups. However only the East End of London, where his anti-semitism had an appeal, provided Mosley with lasting support, and from 1936 British Fascism was virtually dead. A humorist wrote to *The Times* suggesting that Mosley's followers should have B.F. embroidered on the pockets of their black shirts.

Perhaps if parliamentary democracy had been more seriously challenged, its leadership in Britain between the wars might have been more inspiring and its achievements more considerable. As it was, the period 1919–1939 was one of the more depressing in British history. In the face of mounting economic troubles, there was too much caution and too little imagination in high places.

There were three major political parties, Conservative, Liberal and Labour. The Liberal party, the party of Gladstone and of Asquith, had been the main rival of the Conservatives in the late nineteenth and early twentieth century. The Labour party was newest, being founded in 1900 and having first won seats in parliament in 1906. It had begun as a party for industrial workers whose voting strength had been greatly increased by the Reform Act of 1884. In the 1920's it replaced the Liberals to become the main rival to the Conservatives. A major cause of the Liberal decline was the bitter personal quarrel of Lloyd George and Asquith which led to a permanent division of the party. It originated in 1916 when Lloyd George successfully plotted to get rid of Asquith as Prime Minister and ended up himself in Asquith's place. There can be no doubt that the change was in the country's interests but neither Asquith nor many Liberals would work with Lloyd George again. He had to depend on Conservative support to remain Prime Minister and the gulf between him and Asquith was never bridged.

Lloyd George and the Coalition Government 1918–22
In 1919 however, his future and that of his country seemed rosy. The war was won, the Versailles Treaty signed. He was Prime Minister with a huge majority (478) in favour of his coalition and only 139 against. He had high hopes. His task, he said, was 'to provide a country fit for heroes to live in'. For a short time things went well. There was a temporary economic boom and the huge task of demobilizing the victorious army was carried out comparatively smoothly, with the soldiers returning, it seemed, to plenty of jobs. But by 1920 the boom was over and the government, whose parliamentary supporters seemed to Keynes to be made up largely of 'hard-faced men who looked as if they had done well out of the war', cast around for economic policies appropriate to the more depressed situation. The policies it put into effect were those approved of by the traditionally minded business community and in particular by Montague Norman, the influential Governor of the Bank of England. The emphasis was on saving, to reduce both private and public spending, to cut down the state control of basic industries which had been established during the war, and to encourage instead private enterprise and free trade. Typical were the policies suggested by the Geddes Committee appointed by the government in 1921 to investigate possible means of reducing government expenditure. It recommended substantial cuts in educational and defence spending. Such policies were almost certainly wrong. Their effect was to deepen an already deepening depression and significantly increase the number

IRELAND'S FIRST SINN FÉIN BANNER.

of unemployed. At the end of 1920 about 850,000 were unemployed, by May 1921 the number had soared to 2,000,000, and though in 1922 it had dropped back to 1,400,000, it never fell to less than a million again until 1939. So Lloyd George was unable to fulfil his promises to his returning heroes.

The Irish Troubles 1916–22
Left British armoured cars patrol the streets of Dublin after the Chief of Police had been killed in an IRA ambush
Above Sometimes the British must have wished that they had executed De Valera too in 1916

Serious though economic problems were, Lloyd George's major concern was Ireland. In 1914 Ireland had been on the verge of civil war. Asquith's Liberal government had proposed that the whole island should be granted Home Rule. This would have meant that while it would remain part of the British Empire and accept British control of certain economic and defence matters, it would have its own parliament in Dublin to run its own internal affairs. In the northern area of Ulster the Protestants, who were there in the majority, prepared to fight rather than be united to the rest of Ireland which was overwhelmingly Catholic, and thus pass under the rule of a mainly Catholic Dublin parliament. The outbreak of World War I led to the shelving of the Home Rule issue. Nonetheless the situation in Ireland worsened. In 1916, the fiercely nationalistic Sinn Fein movement which was determined to take a united Ireland completely out of the British Empire attempted to seize power in Dublin (the Easter Rising). It was less a serious revolt than 'a sacrifice of blood' for the independence of Ireland. 'The chances against us are a thousand to one,' remarked Connolly, one of the Sinn Fein leaders. A week's street-fighting in Dublin cost the British 500 casualties, the Irish, including civilians, more than a thousand. When it was over, the British stupidly took stern measures, executing fifteen of the Rising's leaders. Connolly, who was dying anyway, was court-martialled in a hospital bed and shot by the execution squad sitting in a chair because he was too weak to stand. Such brutality made martyrs of the Sinn Fein leaders and turned thousands more

RELAND'S FIRST SINN FÉIN BANNER.

Right *British troops in what was left of the G.P.O. building, Dublin, at the end of the Easter Rising, 1916. It had been the Sinn Fein stronghold*

Irishmen into Sinn Fein supporters. In the election of 1918 Sinn Fein won 73 of the 105 Irish seats, and instead of sending its M.P.'s to the Westminster Parliament set up its own Assembly of Ireland (the Dail Eirann) in Dublin. De Valera, who alone of the sixteen leaders of the Easter Rising had survived since he possessed American citizenship and who escaped from prison in the spring of 1919, was declared its first president. He quickly set up a government to rival the British administration and with the help of the Irish Republican Army (I.R.A.) was able to persuade most of Ireland outside Ulster to accept his authority rather than that of the British government and its police force, the Royal Irish Constabularly (R.I.C.). Lloyd George tried tough measures. The R.I.C. was reinforced by ex-soldiers specially recruited for Irish service (the Black and Tans) and ordered to enforce British authority throughout the island. Consequently, for much of 1920, what was effectively a civil war took place, full of terrible atrocities on both sides. 'Things are being done in Ireland', Asquith told the House of Commons, 'which would disgrace the blackest annals of the lowest despotism of Europe'. Lloyd George then decided that the only possible solution was partition, dividing Protestant Ulster from the Catholic south. In December 1920 the Government of Ireland Act set up two parliaments in Ireland, one in Belfast (Stormont) responsible for the six Ulster counties, the other in Dublin for the rest of Ireland. This Act was accepted by most Ulstermen and remained the basis of the Northern Irish constitution until the suspension of Stormont by Heath's government in 1972. The Sinn Fein leaders however were very reluctant to accept partition and only after further violence and long negotiations was Lloyd George able to persuade most of them to accept a treaty in 1921 recognizing all Ireland except Ulster as the Irish Free State, a virtually independent nation within the British Empire. Yet though the Dail Eirann approved the Treaty by a small majority in January 1922, De Valera and some of the I.R.A. would not. A second civil war broke out, this time among the Irish nationalists. In 1922 and 1923 the new Free State government executed more than three times as many Irishmen as the British government in 1920 and 1921, and Michael Collins, its most gifted leader, was killed by a terrorist ambush. Partition did not permanently solve the Irish problem. It did, however, bring an uneasy peace which lasted nearly 50 years before violence broke out on a large scale in Ulster in 1969.

The Irish Treaty was Lloyd George's last great success. By the end of 1922 he was out of office. The reason for his fall was the refusal of the Conservative majority in his coalition to continue to accept his leadership. Basically they distrusted him. He was too clever and too unpredictable. His private life upset many of them, and so did the way he used his position as Prime Minister to sell political honours. (It was not that no other politicians had mistresses or sold political honours, merely that Lloyd George was less secretive about them than most.) The breaking-point came over foreign policy. During the so-called Chanak incident of 1922, the Prime Minister seemed on the point of using British troops against Turkish nationalist forces who were threatening, near the town of Chanak, the international zone established in Turkey by the Versailles Settlement. When the Conservatives met to consider whether they could continue supporting him, a then little-known M.P., Stanley Baldwin, made a decisive speech. Lloyd George, he argued, was 'a dynamic force ... owing to which the Liberal Party had been smashed to pieces. It is my firm conviction', he concluded, 'that in time the same thing will happen to our party.' Conservative support was withdrawn and Lloyd George had to resign. Though he was only 59, comparatively young for a politician, and had another twenty-three years to live, and though he was the outstanding statesman of his generation, he never held office again. The Liberals were split beyond recovery, and without a strong party base no British politician, however gifted, has been able to achieve very much in recent history. It was a dismal end to a brilliant career.

The following year, 1923, was a milestone in the history of the Labour Party because after the election of that year, it formed a government for the first time in its history, with Ramsay MacDonald as Prime Minister. 'We were making history,' recorded J. R. Clynes after his visit to Buckingham Palace to be sworn in as a minister by King George V. 'I could not help marvelling at the strange turn of Fortune's wheel which has brought MacDonald, the starveling clerk, Thomas the engine driver, Henderson the foundry labourer, and Clynes the millhand to this pinnacle beside the man whose forebears had been kings for so many generations.' Some citizens took a

Baldwin and his wife, herself a formidable personality

while to get used to the idea. Snowden, a well-known Labour M.P., was telephoned by a countess who asked if now her throat would be cut and J. H. Thomas, the new Colonial Secretary, always maintained that when he first arrived at the Colonial Office the doorman refused to believe that he was the new minister, but remarked to a colleague, tapping his head as he did so, 'Another case of shell-shock, poor chap!'

The first Labour government lasted only a year. With only 191 M.P.'s, it depended on Liberal support (159) to maintain a majority in the House of Commons over the Conservatives (258), yet it pursued a pro-Russian foreign policy which to the Liberals seemed dangerously soft on Communism. Another election was needed in 1924 which proved a landslide in favour of the Conservatives. They won 419 seats, Labour 151 and the Liberals only 40.

The new Prime Minister was Stanley Baldwin. That this man, who got rid of Lloyd George in 1922 because he was too 'dynamic' and must himself come close to being the least dynamic man to lead this country in the last hundred years, should have been the most dominating figure in British politics from 1922 to 1936, tells us much about the mood of the British electorate between the wars. 'Safety first' was his motto. The trouble was that to the economic problems of the Great Slump and the threats of the Fascist dictators, the apparently safe answers turned out to be no answers at all.

Baldwin was born in 1867, the son of a wealthy Midlands iron manufacturer. His father left him a large fortune and, such was his family's local reputation that Stanley succeeded him as the local Tory M.P. He remained an obscure backbench Conservative M.P. until 1917. Too old to fight in the war, he became parliamentary private secretary to Bonar Law, the Conservative party leader and Chancellor of the Exchequer in Lloyd George's coalition. This was a post of some responsibility, and from it he was promoted first to Financial Secretary to the Treasury and then in 1921 to the Cabinet post of President of the Board of Trade. When ill health forced Bonar Law to resign as Prime Minister in 1923, Baldwin succeeded him. After the brief Labour government of 1923–24, he again became Prime Minister.

He successfully acted the part of the typical unflappable Englishman who smoked a pipe, loved pigs and the countryside and went about his business with little fuss and much commonsense. He was, however, a more complicated character than he liked to make out. Moody, impulsive, sometimes lazy (often in times of crisis), he owed much to his wife who provided the drive and energy which he lacked. And he was a subtle politician. He could not have dominated British politics for so long if he had not been.

One of his earliest and greatest tests came in 1926 when the only general strike in British history took place. Its immediate cause was a bitter dispute between the coal miners and the mine owners which had been rumbling on and off since the end of the war when the mines, which had been under state control during the war, were handed back to private owners. Coalmining was a severely depressed industry and in 1925 the owners demanded that the miners should work longer hours for less pay. Neither side was much prepared to listen to the other. 'It would be possible to say without exaggeration', wrote a member of the government, 'that the miners' leaders were the stupidest men in England, if we had not had frequent occasion to meet the owners.' Negotiations broke down in deadlock. The mine-owners organized a lock-out and the miners went on strike, their demand being 'not a penny off the pay, not a minute on the day'. The miners union also asked the Trades Union Congress (the T.U.C.) for support. Five years before, a miners' strike in very similar circumstances had failed because of lack of support from other unions. This time the T.U.C. prepared to back them. It hoped that the mere threat of a general strike would force the government to intervene in favour of the miners. It did not. At a critical moment in the negotiations between government and T.U.C., Baldwin went to bed instead of waiting up for the T.U.C. negotiators and the strike was on.

The government, however, had been expecting a crisis like this for some months. It had no intention of giving way and was well prepared. It could rely on widespread voluntary, mainly middle class, support. Essential supplies were kept moving and the police and army tactfully organized. The only chance the T.U.C. had of winning was to create a genuinely revolutionary situation, acting itself as an alternative government, and this it had no intention of doing. Baldwin was at his best in the last days of the strike, making it as easy as he could for the union leaders to climb down. He curbed the wilder men of his Cabinet like Winston Churchill who would have happily tried to smash the strike by force. And he made friendly broadcasts on the radio. There would be no victimization once the strike was over, he promised. 'I am a man of peace', he said. 'I am longing and praying for peace.' The General Strike collapsed after nine days and though the miners fought doggedly on for another six months, bitterly attacking their fellow unionists for their spinelessness, they too had to accept defeat. The British trades union movement thus suffered a severe setback from which it did not fully recover until after World War II. In 1921 its membership had been 8,348,000 but the combined effects of the strike and the depression reduced it to 4,392,000 in 1934. The General Strike showed both Baldwin's strengths and his weaknesses. He was cautious and firm yet eager to

Strike-breaking, 1926. Volunteers unload milk at the Hyde Park food depot during the General Strike

1931—the new National Government. Macdonald leads his Cabinet down the steps of 10 Downing Street. Only J. Thomas, just behind him, and Snowden, with the stick, coming down the stairs, were members of his previous Labour government. The rest were mainly Conservatives. Baldwin is in the centre, halfway down the stairs, and Neville Chamberlain above him, just to the left

find a peaceful solution. He kept his government united. But he was short of energy at important times. Strikers were victimized once the strike was over and he allowed his more anti-union followers to carry a Trades Disputes Act through parliament in 1927 which made general strikes illegal and further embittered the trades unions.

Baldwin's government was not inactive in other fields. It managed some social reforms, nationalized the electricity industry and set up the B.B.C. It did not however give the impression of much activity either to its supporters or to the country at large. 'The snores of the Treasury bench [where ministers sit in the House of Commons] reverberate through the Land' a Conservative M.P. commented, and in the 1929 election, campaigning under the 'Safety First' slogan, Baldwin was defeated. Labour won 288 seats, the Conservatives 260 and the Liberals 59.

Ramsay MacDonald and the Labour Party now seemed to have a real chance to prove themselves as an effective governing party. They had the misfortune however to have to grapple with the worst economic crisis in modern history. Only five months after they came to power, Wall Street crashed (see page 73), and by 1931 the effects of the American catastrophe were being felt in Europe. Acute crisis came to Britain in the summer. With banks failing in Europe, money began to drain out of London, one of the world's most important banking centres. Unemployment rose steadily and the government found it more and more difficult to meet its costs. An influential committee, the May Committee, declared that the situation would only get worse unless the government drastically cut down its spending, and New York bankers, from whom the government were seeking a much-needed loan, refused to make the money available unless the recommendations of the May Committee were acted upon. Any government would have found itself in a terrible dilemma but for a Labour government the situation was intolerable. In order to obtain the loan, it would have to cut unemployment pay, directly hurting its traditional supporters and the poor and needy rather than the wealthy. Yet without the loan economic disaster appeared to threaten the whole nation. The party leadership split. MacDonald, the Prime Minister and Snowden, the Chancellor of the Exchequer decided the cuts must be made. The majority of the Cabinet, and most of the parliamentary Labour party and the trades union leaders, demanded that the cuts be avoided and other methods tried. MacDonald then decided to form a National government. The Conservatives joined him, a few Liberals and a handful of his old Labour colleagues. He then called an election in October 1931. The National government won a vast majority, the electorate being convinced that the Labour alternative would mean rapid inflation. The figures were National 554 (of whom 473 were Conservatives), Labour 52. MacDonald was now a captive of the Conservatives, the Labour party utterly demoralized. The cuts were made—in unemployment pay and the salaries of teachers, policemen and civil servants. Though approved of by most economic experts of the time, this economic policy was probably the opposite of the one needed. Keynes, whose ideas were soon, but not soon enough, to revolutionize economic thinking, described the May Committee's recommendations as 'the most foolish document I have ever had the misfortune to read' and the Budget and economic policy of the new National government as 'replete with folly and injustice'. Before long, unemployment increased to more than 3 million. MacDonald, expelled for his 'great betrayal' from the party which he had done so much to create stayed on as Prime Minister for another four years but he was a broken man. He began to lose the thread of his parliamentary speeches and became embarrassing to listen to.

Between 1931 and 1939 the National Government was without inspiration. Baldwin succeeded MacDonald in 1935 and on his retirement in 1937 was succeeded by Neville Chamberlain. That there was some economic recovery—more in fact than in many European countries—owed little to the government. In foreign affairs its record was disastrous (see Chapter 7). Nonetheless it won huge majorities in 1931 and again in 1935. There can be little doubt that its policies reflected closely what the British people in the 1930's actually wanted.

France between the Wars
In France, parliamentary democracy was threatened more severely than in Britain. The French Communist Party was a powerful force on the extreme left throughout the period. In 1936 it held 72 seats in the Chamber (parliamentary assembly) and supported the coalition Popular Front government. It tended to take its orders from Moscow and was, in the long-term, determined to overthrow both capitalism and parliamentary democracy in France. It was also well-supported by trades unionists and had its own union organization (the C.G.T.). There were also parties of the extreme right which held parliamentary democracy in contempt. The oldest of these which dated back to pre-war days was the Action Française which with its own street army—the Camelots du Roi—pre-dated Mussolini's black-shirts. It was reinforced in the 1930's by various leagues of which the Croix de Feu was the largest. They had in common hatred of Communism, extreme nationalism, anti-semitism and contempt for the Third Republic. They had plenty of backing from the French newspapers and big business. In between the Communists on one side and extreme conservatives like the Action Française on the other were many more moderate parties of which the Socialists, the Radicals and the various Catholic parties were the most important. None

of these parties however was strong enough to win an overall majority in the Chamber. Consequently between 1919 and 1939 France was ruled by a succession of coalitions none of which lasted and none of which was strong.

Soon after the war ended Poincaré and Clemenceau, respectively President and Prime Minister of the French Republic, made a triumphant tour of the provinces of Alsace and Lorraine, once more part of France. 'Now I can die,' said Poincaré, feeling that the task of a lifetime was complete. He was wrong. So difficult did the problems of peace prove that he remained a major figure at the centre of French politics until illness forced his retirement in 1929.

Poincaré was born in Lorraine and though his home did not pass into German hands at the end of the Franco-Prussian War, this war, which happened when he was at the impressionable age of ten, bred in him a deep distrust of the Germans. A clever man with a cool precise mind he first became a lawyer then a politician. He was not an immediately impressive person, shy and colourless. As an Englishman who met him many times recalled, 'he was precise and prim . . . completely without any sense of humour. I never heard anyone tell a joke about him and never heard him make a joke.' His seriousness was one of his main assets. With his intelligence, capacity for hard work, his sound financial judgement, and his unusual (for French politicians) honesty, he possessed the qualities the French middle classes most admired and in times of crisis to him they turned.

The World War had cost 1,750,000 French lives and devastated her northern provinces. Naturally enough France's chief concern in the 1920's was to prevent Germany menacing her borders again and to build up her own economic strength, primarily through stern insistence on the regular repayment of reparations.

The first post-war government of France was the firmly right-wing coalition, the Bloc National. Fiercely nationalistic and anti-German, it turned against Clemenceau for being too soft on Germany at Versailles and refused to back him when he stood as President on Poincaré's retirement from that office in 1920. Poincaré became a member of the Senate (the Upper House in the French parliamentary system) and chairman of the important reparations committee. In 1921, reparations were at last fixed at £6,600 million and Germany began paying. Almost immediately she fell behind with her payments. A political crisis in France followed. The Prime Minister, Briand, tended to agree with Lloyd George the British Prime Minister that the Allies should not press the Germans too hard. He was forced to resign and Poincaré replaced him. The new government at once issued this declaration. 'The problem of reparations dominates all others and if Germany, in this essential question, fails in its obligations, we shall be forced to examine the measures to be taken.' Facing an acute economic crisis with some of its population starving, the motto of the German government was 'first bread, then reparations.' It asked the Allies to agree to a delay in repayment. Poincaré refused and sent the French army to occupy the Ruhr, one of Germany's most important industrial areas. 'We have no intention,' he told the French Assembly, 'of strangling Germany or ruining her; we only wish to obtain from her what we can reasonably expect her to provide.' At first, the occupation seemed a success. Despite passive resistance in the Ruhr and condemnation by Britain, French, Belgian and Italian engineers kept the mines and factories working. The German government, overwhelmed by the worst inflation in history, asked to negotiate. But it ended a failure. Poincaré waited too long before negotiating with the Germans whose economic difficulties began to weaken the French economy. The deaths of civilians in the Ruhr and acts of sabotage further isolated France from her allies. Moreover many Frenchmen began to realize that the worse Germany's difficulties, the longer it would take her to pay reparations. In 1924 Poincaré's government was overthrown, the Bloc National disintegrated. In its place came a more left-wing government of Radicals and Socialists, the Cartel des Gauches, which evacuated the Ruhr and pursued a less harsh policy towards Germany.

This new coalition, however, was unable to cope with the financial crisis the Ruhr occupation had begun. The franc plunged in value from 70 to the pound sterling in 1924 to 243 in 1926. This crisis caused the Radicals to switch their support from the Socialists to a government led once again by Poincaré, now viewed as the elder statesman whose financial skill would see the country through in its hour of need. It did. The confidence the business community had in him plus firm conservative financial measures rapidly strengthened the franc and a mild boom took place. Simultaneously, Briand, now Foreign Secretary, was developing a happier relationship with Germany. The years 1926 to 1929 were the most hopeful France knew between the wars.

They did not last. The Wall Street Crash affected France later than other parts of Europe but no less seriously. Unemployment rose, wages were cut and an immigrant population which had entered France in considerable numbers in the 1920's became the target of much ill-feeling. The extreme right-wing Leagues encouraged by the successes of Mussolini and Hitler grew greatly in strength and, in 1934, the year of the Stavisky affair, France came close to civil war.

Stavisky was a Jewish night-club owner and swindler of skill and strong nerves. He moved in the same social circles as many politicians.

*Wall posters in Paris,
1934. 'On étouffe l'affaire
Stavisky'—'The Stavisky
Affair is being hushed up'*

In December 1929 he disappeared when a warrant was issued for his arrest on fraud charges, and when eventually he was found he was dead with a bullet in his head. No-one knows to this day how he died. Officially he committed suicide just as the police burst into his room to arrest him. Most of France however preferred to believe that he had been shot dead by the police to shield senior politicians. The Leagues began to organize huge anti-government demonstrations and street violence became frequent. On the evening of Tuesday 6 February 1934 a massive league demonstration was prevented by the police from marching on the Chamber of Deputies and turned ugly. The police opened fire and after a battle had raged till midnight, fourteen rioters and one policeman were killed, and 1435 demonstrators injured. 'It was a time', a young schoolteacher remembered, 'which recalled the beginnings of the Wars of Religion . . . Armed bands were formed and through the towns marched processions of workers such as had not been seen for decades.' A change of government quietened the Leagues however and the danger of civil war lessened.

Perhaps the most important result of the Stavisky affair was to convince the Socialists and Communists that the danger of a Fascist takeover was so real that, after years of wrangling, they should sink their differences. In 1935 the Rassemblement Populaire or Popular Front was formed, its chief aim being to oppose Fascism. Nationwide the Front aroused enthusiastic support, especially from the trade unions and in the 1936 election it remained united and won a large majority. It formed a government under the leadership of the Socialist Leon Blum.

The situation Blum found himself in was both extremely exciting and extremely difficult. Exciting because the French people had voted for a genuinely reforming government and seemed in the mood to follow its lead, difficult because there was an acute economic crisis and the unity of his Popular Front coalition was at best fragile. Against the advice of the union leaders, workers in many areas celebrated the victory of the Popular Front by occupying their factories and demanding immediate and radical reform. The economic crisis deepened and the middle classes were terrified. And the Communists refused to play an active part in the new government.

Blum's first action was to put into effect the social reforms promised by the Popular Front. By the so-called Matignon agreements, wages of industrial workers were increased by 7%–15%, the working week reduced to 40 hours and paid holidays introduced. Yet simultaneously a costly rearmament programme had to be introduced to meet the menace of Nazi Germany. Only a flourishing economy could have paid for both radical reform and rearmament and the French economy was far from flourishing. Blum had to devalue the

117

LE FRONT POPULAIRE
radical-socialiste-communiste.

① Le radical Herriot, le socialiste Blum et le communiste Cachin sont amis. Mais ...

② ...le radical est toujours mangé par le socialiste...

③ ...et le socialiste, à son tour, est régulièrement ...

④ mangé par le communiste

⑤ Francais! si vous voulez être mangés par les Communistes, marchez et votez avec les radicaux-socialistes ou les socialistes!

Left *An anti-Popular Front poster for the 1936 election. The message is that a vote for the Radicals or Socialists is effectively a vote for the Communists*

franc which both failed to solve the economic crisis yet lost him his moderate support. He then put the brakes on reform and refused to give assistance to the Republicans in the Spanish Civil War (see page 145), thereby turning the French Communists and some of his own Socialist party actively against him. He had to resign in 1937. Blum was a first-class scholar who had also been a journalist before he became a politician. He often asked himself in later life whether he was decisive enough or a good enough judge of people to be able to handle such a challenge as that posed to him by the events of 1936. He failed, but whether anyone else could have succeeded is debatable to say the least.

The riots of February 6, 1934

The failure of the Popular Front left France more divided than ever and totally demoralized. Most political parties seemed to prefer opposition to government. The Communist Party returned to a rigid Moscow line. Most workers, disillusioned by the false optimism of 1936 and the continuing depression, lost interest in political activity. And for many Frenchmen of wealth and property, the Popular Front had so threatened their established way of life that they came to regard Communism/Bolshevism as the greatest threat to France. So Hitler became almost attractive as the chief obstacle to a Bolshevik takeover of Europe.

119

Chapter 7
Dictators

In 1919 there were no dictators in Europe. In 1939 they were the rule rather than the exception. Three in particular, Mussolini in Italy, Hitler in Germany and Stalin in Russia, were determined not merely to prevent any limits in their own personal power but to ensure that their political parties—Fascist in Italy, Nazi in Germany and Communist in Russia—should control the life of their nation in every way, in economic, educational and social matters as much as in purely political ones. Their drive for total obedience and for total control have caused their rule to be described as totalitarian dictatorships.

Italy and Mussolini
The Italy of 1919 was a discontented country. Though she had been on the victorious side in the recent world war, her allies had treated her with little respect at the Versailles conference, and she had gained much less than she had expected. In particular she had not been given any of the colonies confiscated from Germany nor had her gains at the expense of Austria-Hungary included the Adriatic port of Fiume. A tragi-comic event illustrated Italian disgust at the terms of the Versailles Settlement. D'Annùnzio, a fine poet, fiery speaker, daring soldier and extreme nationalist, marched his private army into Fiume in September 1919. The victorious Allies insisted that he withdraw and the Italian government promised to get him out. Nonetheless not until fifteen months had passed and the population of Fiume had suffered real hardship because of D'Annunzio's incompetent rule did they succeed and the Italian troops sent to remove him showed a marked reluctance to carry out their government's orders against him.

Another cause of serious discontent was the economic situation. Italy's war effort had stretched her economic resources to the limit. There was considerable unemployment when the war ended which was made worse by demobilization. Hopes among the peasantry for a redistribution of land in their favour, raised during the war, came to nothing. Powerful socialist and communist groups, encouraged by the 1917 Revolution in Russia, hoped for a similar success in Italy and organized strikes, sit-ins, and riots in both urban and rural areas. During part of September 1920, for example, the steel and engineering industries were paralysed by a sit-in of half a million workers.

Though these tactics failed to bring a revolution, they did bring higher wages to the urban industrial workers. At the same time, the income of most middle class Italians remained stationary while the cost of living rose rapidly. Not surprisingly, the middle classes felt that their government was allowing itself to be blackmailed by the socialists and there was strong anti-government and anti-socialist feeling. The parliamentary government which had to cope with these problems was made up of short-lived coalitions. Between 1918 and 1922 five separate governments came and went, each finding it harder to maintain law and order against extreme political parties in the north and brigand-activity in the more primitive south. By 1922 a political party which could offer some hope of firm government and had a positive anti-socialist and anti-communist policy had a good chance of success. The one which proved most able to build up its support and to profit from the general discontent was Mussolini's Fascist Party.

Mussolini was born in 1883. His father was a village blacksmith who brought up his son to be a keen socialist. For a while the young Mussolini pursued a teaching career but before long he became a journalist and, in the years before World War I, began to make a reputation for himself in Milan, the most important industrial city in Italy. When war broke out, he was editor of the socialist news-paper *Avanti!* (Forward) but he was to find that his socialism and his nationalism clashed. Mussolini became strongly in favour of Italy taking part in the war. Official socialist policy, however, was for neutrality. He was expelled from the party and left *Avanti!* to run his own paper *Il Popolo d'Italia* (the Italian People) which was wholeheartedly in favour of Italy's war-effort. His fierce nationalism led to a complete break with Italian socialism which maintained its disapproval of the war as long as it lasted and when it ended recommended a moderate foreign policy. When he came to found his own Fascist party, it was openly anti-socialist.

The Fascist movement began in Milan in 1919 with the founding by Mussolini of a 'Fascio di Combattimento' or Fighting Group. It began as an emotional and often mindless reaction to the general mood of disillusion, anxiety and hopelessness of post-war Italy. The first Fascists were usually young students or unemployed ex-soldiers or disillusioned revolutionary socialists whose common bond was patriotism and desire for action, preferably violent. They formed themselves into squads (*squadri*), wore a uniform of black shirts, beat up socialists, broke up strikes and showed open contempt for the laws and customs of traditional Italian society which, they were convinced, was doomed. 'Me ne frego,' (I don't give a damn) was a favourite saying. To begin with, there seemed little reason to take them seriously. In the November election of 1919 Mussolini

managed to win only 2% of the poll in a Milan constituency. 'A corpse in a state of decay was found today floating in the Naviglio [a canal which runs through Milan]. It seems that it is that of Benito Mussolini' crowed *Avanti!*, his former newspaper now his deadly enemy. Its joy was premature. The Fascist street armies continued their violence. Rich landowners, determined to end the socialist-organized strikes in the countryside, began to make use of them. In November 1920 a Fascist-inspired campaign of terror, financed by landlords and owners of sugar factories, was let loose in the Po valley. The attempts of the government to suppress the blackshirt squads came to nothing. The campaign was a success and the Fascist movement began to attract more support. Violence continued. Communist versus Fascist street-battles in Florence and Pisa caused several deaths. Fascism throve on violence and its anti-Communism made it attractive to many sections of Italian society. In the elections of 1921 it gained a foothold in parliament, Mussolini and 34 others being elected. Since many senior army officers and police chiefs favoured Fascism, blackshirt violence continued unchecked. In May 1922 the Bologna city government was taken over by force, and the squads then marched on Ravenna. 'We passed through the province destroying and burning all the offices of the Socialists and Communists,' one of their leaders recalled. 'It was a terrible night. Our march was marked by tall columns of fire and smoke.' In October 1922 Mussolini felt strong enough to demand

Waiting for Mussolini. A black-shirt guard of honour, 1922

power openly. Reviewing 40,000 blackshirts in Naples on 24 October, he challenged the government to solve the country's problems or make way for the Fascists.

Three days later, the famous 'March on Rome' took place. While blackshirt squads occupied public buildings all over northern and central Italy, others, in their tens of thousands, marched on Rome. A determined government could have dealt with the crisis. It had 12,000 well-armed troops at its disposal. The marchers on Rome were ill-armed and poorly organized (Mussolini was still in Milan). The heavy autumn rain had dampened their fighting spirit. But Facta, the Prime Minister, hesitated. When at last he decided to declare a state of emergency and use the troops, King Victor Emmanuel III refused to sign the necessary emergency decree. Why the king made this fateful decision is not clear. He may have feared civil war, he may also have thought that his younger brother, the Duke of Aosta, might seize the throne with Fascist support. Whatever his reasons, he gave himself little choice but to sack Facta and invite Mussolini to became Prime Minister. Arriving by train from Milan, the Fascist leader accepted the king's offer. The squads then marched in triumph to the centre of the city. Yet it was not so much their actual strength which brought the Fascists to power as the lack of nerve of the politicians and the king. The March on Rome was a gigantic bluff which paid off.

Four years later, Italy had been transformed from a multi-party parliamentary democracy to a one-party dictatorship with Mussolini as *il Duce* (the Leader). In 1923, a Fascist Grand Council was established alongside parliament and took over some of its duties. In contrast to members of parliament who were elected by the Italian people, members of the Grand Council were appointed by Mussolini. In the same year, the blackshirt squads were converted into the Volunteer Militia for National Security (M.V.S.N.) so Mussolini now had his own official private army. In 1924 he persuaded parliament to allow a general election and to agree that the party which gained most seats as a result of this election should automatically be awarded two-thirds of the seats in parliament. In this way, he argued, the most successful party would possess a firm parliamentary majority and the evils of a weak multi-party coalition would be avoided. There could be only one result of the election of 1924. Mussolini persuaded the more conservative political parties to co-operate with the Fascists during the election campaign and the Fascist militia terrorized the opposition. Consequently, the Fascists took control of parliament.

Not long after the elections, Matteotti, one of the leading socialist M.P.'s still in parliament, delivered a hard-hitting attack on the Fascists, giving many examples to prove how Mussolini's followers had used violence to influence the election results. Ten days later, he was kidnapped in broad daylight and disappeared without trace. After many months his remains were found in a shallow grave just outside Rome. There is no doubt that Fascists murdered him, whether on Mussolini's orders is unclear. So great was public disgust that, for a short time, it looked as if Mussolini might be overthrown. None of his opponents however was prepared to counter his violence with violence of their own. The Socialists contented themselves with walking out of parliament (the Aventine Secession), an action which won them temporary publicity but in the long run ended what little political influence they still possessed. And no individual came forward round whom the opposition to Fascism could unite. Mussolini survived and Matteotti's fate acted as an awful warning to other would-be critics of his government. In 1925 and 1926 he felt strong enough to pass laws which banned political parties opposed to him, censored the press, postponed indefinitely provincial and municipal elections and further strengthened the Fascist Grand Council at the expense of parliament. By the end of 1926 his rule was to all intents and purposes a dictatorship and many of his political opponents went into exile.

At this stage, it is worth asking what Fascists believed in. In fact they were much more clear about what they were against than what they were for. They were against Communism. They were also against democracy. 'Italian Fascism,' declared Mussolini, 'represents a reaction against Democrats who would have made everything mediocre and uniform . . . Democracy has taken elegance from the lives of the people but Fascism brings it back.' They were also contemptuous of the 'comfortable' life. In contrast Fascism was austere, heroic and holy. On the subject of what they were for rather than against, they were rather vague. Mussolini's article in the Italian Encyclopaedia about Fascism is a masterpiece of saying very little in a large space. They glorified the state. 'For the Fascist everything is in the state and nothing human or spiritual exists, much less has value, outside the state,' he wrote. Their aims were absolutely totalitarian. 'Fascism wants to remake not the form of human life but its content, man, character, faith.' The intellectual content of Fascism was basically shallow. It was less a well thought-out political philosophy than an attitude to life—a craving for power, excitement, glory and violence. Mussolini himself typified this attitude. 'I want to make a mark on my era, like a lion with its claw', he once said. Favourite Fascist slogans were very similar. 'Better to live one day like a lion than a hundred years like a sheep.' 'A minute on the battlefield is worth a lifetime of peace.' Perhaps the most important characteristic of Fascism was its extreme nationalism. Expanding the power of one's nation by an aggressive foreign policy provided excitement and the chance to be leonine, heroic and aggressive in a way which was quite impossible in domestic matters.

Not surprisingly, Mussolini's foreign policy was much more decisive than his domestic policy. In home affairs, for all its slogans and apparent activity, Fascism made surprisingly little impact. In theory, Italian economic life was revolutionized by the new Corporate State. Trades unions passed under Fascist control, first 13 and then 22 corporations being set up, each representing particular industries or groups of industries. They were all supervised by a Minister of Corporations, originally Mussolini himself, and they were intended to provide the means whereby employers and workers, encouraged by the state, could settle in harmony wage rates, working conditions and future industrial development, avoiding the industrial conflicts which had so weakened Italy in the past. Lockouts and strikes were forbidden. The chief effect of the Corporate State was to strengthen the position of the employers at the expense of the workers. Not surprisingly, Mussolini could count on the support of the majority of Italian businessmen for most of the period he held power.

How economically effective the Corporate State was is hard to say since first the Great Slump and then the Abyssinian war placed strains upon it which any system would have found hard to bear. The average Italian seems to have become neither worse nor better off—despite the optimistic claims of Fascist officials—and the Italian economy proved no better able to withstand the strains of World War II than it had World War I. The government's greatest success was at the propaganda level. There was the 'Battle for Grain' which had some success in making Italy more self-supporting. Schemes like the draining of the Pontine marshes, building new roads and public monuments were given huge publicity but their effect on unemployment—which remained obstinately high—was insignificant. Foreigners were particularly impressed by the fact the Italian main-line trains at last ran on time. Like so much of il Duce's Italy, this was mere window-dressing. The branch-lines continued as before.

The greatest success of Mussolini's home policy was the Lateran Treaties, signed with Pope Pius XI in 1929. They began a new period of co-operation between the Roman Catholic Church and the Italian state after nearly sixty years of serious disagreement. When Italy had been unified in 1860, the then Pope, Pius IX, had bitterly opposed unification and refused to allow the city of Rome to become part of the new nation. Ten years later, Rome was united to the rest of Italy but only after fighting between papal and government troops. Pius IX shut himself away in the huge Vatican Palace, declared himself a prisoner of the Vatican, forbade Italian Catholics to co-operate with the Italian government and refused himself to do so. Though his successors were a little less uncompromising, the basic dispute remained. In the 1920's, however, Pius XI was anxious to heal the split, and he feared a Fascist campaign against the Church if no agreement was reached. Mussolini also hoped for an agreement. He himself was no Christian—indeed his father had brought him up to hate priests—but he realized that most Italians were practising Catholics and hostility between Fascism and the Church might weaken his position. By the Lateran Treaties, the Pope at last accepted that the lands lost by the Church during the unification of Italy were gone for ever. In return, the Church received £30 million pounds worth of compensation and the area round the Vatican Palace, the Vatican City, as an independent state under the Pope's rule. Included in the treaties was a Concordat (Agreement) which regulated the relations between Church and State. All bishops and priests were to be paid by the State. Bishops were to be appointed by the Pope on condition that the government was notified well in advance of the date of the appointment so that it could make sure that the candidate was politically sound. Moreover all bishops were to take an oath of loyalty to the king and churchmen generally were forbidden to play an active part in politics. 'We have given back to God to Italy and Italy to God,' said the Pope when the treaties were signed. Both sides had reasons for satisfaction. Mussolini had ended a long-standing dispute and won the co-operation of one of the most influential institutions in Italian society. He had also ended the danger that a specifically Catholic party might develop to challenge Fascism. For his part, the Pope had won a secure well-defined and reasonably independent position for himself and his Church within Fascist Italy. Though he had no political power, his Church's existence was in no danger and as long as it existed and kept the loyalty of millions of Italians, it was a major obstacle to Fascist Italy becoming a genuinely totalitarian state. Problems did arise in the 1930's, especially over education which the Fascists wished to take over completely. The Church protested, with some success. The Lateran Treaties and the Concordat are among the very few of Mussolini's achievements which have survived through to the present day.

Perhaps the most attractive thing about Italian Fascism was its inefficiency, which prevented its essentially barbaric attitude to life from doing much harm to many Italians. It was easily satisfied with appearances rather than reality. In comparison with Stalin or Hitler, Mussolini seems an almost comic figure and his violence mild. Some political opponents, like Matteotti, were murdered. There was a secret police. Thousands were exiled and thousands more imprisoned or temporarily banished to small Mediterranean islands. But there were no forced labour or concentration camps, no purges or firing squads. More typical of Italian Fascism was the forcible feeding of political opponents with castor oil. From 1937, under Nazi influence, the government passed some anti-semitic laws. They were for the most part ignored.

The people who suffered most from Italian Fascism were not Italians at all, since much of the viciousness of Fascism was concentrated in Mussolini's foreign policy. The first to suffer were the weak—Abyssinians, Greeks, Albanians. Before long, however, il Duce's unrealistic belligerence and his alliance with Hitler caused Italy to attack much more dangerous opponents. Thousands of Italians then marched to their death and Italian Fascism to its destruction.

Germany and Hitler
Parliamentary democracy lasted a decade longer in Germany than in Italy. However, when it was finally overthrown in 1933, the dictatorship which replaced it was much more terrible and threatened the freedom of Europe and the rest of the world far more dangerously than Mussolini's.

The parliamentary democracy which was created in Germany at the end of World War I is usually known as the Weimar Republic since its constitution was drawn up in the small and peaceful city of Weimar in Central Germany. The first five years of its existence (1918–1923) were chaotic. Its inheritance from the Kaiser—who fled into a Dutch exile—was a defeated army, a mutinous navy, a humiliating peace-treaty to sign, major cities in revolution and an economy in ruins. Though it had firm support from the powerful socialist party, the S.P.D. (its first President, Ebert, was a socialist), it was attacked from all sides and had a desperate struggle even to survive. The first serious threat came from the Communists who tried to seize power in Berlin in December 1918. They were led by Liebknecht and Rosa Luxemburg and called themselves the Spartacists after the rebellion of slaves led by Spartacus against the government of Ancient Rome. In the spring of 1919, moreover, a Soviet republic was established in Bavaria. To end these Communist revolts, the government had to call in the army. The Spartacists were brutally crushed—Liebknecht and Rosa Luxemburg being murdered without any kind of trial—and by the summer of 1919 some sort of peace was restored. It was a bad start. Henceforward, the German Communist Party (the K.P.D.) which held the loyalty of a considerable number of German workers was the unforgiving enemy of the Weimar Republic. More seriously, the government was seen to be powerless without the support of the German army. The German army chiefs had political ambitions of their own. They never became the faithful servants of the Weimar Republic which they despised. They saw themselves as the survival of all that was best in Imperial Germany and looked forward to the time when Germany would be strong once more and the army recover its honoured place in German society. Though the Versailles Treaty had stated that the German army should not number more than 100,000, it had secretly been ignored. The army's morale was high and among its officers the percentage of aristocrats was higher in 1921 than in 1913. Its independence and lack of loyalty proved to be one of Weimar's most serious weaknesses.

The government leaders then had to go through the agony of accepting the humiliating peace-terms dictated to their representatives at Versailles. They considered refusing to sign but, in the end, they had no choice. Nonetheless, millions of Germans never forgave them, holding them responsible, though they were not the real villains, for both the defeat and the miserable peace. In March 1920 a former civil servant, Kapp, tried to seize power in Berlin. He and the army officers who supported him hoped to use the patriotic disgust with the Versailles Treaty against the government. The Kapp putsch (attempt to seize power) ended after a few days, a general strike organized by Berlin workers in favour of Ebert's government being a major cause of its failure. But while it lasted, Berlin experienced some terrible events. Army units supporting Kapp executed at least 2,000 workers. Kapp managed to escape to Sweden and the other conspirators got off lightly. Both the army leaders and the high-court judges, who were happy to take stern action against the Communists, were much more reluctant to do so against Nationalists like Kapp. Disorder continued. Bands of de-

Inflation 1922–3. Checking the exchange rates, Berlin, 1922

mobilized and unemployed soldiers (the Freikorps or Free Corps) roamed the streets and countryside looking for trouble. They were mostly extreme nationalists and particularly ready to harass socialists and communists. In 1920 there was an attempted communist revolt in the Ruhr, and another in 1921. Both failed. In 1921 Erzberger, a politician who had argued strongly that the Versailles Treaty must be signed, was assassinated. The following year, Rathenau, the Foreign Minister—whose main crime seems to have been that he was born a Jew—was shot dead in a Berlin street. Both these murders were the work of extreme nationalist groups.

Economic conditions went from bad to worse. In 1922 and 1923 Germany suffered the worst inflationary crisis in history (see page 72). The mark became worthless and millions of thrifty Germans who had carefully banked their savings over the years found that they were ruined. Thus another important section of German society lost all confidence in the Weimar government. In the autumn of 1923 there was another attempted revolt. In Munich an extreme nationalist party, the N.S.D.A.P. (generally known as the Nazis), led by Adolf Hitler, attempted to seize control of the city. The revolt was bungled and Hitler sentenced to five year's imprisonment.

After 1923, however, the Weimar Republic moved into calmer waters. Economic conditions improved, so too, with the Locarno Pact, did Germany's relations with the rest of Europe. Moreover a statesman of the first rank emerged in the person of Stresemann, Chancellor and Foreign Minister from 1923 to his death in 1929. The calm was brief. Three weeks after Stresemann's death, Wall Street crashed (see page 73) beginning the Great Depression which

A 5 million mark stamp of the early 1920s

Hitler, on the extreme left, with Nazi companions during his far from rigorous imprisonment in the Landsberg fortress, 1924

affected first the U.S.A. then Europe. In the political storms which accompanied the Depression, the Weimar Republic was eventually engulfed. Unemployment in Germany began to rise in 1930. In 1931 it reached serious proportions, and in 1932 the previously unheard of figure of 6 millions, the worst in Europe. In such bleak conditions, the extreme political parties began to win support, on one hand the Communists, on the other the Nazis. In the election of 1930 the Communists won 77 seats in the German parliament (Reichstag), the Nazis 107. Hitler, the Nazi leader, was now a man of real importance in German politics.

That a man like Hitler could become the leader of a major political party was itself an indication of how deeply disturbed German society had become by 1930. He was born in Austria, close to the German border, in 1889. His father, a customs official in the Habsburg civil service, was only moderately well-off but he was ready to pay for his son to attend Linz High School so that he too might qualify for a civil service career. Adolf, however, had a poor school record, mainly because he was incapable of sustained hard work. 'Hitler was certainly gifted,' one former teacher remembered, 'although only in particular subjects, but he lacked self-control and, to say the least, he was considered argumentative, autocratic, self-opinionated and bad-tempered and unable to submit to school-discipline. Nor was he industrious, otherwise he would have achieved much better results.' The only teacher Hitler remembered with affection—the rest he dismissed as 'erudite apes'—was his history teacher who was a passionate German nationalist and aroused in the young Hitler an intense interest in the glories of Germany's past. So poor was his progress that he had to leave Linz for the local school but did no better there and left without any qualifications. His father had died when he was thirteen and he gave up any idea of a civil service career, intending instead to become a painter since at school he had showed a little talent at drawing. Between the ages of 16 and 19, however, he preferred to live off his mother who was always ready to pamper him and managed to maintain herself, her son and his sister Paula on a widow's pension. He seems to have had only one close male friend and no female ones. He spent his time wandering along the banks of the Danube or reading, especially German history. He did once try to enter the Vienna Academy of Fine Arts but was turned down for lack of ability. In 1908 his mother died and he had no choice but to try to earn a living.

For the next few years he lived in Vienna and seemed neither willing nor able to keep a regular job. He made a little money painting picture postcards of Viennese scenes to be sold by street traders. Otherwise he worked on building sites, shovelled snow and beat carpets. He lived in acute poverty, unable to afford regular lodgings, and survived in doss-houses and charity kitchens. Though he played no active part in Viennese politics, he was fascinated by them and permanently influenced by the anti-semitism that flourished in the Austrian capital before World War I. He was also impressed by the desire of many German-speaking Austrians (a desire which he himself shared) to see the end of the Habsburg Empire and the union of all German-speaking people in a greater German Empire. Otherwise his main interest was reading German history. In 1913, at the age of 24, he was an eccentric book-reading tramp in the Viennese gutter, a man without a future.

Then came the war. For millions of Europeans World War I was a catastrophe. For Hitler it was marvellous. 'It came as a deliverance,' he later wrote, 'from the distress that had weighed upon me during the days of my youth . . . I thanked Heaven out of the fullness of my heart for granting me the good fortune to live at such a time.' The war gave him a new chance. He was in Germany when war was declared and persuaded the German authorities, Austrian though he was, to let him join the German army. He was a good soldier. He fought through some of the most terrible battles on the Western Front and was twice decorated for bravery. His fellow soldiers remembered him as unusual—one who never grumbled about conditions at the Front yet went on and on about the Jewish conspiracy inside Germany that might lose the war. As far as he was concerned, Germany's collapse confirmed his worst suspicions. The army had not really been defeated, it had been stabbed in the back by Jews and Communists. 'Did all this happen,' he said when the news of Germany's surrender reached him, 'that a gang of criminals could lay hands on the Fatherland?' In the chaos of postwar Germany, he soon became involved in politics. The army appointed him 'educational officer' in Munich to act against dangerous ideas like socialism. His work brought him in contact with a tiny extremely anti-semitic party, the German Workers Party. This he joined, soon dominated, and in 1920 rechristened it the National Socialist German Workers Party (N.S.D.A.P.). The Nazi Party was born.

Hitler kept his connexion with the army and also built up his own private army of thugs (the S.A.) under the command of a close friend and fellow-Nazi, Captain Röhm. At the end of 1923 he persuaded some army officers, including the famous General Ludendorff, to join him in an attempt to seize power in Munich, the capital of the province of Bavaria. The attempt—which is often known as the Beer-hall Putsch since it began in a Munich beer-hall—was a fiasco. The police stood their ground and fired on the Nazi demonstrators, most of whom fled including Hitler who dislocated his shoulder in the tumult. He was arrested two days later, put on trial

131

and though lives had been lost in the revolt, received the lenient sentence of five years imprisonment which was later reduced to nine months. Hitler made sure that he gained plenty of publicity at his trial. He proudly claimed complete responsibility for the revolt and expressed his political views at some length. He also put his time in the Landsberg prison to good use, writing *Mein Kampf* (My Struggle), a book of 782 pages which described his life (briefly) and his political ideas (at length). When Hitler eventually became Chancellor in 1933 more than a million copies of *Mein Kampf* were sold that year. It was the bible of the Nazi Party and most prudent Germans made sure that they had a copy in their library. Between 1925 and 1929, however, sales never reached 10,000 per year, and in 1928 they were down to a mere 3,000. These figures reflect quite accurately the position of the Nazi Party in these years. After his release from prison, Hitler could do little but hold his party together and wait. In the 1924 election the Party had won 14 seats in the Reichstag; in 1928 the number dropped to 12. The Great Depression, however, transformed its fortunes. The Nazi leader and his political ideas turned out to have something which, in difficult times, attracted Germans other than those on the lunatic fringe.

Many of these ideas which became central to Nazi policy between 1933 and 1945 were to be found in *Mein Kampf*. Among the most significant were those which concerned 'race'. Hitler was a racialist. He believed that some races were superior to others and racial purity was a quality to be preserved at all costs. The most superior race was the German or, as he preferred to name it, the Aryan. 'All human culture, all the results of art, science and technology which we see before us today,' he wrote, 'are almost exclusively the creative product of the Aryan . . . He is the Prometheus of mankind from whose strong brow the divine spark of genius has sprung at all times forever kindling anew that fire of knowledge which . . . caused man to climb the path to mastery over the other beings of this earth.' As long as the Aryan maintained his racial purity, he must dominate. Germany therefore could not remain a minor power, nor indeed one among the major powers of Europe. She must dominate the continent. This must mean—and Hitler was quite explicit about this—a fight to the finish with France and then a massive eastward expansion of the German race at the expense of the inferior Slavs. 'We (Nazis) stop the endless German movement to the south and west and turn our gaze to the land in the East. If we speak of soil in Europe today, we can have in mind only Russia and her vassal border-states.' There the living-space (*lebensraum*) vital for a growing Germany must be found.

How this new expanded German nation was to be run, Hitler does not make very clear, except that it would be a one-party dictatorship with himself as Leader (*Führer*) ruling in the interest of the *Volk* (Folk or national community). The unity of the nation would be stressed and the total obedience of all citizens to the state be demanded. National unity would be strengthened by the elimination of two undesirable groups, the Communists and the Jews. Hitler hated Communism, partly because Marx the founder of Communism was a German Jew, partly because it was international rather than national in outlook, and partly because a Communist government had successfully seized power in Russia and placed a major obstacle in the way of his dreams of eastern *lebensraum*. As for the Jews, they were responsible for most of the serious problems of mankind. Moreover there was an international Jewish conspiracy to keep Germany weak. Among the many appalling things for which he held the Jews responsible was international prostitution and the white slave traffic. This 'truth' dawned on him one day in Vienna before World War I. 'When for the first time I recognized the Jew as the cold-hearted shameless and calculating director of this revolting vice,' he wrote in *Mein Kampf*, 'a cold shudder ran down my back.' Hitler's antisemitism was an obsession which grew worse rather than better as he grew more powerful. As the Russians closed in on Berlin in 1945 and he prepared to take his own life almost his last order to Doenitz, his appointed successor, was 'to uphold the racial laws to the limit and to resist mercilessly the poisoner of all nations, international Jewry'.

The trouble with Hitler's ideas was that ridiculous though they were in their extreme form they did reflect in an exaggerated way what millions of Germans felt in the 1930's. They were very patriotic and did feel that the Germans were a rather special race. They felt deeply humiliated by their defeat in 1918 and by the peace-terms of 1919, and they wished to see their nation strong and feared once more. They were most suspicious of Communism and regarded Bolshevik Russia as a most dangerous enemy. And antisemitism, which had a long history in Europe, had been a powerful force in Germany as well as Austria since the 1880's. As conditions worsened after 1929, more and more Germans came to see sense rather than nonsense in Nazism.

Between 1930 and 1934, the Weimar Republic was destroyed and Hitler emerged as dictator. The sequence of events was as follows.

As the economic crisis worsened throughout 1931 and 1932, law and order began to break down. The unofficial armies of various political parties, among whom the Nazis and Communists were the most prominent, fought each other in the streets. An English writer, Christopher Isherwood, was living in Berlin at this time. 'The city,' he wrote, 'was in a state of civil war. Hate exploded suddenly without warning out of nowhere . . . In the middle of the street a young man would be attacked, stripped, thrashed and left bleeding on the pavement; in 15 seconds it was all over and the assailants had disappeared.' The coalition governments of the Weimar Republic were unable to cope with this turbulence. Since 1930 the S.P.D., whose support was essential for an effective coalition, had refused to co-operate with any other party, and the various Chancellors, first Bruning, then von Papen, and, in December 1932, General von Schleicher had to govern by decree because they had no reliable majority in the Reichstag. Since the decrees had to be approved by the President before they became law, President Hindenburg, the aged war-hero who was now becoming senile, was no figurehead. He had a vital political role to play. Meanwhile the Nazis were busy. While the S.A. strutted in the streets, the leaders worked hard to persuade businessmen and army officers that they were worth supporting. They had some success. The numerical and financial strength of the Party increased swiftly. In the spring of 1932 the presidential election was due and Hitler stood against Hindenburg. He won 37% of the votes against Hindenburg's 53%. Though he lost, no-one doubted that his was a considerable political success. In July the Reichstag elections were held, and during the election campaign violence reached new heights. In the province of Prussia 461 pitched battles were recorded, as a result of which 82 people died and 400 were seriously wounded. The July elections were a Nazi triumph. They polled 13,745,000 votes and won 230 seats which made them the largest party in the Reichstag. The Communists also gained, winning 89 seats. Hitler now expected to be invited to join the government and began discussions with von Papen and with Hindenburg. These came to nothing. Hitler demanded that he should be Chancellor but in the course of a very cool interview Hindenburg refused him the Chancellorship and at the same time gave him a stern lecture on the lawlessness of the S.A. Further elections were held in November which proved a serious setback for the Nazis. They lost 34 seats which left them with 196 while the Communists secured 100 (a gain of 11 on July). For a moment it seemed as if the Nazis had passed their peak. However, in a situation of acute political crisis, the politicians in Berlin were continually plotting and intriguing. The Chancellor of December 1932 to January 1933, Schleicher, himself a skilful plotter, remarked, 'I stayed in power 57 days and on every one of them I was betrayed 57 times!' On 28 January, unable to maintain a stable government, he resigned, and two days later a coalition organized by von Papen, in which Hitler was given the Chancellorship and two other Nazis held ministerial posts (Von Papen was to be Vice-Chancellor), was approved by President Hindenburg. Why Hindenburg and von Papen were ready to co-operate with Hitler in 1933 when they had refused in 1932 is something of a mystery. Perhaps they felt that the

'*A vote for the Führer is a vote against unemployment*'. *This Nazi election poster of 1933 plays cleverly on the anxieties of millions of Germans, as the depression grows worse*

March 1933, members of the S.A., the Nazis' private army, terrorize Communists in Berlin

November election results showed the Nazis to be on the decline and that Hitler was a simple man who could be managed by more experienced politicians. 'In less than two months,' von Papen commented, 'we will have pushed Hitler so far into a corner that he'll be squeaking.'

The Creation of the Nazi Dictatorship

Theirs was one of the most costly miscalculations of the twentieth century. Like Mussolini, but with much more speed and ruthlessness, Hitler was able to transform himself from being Chancellor of a multi-party parliamentary democracy to Führer of a one-party dictatorship. The first step was to hold new elections. The S.A. stepped up its violence in the weeks before polling day which was to be 5 March 1933. On 27 February the German people were shocked to learn that the Reichstag building had been burnt to the ground. A Dutch Communist half-wit by the name of Van der Lubbe was found wandering near the scene of the crime. He was arrested, tried and found guilty. However there are good reasons for believing that the Reichstag fire was organized and Van der Lubbe framed by the Nazis to discredit the Communists on the eve of the election. The building burnt so fast that there must have been far more explosives and chemicals than Van der Lubbe could have placed on his own and the basement of the Reichstag was connected by underground passage to the residence of Goering, a leading Nazi. Whoever was responsible, the Communists suffered. Between four and five thousand were immediately arrested.

135

Despite S.A. violence, the Reichstag fire and the arrests, the elections failed to give Hitler the massive majority he hoped for. The Nazi Party won 44% of the votes and gained 288 seats. It was the largest political party but more Germans were still against it than for it. Hitler however had the politicians mesmerized. Playing on their general fear of Communism and widespread disorder, he persuaded the Reichstag to pass an Enabling Act which transferred political power for a period of four years from parliament to Hitler and his ministers. Only the S.P.D. members opposed him. Hitler turned on them in a rage. 'You are no longer needed,' he shouted. 'Your death-knell has sounded. I do not want your votes. Germany will be free but not through you.' The final voting figures were 441 in favour, 84 against. The Weimar Republic ceased to exist.

Hitler had come to supreme power legally with the active or passive approval of many of the most influential sections of German society. Two days before the Enabling Act was passed, the Nazis organized a great ceremony at Potsdam, the burial place of the heroic warrior-king of Prussia, Frederick the Great. The theme was national revival. The army, the churches, civil servants and party politicians were all represented. After Hitler ended his speech, President Hindenburg shook him by the hand, and units of the army, the S.A. and Hitler's personal bodyguard, the S.S., goosestepped past them both. A few days later, the Roman Catholic bishops made a statement publicly withdrawing their earlier criticisms of the Nazi party. Hitler had managed to deceive almost every German who mattered that their future was best placed in his hands.

Now firmly in command, he began to show his true colours. The Communist Party was already banned, the S.P.D. and Centre Party soon followed. In July, Germany was declared to be a one-party state, trade unions abolished and strikes forbidden. Henceforward his power could only be destroyed in two ways—either by the army, or by a revolt within the Nazi party. In 1934, both seemed possible. Röhm, the leader of the S.A., had nearly two million men under his command and wished to take over the German army. He also had his differences with Hitler whom he felt was not interested enough in revolutionizing German society. The army high command were very suspicious of Röhm's ambitions and hated the S.A. leaders. 'Thieves, drunks and sods' was how one general summed them up with some accuracy. In April 1934 Hitler made a secret deal with the army. If he broke Röhm, the army would back him as Head of State when Hindenburg died—an event which could not be far away as the President was 87 and in poor health. In June, Hitler invited the S.A. leaders to a conference. In a night purge, they were arrested and shot without trial. Hitler himself arrested Röhm and arranged that a pistol be left so that his old friend could have the privilege

Newly appointed Chancellor, Hitler shakes hands with President Hindenberg

Right *David Low, cartoonist of the London* Evening Standard, *makes his pictorial comment on the Night of the Long Knives, 1934*

of shooting himself. 'If I am to be killed,' said Röhm, 'let Adolf do it himself.' Adolf left it to two S.S. lieutenants. Simultaneously between three and five thousand Germans were arrested in an S.S. countrywide swoop and were generally murdered then and there. Most were S.A. men, others people who had annoyed the Nazis for one reason or another. Some died simply by mistake: Dr. Willi Schmidt, music critic of a Munich newspaper, married with three children, was arrested and shot by the S.S. before it was discovered that he was not Willi Schmidt, an S.A. leader. After this 'Night of the Long Knives', opponents of Nazism tended to keep quiet.

There had been an S.A. plot to overthrow the government, Hitler told Germany and the world, and he was congratulated by Hindenburg 'for his gallant personal intervention which had nipped treason in the bud'. Shortly afterwards the old President died. Hitler proclaimed the post of President abolished and himself Führer and Reich Chancellor. The army kept its part of the bargain, every soldier taking an oath of personal loyalty to the Führer. He then asked the German people whether or not they approved of his new position. Of the $42\frac{1}{2}$ million who voted, 90% were in favour.

How it was that an eccentric Austrian drop-out whose political ideas were half-baked or vicious could, in the twentieth century, become dictator of one of the most powerful, best-educated and civilized nations of the world is an important question which needs a lot of explaining. Certainly there is no easy answer. Partly it was for reasons already outlined—Germany's sense of humiliation after 1919, the terrible economic crises of 1923 and 1929–33, a deep

THEY SALUTE WITH BOTH HANDS NOW

137

fear of Communism, a prevailing sense of insecurity and lack of confidence in the kind of government provided by the Weimar Republic. For reasons such as these, Germany was looking for new policies and new leaders. Moreover, in the exceptional circumstances of the early 1930's, Hitler made extremely skilful use of his opportunities. He was an electrifying public speaker with a remarkable ability to sense the mood of his audience and, having won their attention, to carry them away. As a leader, he knew exactly what he wanted and pursued it with an abnormal intensity of purpose. Power was the one interest, indeed passion, of his life. Moreover there can be no doubt that as a leader he inspired both his close associates and his mass of followers. He also developed methods of propaganda, of swaying public opinion in favour of Nazi policies, well beyond anything previously attempted. Here he was ably assisted by Goebbels —a Doctor of Philosophy of Heidelberg University, unsuccessful writer and journalist before he became a Nazi. Goebbels was in many ways as effective a public speaker as Hitler. Earlier than anyone he realized the immense power of radio to mould public opinion. He also was a great organizer of marches and demonstrations with thousands of Nazi stormtroopers singing their marching songs and bearing their swastika banners or, after dark, their flaming torches. These ceremonies projected an image of power, purpose and excitement in a country otherwise miserable and purposeless. And against those who refused to be impressed, Hitler had no hesitation in using the S.A.

Hitler was also a master of deception. When he wished he could be both reasonable and charming and, with experience, he learnt to play convincingly the part of the sober, moderate statesman. Before 1934 he managed to deceive most leading Germans, and between 1934 and 1939 he was equally successful in deceiving foreign statesmen. But when all is said and done, the key to Nazism's success was its ability to respond to the mood of Germany better than any other political party. 'All over the land,' wrote Ludecke, one of the more thoughtful Nazis, 'young spirits were rearing up in defiant protest against the wretchedness of a life which their fathers seemed to have spoiled for them. Whether they marched with the Nazis, as they did in increasing numbers, or shouted the battle cry of the Communists, they were resolved on change, on a new order.'

The Nazis were as totalitarian in intention as the Italian Fascists and much more efficient in making a practical reality of their intentions.

Their first aim was to make Nazi opinion German opinion. Goebbels became Minister of Propaganda and took control of German radio. He also introduced strict political censorship of newspapers and films. Two of the most famous newspapers, the *Vossische Zeitung* and *Berliner Tageblatt*, were forced out of business. The censored

A master propagandist . . . Goebbels, speaking in Berlin in 1935

. . . and his most effective instrument, government controlled radio. The text reads 'The whole of Germany listens to the Führer on the radio'

newspapers remaining were so boring that their circulation dropped considerably. Cinema audiences were not much impressed by government sponsored propaganda films. Hissing became so common that the government had to issue a stern warning 'against the treasonable behaviour of cinema audiences'. Books by authors of whom the Nazis disapproved were banned and great book-burning demonstrations were held by party members to show their contempt for the work of men of genius like Thomas Mann, Alfred Einstein and Sigmund Freud.

Education was given most careful attention. 'This new Reich (Empire) will give its youth to no-one,' wrote Hitler, 'but will itself take youth and give to youth its own education and upbringing.' Every German teacher had to join the National Socialist Teachers League and to promise to be ready at all times to defend without reservation the National Socialist state. All schools came under the direct control of the Ministry of Education. Textbooks were rewritten to fit in with Nazi ideas and Hitler's lunatic racial theories were taught throughout the educational system up to and including universities. While many outstanding German scholars, especially those with Jewish blood, left Germany forever the resistance of the German teachers to the nazification of education was limited. Alongside the schools was established another organization for young people, the Hitler Youth. All other youth organizations were banned and parents strongly advised to see that their children between the ages of 6 and 18 belonged to the Hitler Youth. Its aim was to produce citizens strong in body and Nazi in mind. From the age of 6 to 10, the member served a period of apprenticeship. If, at the age of 10, he passed tests in athletics, camping and history (Nazi version) he was promoted to the *Jungfolk* (Young Folk). He then took an oath of loyalty to Hitler which ended 'I am willing and ready to give up my life for him. So help me God.' For males between the ages of 14 and 18 military training was added to their other activities. For girls over 14 there was the *Bund Deutsche Maedel* (the League of German Maidens). Here the emphasis was on becoming healthy mothers of healthy children, devoted to the Fatherland. By the end of 1938 the Hitler Youth numbered 8,000,000.

The Nazis realized that in their struggle for the hearts and minds of ordinary people, the churches were a serious rival. Hitler had been brought up a Catholic but had lost his faith when a young man and hated the Christian churches. Nonetheless, he realized their strength and moved against them cautiously. Since many leading German churchmen had welcomed his rise to power he first tried to keep their support by seeming friendly. In July 1933 he reached an agreement with the Pope which promised the Catholic Church in Germany the right 'to regulate its own affairs'. It was however an

empty promise. Nazi educational and racial policies made co-operation between convinced Christians and Nazis impossible. The Catholic Youth League was destroyed, Catholic publications banned and the leader of Catholic Action in Germany murdered. By 1937 the Pope publicly accused Hitler of breaking his word and of 'open hostility to Christ and his Church'. His protest was ignored. The German Protestant churches had welcomed Hitler to power even more whole-heartedly and some priests were ready to form a new Reich Church which would be both Christian and Nazi. Ludwig Muller a former army chaplain and associate of Hitler became the first Bishop of the Reich. His attempts to unite all Protestants into his church failed. In 1934 a large number of Protestants, led by Pastor Niemöller, broke away and formed the Confessing Church. They were appalled by Muller's attempts to nazify the Protestant churches and in particular to forbid men of Jewish blood to become priests. Moreover they publicly criticized the totalitarian nature of Nazi policies. Such defiance could not be allowed. In 1937 Niemöller and 806 other leaders of the Confessing Church were arrested and many ended up in concentration camps.

Dr Schacht stands immediately on Hitler's left, a piece of paper in his hand

Setting a good example in the 'Battle for Work' Hitler does some digging on the Frankfurt–Heidelberg autobahn, 1933

The most pressing problem Hitler faced when he came to power in 1933 was economic, in particular how to lower the number of unemployed. However, he was not interested in economics, and had little time for economic experts. Most of their problems, he believed, were not really problems at all. They could be overcome by will-power. The economic policies of the Third Reich, therefore, turned out to be something of a muddle! The Propaganda Ministry at once declared a 'Battle for Work'. Temporary jobs were quickly found which almost at a stroke cut a million off the unemployment figures. An expanding armaments industry provided more permanent jobs for others as did the slow recovery of German industry and commerce in step with the recovery of world trade. By the end of 1933 the official number of unemployed was down to 4,000,000, by October 1934 2,500,000, and by August 1935 1,700,000. Moreover between 1932 and 1937 the German G.N.P. rose by 102%. This remarkable achievement was chiefly the work of Dr. Schacht, an economic wizard whom Hitler, for all his suspicion of economic wizardry, appointed as Minister of Economics. Schacht had been a successful banker during the Weimar Republic. In 1930 he became an enthusiastic Nazi and played an important part in bringing big business to support Hitler. Later his enthusiasm was to turn to disillusion and he, like so many others, ended up in a concentration camp. Before that however his contribution to the economic success of Nazism was immense. What he did from 1935 to 1936 was to juggle with exchange rates and fix the import and export quotas of particular goods so skilfully that the industries vital to Germany's military strength expanded fast. Profits increased and so did business confidence. Businessmen also appreciated the Nazi attitude to labour relations. While employers were generally left in peace, trade unions were banned and what little labour unrest showed itself was quickly suppressed by the police. In 1936, however, Schacht resigned. Hitler was demanding that Germany be ready to fight a major war within four years yet forbade any unpopular measures like heavier taxation or the reduction in the production of consumer goods to make way for armaments. Schacht insisted that he was asking the impossible. The Führer was unimpressed. He replaced Schacht by General Thomas and ordered that a Four-Year Plan be put into effect. Government loans were used to encourage a massive drive towards greater economic self-sufficiency and an unexpected revival in world trade helped the Plan along without the need for any unpopular measures. Like Schacht, General Thomas and businessmen generally were shocked by Hitler's policy of 'wrapping the workers in cotton-wool' and by the enormous financial risks caused by his demands. But he got away with it. In 1939 Germany went to war against Poland with limited economic resources and maintained her war-effort by plundering the nations she defeated. Not until 1942 did the Nazi government begin to demand significant economic sacrifices of the German people.

In contrast to economic affairs, Hitler was both interested in and had a clear policy for the Jews. They were to be driven from every position of influence and, if possible, out of the country altogether. In 1935 the Nuremberg Laws ended the right of a Jew to be a German citizen and marriage between Jews and Aryans was forbidden. Thousands lost their jobs. By 1938 the civil service, journalism, farming, teaching, entertainment, the stock exchange, law and medicine were all closed to Jews. Furthermore shopkeepers all over the country, including essential ones like grocers and chemists, refused to serve them. And taunting signs like 'Jews enter this place at their own risk' or 'Drive carefully! Sharp curve! Jews 75 m.p.h.' sprang up everywhere. In 1938 legal and social harassment turned to open violence. On 7 November a Nazi diplomat in Paris was murdered by a teenage German Jewish refugee. In revenge, Goebbels and Himmler, Head of the S.S., set the secret police to work at organizing 'a spontaneous demonstration'. The result was, as a *Times*

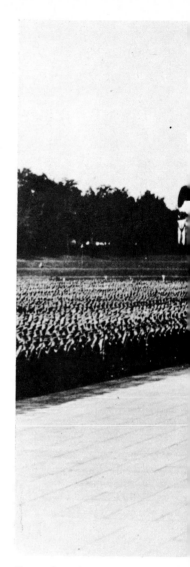

Power through pageantry: a party rally in Nuremberg, 1936. Eagles, banners, thousands of party members in their uniformed serried ranks with their leader in the middle of it all.

reporter put it, 'scenes of systematic plunder and destruction which have seldom had their equal in a civilized country since the Middle Ages'. At least 7,500 Jewish shops were looted, 200 synagogues destroyed and 20,000 Jews arrested; 70 were killed or seriously injured. This event became known as Crystal Night because so much glass was smashed. The Jewish community was then fined one billion marks to pay for the damage, rather than the Aryan insurance companies who were legally liable!

It is very difficult to estimate with accuracy how most Germans felt in 1939 after six years of Nazi rule. In a police state of such ruthlessness, wise men kept their opinions to themselves. Nor is it clear how much was generally known about Nazi violence. The existence of the concentration camps, in which there were between twenty and thirty thousand prisoners in 1939, was kept secret and, apart from Crystal Night, violence against the Jews was not given much publicity. If most Germans had lost much of their freedom they had little else to complain of. Unemployment was gone, their standard of living was rising. Furthermore, by a series of brilliant diplomatic moves, Hitler had made meaningless the hated Versailles Treaty and Germany was once again the most powerful and feared nation of Europe. There can be little doubt that he enjoyed even greater popular support in 1939 than in his first days of power.

Spain and Franco

Meanwhile in Spain the first forty years of the twentieth century were among the most unhappy in her history. The country was bitterly split. On the one hand, conservative groups like the landowners, the Catholic Church and the army were privileged, politically power-ful and strongly opposed to any change, while the growing working classes whose conditions were as bad as anywhere in Europe tended to anarchism and anti-clericalism. Moreover the Basques and Catalans in the north resented being ruled from Madrid. Between 1900 and 1920 assassination and terrorism were commonplace. In 1921 the Spanish army suffered a disastrous defeat at the hands of the Riff rebels in Morocco. This caused a political crisis which led, in 1923, to the end of the Cortès, the representative assembly, and to the dictatorship of General Primo de Rivera. Overwhelmed by the Great Depression, Primo resigned in 1929, old and weary, and in 1931 King Alfonso XIII retired into exile, conscious of a rapid increase in popular support for a republic.

A new constitution was drawn up in December 1931. Spain, it declared, was 'a workers' republic.' Radical reforms were begun which were intended to destroy for ever the privileged positions of the Church, army and landowning class. However the republicans tried to do too much too quickly. They also promised more to their worker and peasant supporters than they could possibly achieve in the economic conditions of the time. Spain slipped into chaos. There were peasant riots, innumerable strikes, military plots and anarchist terror. In 1934 a miners' strike devastated the city of Oviedo and cost 3000 lives before it was suppressed. The Falange, a fascist organization, built up support and the army began to plot a military takeover. In 1936 the anti-conservative parties—moderate Republicans, Socialists, Communists, Anarchists and Basque and Catalan separatists—formed a Popular Front government to continue the radical reforms begun in 1931. The army decided the time had come to take action. On 18 July 1936 General Franco in Spanish Morocco raised the flag of revolt against the Republican government

144

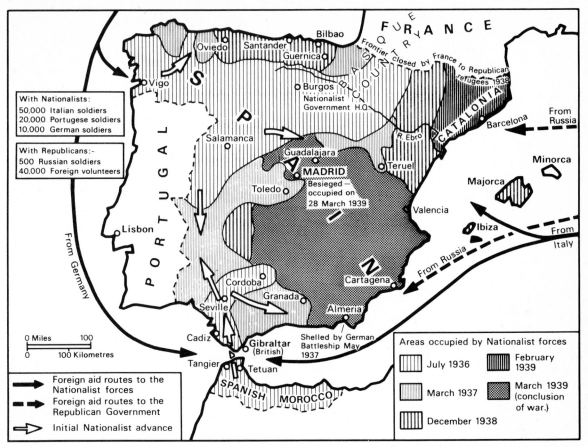

With Nationalists:
50,000 Italian soldiers
20,000 Portugese soldiers
10,000 German soldiers

With Republicans:-
500 Russian soldiers
40,000 Foreign volunteers

MADRID
Besieged –
occupied on
28 March 1939

Shelled by German
Battleship May
1937

Areas occupied by Nationalist forces

July 1936
March 1937
December 1938
February 1939
March 1939 (conclusion of war.)

Foreign aid routes to the Nationalist forces
Foreign aid routes to the Republican Government
Initial Nationalist advance

Map 6 The Spanish Civil War 1936–39

and ten days later invaded the mainland with his Moroccan troops. So began a civil war which was to last three years. Supporting Franco's Nationalist cause were the Catholic Church, the army (though not the navy and air force), the Fascists and the monarchists. From outside Spain he could count on the active support of both Mussolini and Hitler. For its part, the Republican government received some assistance from Russia and from individual sympathizers from all over the world who, in what seemed to them a crusade against Fascism, formed themselves into international brigades to reinforce the Republican armies.

When the war began, the southern and eastern parts of Spain, including the two main cities of Madrid and Barcelona, were Republican; the north, except for the Basque country, was Nationalist. In the first (1936) phase of the war, Franco tried but failed to win Madrid; in the second (1937), the Basques were overrun and the fragile unity of the Republicans was temporarily shattered when Anarchists fought Communists in the streets of Barcelona. The turning-point of the war came in the winter of 1937–38. The

145

Nationalists fought off a Republican assault on the strategically vital town of Teruel. Then they took the offensive and won control of a wide band of land from Teruel to the Mediterranean, thus cutting off Barcelona from the rest of Republican Spain. A desperate counter-attack by the Republicans in the Ebro valley failed in the summer of 1938 and what was left of the Republican army marched into a French exile. Barcelona fell in January 1939 and Madrid two months later. Franco established himself as a military dictator.

This terrible war cost the lives of about 600,000 people of whom probably half were killed on the battlefield and the rest in the mass executions and other atrocities which each side inflicted on the other. A major reason for the Nationalist victory was the support Franco received from Mussolini and Hitler. Italy provided infantry, planes and money, Germany a smaller but more efficient force of fighter-bombers and tanks. Foreign aid to the Republicans was much less effective. The Russians provided planes and tanks but on a smaller scale and the international volunteers could not make up by their enthusiasm what they lacked in weapons and training. The British and French governments' chief anxiety was to avoid getting involved. They set up a Committee of Non-intervention the aim of which was to persuade the nations of Europe to leave the Spanish to settle their differences on their own. This committee, which included representatives of Germany, Italy and Russia, was remarkable for its ineffectiveness.

Thirty-five years after the civil war ended, Franco was still dictator of Spain. He was often described as a fascist, without much accuracy. He was, rather, an extreme conservative whose main achievement was to keep Spain at peace after 1939 and to restore the traditional conservative institutions, especially the Catholic church, to their former powerful position in Spanish society. He was careful to avoid becoming entangled in a Nazi alliance during World War II and he planned that after his death the Spanish monarchy should be restored.

Chapter 8
Hitler's Triumphs

Aggression and Appeasement

'Better a minute on the battlefield than a lifetime of peace'.
Italian Fascist slogan.

'There must be a final active reckoning with France . . . a last decisive struggle.'
Hitler, Mein Kampf, 1925.

'The whole teaching of Hitlerism is to justify war as an instrument of policy and there is hardly a boy in Germany who does not view the preparation for ultimate war as the most important aspect of his life.'
S. H. Roberts (an Australian in Germany), 1936.

The main interest of both Hitler and Mussolini was foreign affairs. Not only did they wish to make their nations respected once again but they were determined to conquer other countries. Both were happy to use the threat of war and war itself to do so. And so successful were they that in 1942 Fascist Europe stretched from the Pyrenees and Belgium in the west of Europe to the suburbs of Leningrad and to the Caucasus in the east. Between 1933 and 1936 Mussolini had looked the stronger, more aggressive leader, but by 1942 he was virtually Hitler's puppet. The Europe of 1942 was Hitler's Europe. He had conquered more of the continent than any man in history, more even than Alexander the Great, or Julius Caesar, or Napoleon. His success was due to a combination of characteristics which amounted to evil genius. His aims were clear and simple. They were pursued with a fanatical determination and uncanny sense of timing. He possessed moreover a capacity to inspire the most misplaced trust in even the most experienced diplomats.

He had two main aims, the first of which he never made any attempt to conceal. As Point 2 of the Nazi political programme stated as early as 1920, 'equality of rights for the German People in its dealings with other nations and the abolition of the Peace Treaties of Versailles and St. Germain must be secured'. The second, which he talked much less about in the years before 1939 but had expressed clearly in Mein Kampf, was to win 'land and territory for the nourishment of our people and for settling our superfluous population'. This living-space or lebensraum would be won in the east from the Slav races and would also mean a war with France. Between 1933 and 1939 he had effectively abolished the Versailles Treaty. Between 1939 and 1942 France was smashed and 'living-space' in the east seemed to have been won to an extent far beyond his wildest dreams.

His first defiance of Versailles took place in 1933 almost as soon as he became Chancellor. He took Germany out of the League of Nations, arguing that France was being unreasonable about disarmament, and then accelerated the secret German armament production which had been going on since 1919. In March 1934 he publicly announced that the German army would be expanded to 600,000, six times the number laid down at Versailles. He also announced that Germany would have an air force (forbidden at Versailles) and a far larger navy. Britain, France and Italy— Mussolini at this time regarded himself as the champion of Austria against Germany—met together and condemned his action. So did the Council of the League of Nations. Hitler was then all sweetness and light. In a much-publicized speech to the Reichstag in 1934 he insisted that Germany's only desire was for peace. Her rearmament was only to secure her defences against her neighbours. 'Whoever lights the torch of war,' he said, 'can wish only for chaos. We however live in the firm conviction that in our time will be fulfilled . . . the renaissance of the West. That Germany may make an imperishable contribution to this great work is our proud hope and our unshakable belief.' Such words were what Europe desperately wanted to hear. Said *The Times*, 'It is to be hoped that the speech will be taken as a sincere and well-considered utterance meaning precisely what it says.' As Hitler expected, no country did anything more than protest and German rearmament continued. Furthermore Britain, who felt that France was unnecessarily bloddy-minded about Germany and that Europe would be a safer place if Germany ceased to feel so bitter about Versailles, signed a naval agreement with Hitler in 1935 which allowed Germany to build to up to 35% of the tonnage of the British navy. This agreement was made without consulting either France or Italy. It undermined the League of Nations, made nonsense of the Versailles Treaty and played directly into Hitler's hands.

Mussolini then took the limelight, conquering Abyssinia in open defiance of the League of Nations (see page 61). Britain and France only half-heartedly imposed sanctions on Italy. Fearing Hitler more than Mussolini, they were anxious to stay friendly with Italy if they possibly could while saving Abyssinia. Their policy could not have been less successful. Abyssinia fell and Mussolini left the League. He also became more friendly with Hitler who, for his part, could not fail to notice the feebleness of Britain, France and the League when faced with open aggression.

On 3 March 1936, therefore, Hitler took his first major gamble. He sent troops to occupy the Rhineland, that is the German territory

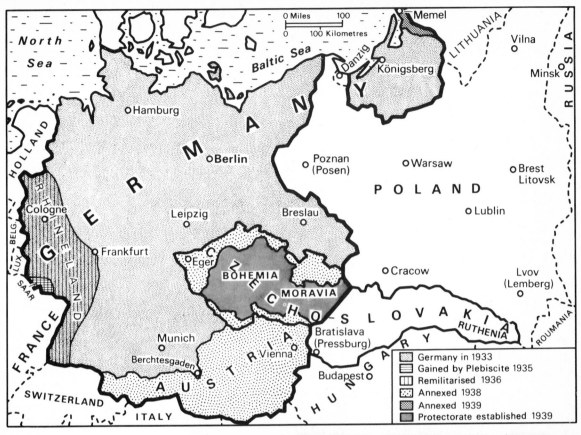

Map 7 German expansion
1936–39

Legend:

- Germany in 1933
- Gained by Plebiscite 1935
- Remilitarised 1936
- Annexed 1938
- Annexed 1939
- Protectorate established 1939

1936—German troops crossing
the Rhine into the de-militarized
zone, 1936

west of the Rhine with a fifty kilometre zone on the eastern bank which the Versailles Treaty had declared demilitarized. This was not just another defiance of Versailles, it was a direct challenge to France who had insisted, for her security, on the demilitarization of the Rhineland. It was a gamble because the German army simply was not strong enough to stand its ground if the French had threatened war. The troops would have had to make a humiliating withdrawal. The German general staff were appalled by the risks they were being made to take. Only Hitler kept his nerve throughout the crisis. His bluff was not called. The French were on the eve of a general election and though some ministers were in favour of immediate military action, the government decided to ask first for British support. The British however were not prepared to act. The comment that the Germans 'were after all going into their own back-garden' summed up the prevailing British attitude. Without British support, the French lacked the confidence to act alone. They protested but German troops stayed in the Rhineland. This was a decisive episode. In 1936 Hitler could have been stopped without bloodshed. He was not, and his confidence, his ambitions and his popularity within Germany all increased. He had gambled once and won. He would gamble again. Next time he would be harder to stop.

His successes attracted allies. In the autumn of 1936 Hitler and Mussolini decided to pursue a common foreign policy. They formed what il Duce termed 'the Axis' round which the rest of Europe would in future revolve. Soon afterwards Japan and Germany signed the 'Anti-Comintern Pact' to defend themselves against the menace of Communist Russia. It was a measure, Hitler told the world, to safeguard the whole of civilization against the Bolshevik menace. The new Fascist partnership was seen in action in Spain during the Civil War of 1936 to 1939 (see page 145). The Axis—especially Italy—supplied Franco's Nationalists with men and arms, including aircraft. Russia supplied the Republicans. Though the Spanish Civil War gave Axis forces useful battle practice, Hitler was careful not to become too involved there. He was sure that a much more important war was close at hand. As he pointed out to the Italian foreign minister in 1936, 'German and Italian rearmament is proceeding much more rapidly than rearmament can in England . . . In three years Germany will be ready'.

Two years later, in 1938, he pushed Britain and France much harder, again with considerable success. The Versailles Treaty had forbidden the unification of Austria and Germany. Since 1919 Austria had been a small country most of whose inhabitants were German-speaking and many of whom were passionately in favour of a union with Germany. In 1934 the Austrian Nazis assassinated Dollfuss, Chancellor and virtual dictator of Austria, but failed in their attempt

Something of a homecoming—Hitler's triumphant entry into Vienna in 1938, after the Auschluss

to seize power. At that time, Hitler was reluctant to give them active support because Mussolini seemed ready to defend Austrian independence. With the formation of the Axis, however, Hitler was able to interfere more and more in Austrian affairs. The Austrian Nazis redoubled their demonstrations, many of which were violent, against the government and early in 1938, Hitler summoned Schuschnigg, the new Austrian Chancellor, for an interview at which he demanded that two Nazis be appointed ministers, one of whom must command the police. Well aware that a German take-over was planned, Schuschnigg ordered a plebiscite to be held to see if a majority of Austrians really did wish to be united with Germany. This the Nazis were not prepared to risk. Arguing that law and order had broken down (in fact it was the Nazis themselves who were responsible for most of the rioting!) German troops marched across the border. On 13 March 1938 the union (*Anschluss*) of Austria with Germany was proclaimed, and Hitler entered Vienna in triumph the following day. There was no resistance. When he entered the city where just twenty-five years earlier he had been a starving vagabond, Hitler was, the German ambassador noted, 'in a state of ecstasy'. The S.S. then got to work. 80,000 possible political opponents were arrested and Jews publicly humiliated. Then a 'free and secret' plebiscite was held under Nazi supervision which produced a 99% vote in favour of the Anschluss! The British and French reaction was that there was nothing they could do.

Hitler's second move in 1938 was even more serious. His target this time was Czechoslovakia. Czechoslovakia had been created in 1918, and, though a mainly Slav country, contained a German-speaking population of about three million, most of whom lived in the Sudetenland, an area which bordered on Germany and Austria. In the 1930's a Sudeten German Party won considerable support in the Sudetenland; it was led by Henlein and demanded greater independence within Czechoslovakia. The Nazis were quick to support Henlein, and in the spring of 1938 Hitler told the Reichstag in Berlin that it was his intention to give the Sudeten Germans the same right of self-determination that every other race, except the Germans, had been granted by the Versailles Treaty. Simultaneously he told his generals to prepare for war. The Czech government, while ready to consider Henlein's demands for greater freedom within Czechoslovakia, refused to consider the handing over to Germany of the Sudetenland. Such a change in the border would drastically weaken her defences against her most dangerous neighbour as well as deprive her of some of her most valuable industry. Though a small country, Czechoslovakia possessed a most modern and capable army. She was also convinced that she could count on the support of Britain and France and therefore prepared to defend herself.

Britain and France however were most reluctant to give Czechoslovakia their whole-hearted support in this crisis. Chamberlain, the
British Prime Minister who had made foreign affairs his particular
concern, believed that many of Hitler's demands were reasonable.
He had no doubt that the eastern borders of Germany would have to
be revised in Germany's favour if peace in Europe was to be preserved.
He believed that the Sudeten demand for the right of self-determination was just, and, perhaps most crucially, he believed he
understood the Führer and that when he gave his word that he had no
intention of destroying Czechoslovakia, he could be trusted to keep
it. In September, serious trouble erupted in the Sudetenland.
Henlein fled the country and the Czechs proclaimed martial law.
War seemed very close. Chamberlain went into action. On 1
September he flew to Berchtesgaden and met Hitler, who demanded
immediate self-determination for the Sudetenland which in turn
would mean its takeover by Germany. Chamberlain accepted these
terms and managed to persuade the French and the Czechs (who
believed that they had no chance on their own) to accept them too.
He then met Hitler again at Bad Godesberg on 22 September.
To his dismay, he found that the Führer had upped his demands.
The Sudetenland should be handed over to Germany without delay.
Moreover plebiscites were to be held wherever else in Czechoslovakia
there were significant proportions of German-speaking people. Polish,
Hungarian and Slovak claims against the Czechs should also be
satisfied—and by 25 September at the latest. Chamberlain returned
to London a worried and exasperated man. He was ready to accept
Hitler's new terms but had trouble in persuading his cabinet and
the French to go with him. The Czechs now made it clear that
Hitler's demands were impossible. 'They would deprive us,' they
said, 'of every safeguard of our natural existence. We rely,' their
government's statement ended, 'upon the two great Western Democracies whose wishes we have followed much against our own
judgement, to stand by us in our hour of need.' The French seemed
to be standing firm. They backed the Czech stand and began
mobilizing their army. With the German generals not at all keen on
the prospect of fighting, Mussolini getting cold feet and the combined firmness of the French and Czechs, Hitler paused. He wrote to
Chamberlain a charming moderate-sounding letter asking him to do
his best to bring the Czechs 'to reason at the very last hour'.

It was an inspired move. Instead of war, Chamberlain was able to
arrange a conference at Munich at the end of September attended
by Hitler, Mussolini, Daladier for France and himself. The Czechs
were not invited. Hitler would not have it. The nearness of war had
unnerved Chamberlain and Daladier. 'How horrible, fantastic, in-
credible it is,' Chamberlain had said to the British people on a

B.B.C. radio broadcast a few days before, 'that we should be digging trenches because of a quarrel in a faraway country between people of whom we know nothing.' They agreed that the Sudetenland was not worth a major war and worked to maintain peace at almost any price. Consequently, at Munich, Hitler got almost all he wanted. German troops were to occupy the Sudetenland on 10 October. There should be further plebiscites and Polish and Hungarian claims met by the Czech government. Then the Czech representatives were informed by British officials what the terms were. Their bitter protests were met with the reply 'If you do not accept, you will have to settle your affairs with Germany absolutely alone'. They had no choice but to agree. 'We were abandoned. We stand alone', said their President Benes when he broadcast the news to his people. Chamberlain, however, returned to his people a hero. At one stroke, he seemed to have banished the terrible fear of war which had loomed so darkly over Europe before the Munich conference. 'My good friends,' he said to the cheering crowd outside 10 Downing Street, 'this is the second time in our history that there has come back from Germany to Downing Street peace with honour. I believe that it is peace in our time.' Daladier, who had a much clearer understanding of the extent to which he and Chamberlain had sold out the Czechs, thought the crowds waiting for him at the airport were there to lynch him. He was both relieved and surprised to find they too had come to cheer.

Once again Hitler had triumphed by threat of war, strong nerves and a skilful policy of playing upon the hesitations and divisions of his opponents. Once again he was bluffing. Almost certainly the combined Czechoslovak, French and British armies would have been too strong for a German force not yet fully ready for a major war.

He now bothered less about cloaking his aggression in reasonable-sounding arguments about the right of German minorities to self-determination. In March 1939 he occupied the rest of Czechoslovakia. For Chamberlain, and for British public opinion, this was, at long last, the turning-point. Hitler could not be trusted and any further Nazi aggression must be resisted by force. The British rearmament programme was speeded up, conscription introduced for the first time ever in peacetime and the U.S.S.R. approached in the hope of signing an anti-Nazi agreement.

Hitler, however, was convinced that he could now ignore Britain and France. Ten days after occupying Czechoslovakia he occupied Memel, a Baltic port hitherto part of Lithuania. In April 1939 Mussolini invaded Albania and the two signed the Pact of Steel, promising each other assistance in the event of war.

Hitler then set about treating Poland like Czechoslovakia. The circumstances were similar. In the area round the great Baltic port of Danzig and in the Polish Corridor a large proportion of the population were German-speaking. Many of them—as in the Sudetenland—wished to be united with Germany. Demands that this German minority should be granted self-determination, coupled with the threat of war if this self-determination was not allowed by the Polish government, should be enough to bring about the downfall of Poland in 1939 as they had done for Czechoslovakia in 1938. Hitler's confidence was boosted by a considerable diplomatic success in August 1939. He persuaded his arch-enemy Russia to sign a non-aggression pact (the Nazi-Soviet Pact). Stalin had hoped for some kind of anti-Nazi agreement with Britain and France. These two powers, however, moved very slowly. Their chief aim was to safeguard Poland and the Polish government regarded Stalin as no less a menace than Hitler. Stalin therefore sided with Hitler (see page 103). The terms of the Pact were that Estonia, Latvia and part of Poland should go to Russia, Lithuania and the rest of Poland to Germany. The only power therefore that could have effectively helped Poland was bought off.

Britain and France, however, backed Poland much more firmly than they had backed Czechoslovakia. Hitler was not impressed. 'They are little worms,' he said. 'I have seen them at Munich.' When the Poles refused to consider his demands concerning the German minority, he threatened war. When this had no effect and Britain and France stood firm behind the uncompromising Polish government, he sent his troops into action. On 1 September Germans in Polish uniform faked an attack on a German radio station and the German invasion of Poland began. After a day's wait in the hope of peace negotiation, Britain declared war on the morning of 3 September and France followed her example the same afternoon. World War II had begun.

The policies of the British and French governments towards Hitler between 1933 and 1939 are and have been easy to criticize. They are usually described as the policies of appeasement, which, meaning the 'soothing of threatening enemies by satisfying their demands', is not a complimentary term. They look spineless, badly thought-out and their consequences were catastrophic. Nonetheless, they were the thoughtful policies of intelligent men and until 1939 they had the overwhelming support of public opinion in both Britain and France. Fundamental to British policy was the belief that the Versailles Treaty had been unnecessarily hard on Germany and that Hitler's determination to abolish the Treaty was not unjustified. Coupled with this was the memory of the horrors of World War I and a reluctance among all sections of the population to consider the need for another war. Consequently German rearmament, the occupation of the Rhineland, the Anschluss and the demand for self-determination of the Sudetenland were not regarded as particularly dangerous. Hitler's methods of bargaining might be a trifle heavy-handed but they certainly did not justify counter-threats. Chamberlain was a tough and experienced politician. He was convinced that Germany had justifiable complaints against the Versailles Treaty. His policy of appeasement was designed to end these complaints and maintain the peace of Europe through reasonable negotiations by reasonable men. The fatal flaw in this otherwise logical policy was Hitler's lack of reasonableness. 'In spite of the hardness and ruth-lessness I thought I saw in his face,' Chamberlain commented about Hitler in 1938, 'I got the impression that here was a man who could be relied upon when he had given his word'. He was wrong, but he should not be too harshly judged for that. Hitler deceived literally thousands of Germans and non-Germans in the course of his career. For Chamberlain, the events of 1939 and 1940 were a personal

In the early days of the Munich Crisis, Chamberlain flew to see Hitler in his mountain retreat near Berchtesgaden. This Punch *cartoon reflects the profound but vain hopes most Britons had in Chamberlain's dramatic efforts to keep Europe at peace*

STILL HOPE

tragedy. When war was finally declared, he made this pathetic broadcast to the British people. 'This is a sad day for all of us and to none is it sadder than to me. Everything I have worked for, every-thing I have believed in during my public life has crashed into ruins.'

Though French policy worked out much the same as the British, the reasons behind it were different. For the French, the Versailles Treaty was too soft not too hard on the Germans whom they never trusted either before or after Hitler came to power. After the failure of the Ruhr occupation, however, they never felt strong enough to act against Germany on their own. Without British support, they felt they could only stand and watch while Hitler rearmed and occupied the Rhineland. Furthermore, from 1934 onwards, political disunity within France became so serious as to paralyse French foreign policy. Some conservative Frenchmen came to think that Nazi rule in France might be preferable to a French socialist government. The country grew defeatist and, hopeful that the newly-built Maginot line of fortification was an effective defence against German attack, proved ready to follow Chamberlain's 'appeasement' approach. As in Britain, pacifism was deep-rooted. Just after the Spanish Civil War began, the novelist Roger Martin du Gard wrote to a friend: 'I am hard as steel for neutrality. My principle: anything rather than war! Anything, anything! Even fascism in Spain . . . even fascism in France! Anything: Hitler, rather than war!' Both Britain and France believed that Germany was militarily much more powerful than in fact she was; they also were too ready to treat Mussolini with kid gloves in the unreasonable hope of diverting him from his German alliance, and they were unnecessarily suspicious of Russia as a possible ally. All in all the two western democracies were both deceived and outmanoeuvred by the Axis dictators.

The Phoney War

From the military point of view, Britain and France could not have chosen a worse issue on which to fight Hitler than the defence of Poland which simply could not be defended from a German attack by the two Western nations. The German attack was swift, overwhelming and deadly. The Polish frontier was long, most of the countryside ideal for the mechanized warfare the German army had perfected. Waves of dive bombers and heavy bombers disrupted Polish communications, harassed the infantry and spread panic in the big cities. The German armoured columns swept all resistance aside. In less than a fortnight the bulk of the Polish army was destroyed and the Russian army, marching in from the east, was able to take its share of the country with hardly any fighting. This did not however prevent it from committing the first of the all too many atrocities of World War II. Thousands of Polish officers surrendered to the Russians rather than the Germans in the hope of better treatment. They were cold-bloodedly murdered and buried in a mass grave in the Katyn forest. Warsaw surrendered at the end of September, and the last Polish army a week later. Then Poland ceased to exist. When the war began, Britain and France had no plans of how to help their unfortunate ally and the German victory came so swiftly that in fact they did next to nothing. The French began a tiny attack against the western frontier of Germany which they stopped once it became clear that Polish resistance was collapsing. Thereafter, they concentrated on strengthening their defences. The British were even feebler. For fear of provoking the Luftwaffe (the German Air Force) into bombing Britain, British bombers were careful not to bomb Germany except with propaganda leaflets. Neither the Germans nor the Poles were much impressed.

Then followed a curious period through the autumn and winter of 1939–40 when both sides were officially at war but there was next to no fighting. The Germans called it the 'Sitzkrieg' (the Sitting-War), the British 'the phoney war'. It happened because Hitler had attacked Poland convinced that Britain and France would not have the nerve to fight and his armies were not ready to fight France in 1939. And Britain and France, while having the nerve to declare war, had no plans to attack Germany. They had assumed that Hitler and Mussolini would fight together and planned an attack on Italy. Mussolini, however, had stayed neutral! The only serious fighting was in Finland between Russia and the Finns. Stalin's aim was to seize land from Finland to strengthen the northern frontier of Russia near Leningrad. He assumed that he could get what he wanted without fighting because Finland after all was very small and like Poland quite without effective allies while Russia was very large. The Finns however fought, and fought heroically. Though eventually they had to accept defeat and sign away all the land which Stalin wanted by the Treaty of Moscow (1940), they kept their independence. Moreover, their successes against the Russian army persuaded Hitler that the German army could make mincemeat of the Russians whenever he wished.

The only active policy Britain and France pursued against Germany at this stage was an economic blockade and Britain's attempts to make it more effective ended the phoney war. In April 1940 Chamberlain's government decided to mine the coast of Norway and disrupt the supply of vital Scandinavian iron ore to German industry. Hitler, however, was yet again a step ahead. Denmark was occupied on 9 April and Norway then invaded in such force and with such speed that the Norwegian army and the small British force backing it up were crushed. Southern and central Norway were in German hands by the end of April and the northern areas finally conquered in June. The fall of Norway was a bad blow for Britain. Once again she had been out-thought and outfought by Germany. Not long before, Chamberlain had commented, with a typically unfortunate choice of phrase, that 'Hitler had missed the bus'. Now Hitler had proved him wrong again and his leadership came under increasing criticism. The Norway defeat caused a passionate debate in the House of Commons. Amery, a leading member of Chamberlain's Conservative party, turned on the Prime Minister using Oliver Cromwell's famous words to the Rump Parliament, 'Depart, I say, and let us have done with you. In the name of God, go!' And Lloyd George, the great war-leader of World War I and now in his 70's, ended his last important speech by telling the Prime Minister that 'there is nothing which can contribute more to victory . . . than that he should sacrifice the seals of office'. Realizing that he was fast losing the confidence of his party and of the nation, Chamberlain resigned and King George VI asked Winston Churchill to form a national government representing all the major political parties in the country.

Churchill was 66, an experienced politician with a stormy record. He was the son of Lord Randolph Churchill, a famous Conservative politician and his brilliant American wife Jennie. He entered politics as a Conservative but switched to the Liberals, for whom he was a successful cabinet minister from 1906 to 1915. The Gallipoli campaign (see page 33) brought temporary disgrace but after the war he rejoined the Conservatives and was Chancellor of the Exchequer from 1924 to 1929. He won himself a reputation as the scourge of striking workers and as an extreme imperialist. In the 1930's he quarrelled with the party leaders over Indian affairs and foreign policy and was for most of the decade a backbencher of little influence. Earlier than most, he realized how dangerous Hitler really was and constantly warned the country against the policy of appeasement. When Chamberlain returned from Munich in 1938, talking of 'peace with honour', Churchill told the House of Commons, above the shouts of protest from Chamberlain's angry supporters, 'We have sustained total and unmitigated defeat . . . We are in the midst of a disaster of the first magnitude . . . And do not suppose this is the end. It is only the beginning.' When it became clear that he had been right all along, his fortunes improved. On the outbreak of war he became First Lord of the Admiralty and immediately displayed a fiery energy and determination which made his colleagues seem pale shadows in comparison. Churchill was a fighter by temperament, a man of tremendous passions. In times of peace he seemed impetuous and lacking in judgement, but in a fight to the finish against Nazism, no country could have wished for a better leader, especially in the dark days of defeat. He was an orator of spell-binding power and his speeches in the House of Commons and over the radio were a constant source of hope and inspiration not only to Britain but to the whole free world. And he had complete confidence in himself. 'I felt that I was walking with Destiny,' he recalled after he had just been appointed Prime Minister, 'and all my past life had been a preparation for this hour and this trial. I thought I knew a good deal about it all and I was sure that I should not fail.'

The Fall of France

The events of the rest of 1940 were to stretch his powers of leadership to the utmost. On paper, the position of France and Britain seemed quite strong, in reality it was desperate. Though their combined population outnumbered Germany's (89 million to 68 million), and though their armies were slightly and their navies considerably larger, in vital matters they were much weaker. They could only put 2,700 aircraft in the air against 4,000, and though they had as many tanks as the Germans, they were scattered and not massed as the Germans used them. In fact the German army was incomparably the best in the world. It had perfected a new type of warfare, as the campaigns in Poland and Norway showed, in which air attack, armoured units and infantry were blended together in a campaign of rapid movement. The 'blitzkrieg' tactics had been recommended by both French and British military writers but only the Germans had put them into practice. By 1939 their armour, including excellent tanks taken from the Czechs, was superior in quality to the French and British; so too were their supporting aircraft. Their front-line generals were also better. Men like Guderian, Manstein and Rommel were not only masters of the 'blitzkrieg' but commanders of individual flair and courage. The German troops were well-trained. Many of them had already had experience in Spain and Poland. Their morale was high and they had a zest for war. The main weight of any German attack was bound to fall on the French army. Compared with the German, the French army's thinking was completely out of date. With their right flank secured by the massive fortifications of the Maginot line, their leaders assumed that the Germans must attack as in 1914 across central and western Belgium and could be held there by a French counter-attack. But the slow movement of the army, its unimaginative use of armour and its slow-thinking commanders, first Gamelin, then Weygand, left it quite unprepared to cope with the blitzkrieg when it came.

On 10 May 1940, the day Churchill became Prime Minister, the German attack on the west began with German paratroops pouring out of the sky over neutral Belgium and Holland. German troops moved swiftly across the borders to join them. Thinking this to be the main attack and their right flank secure, the French army lumbered into Belgium where it was joined by the British Expeditionary Force. The main German attack, however, was launched through the

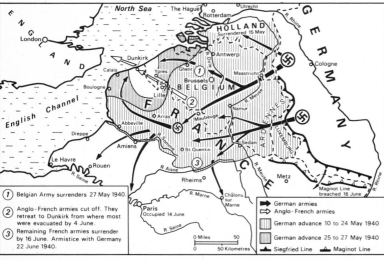

Top left *Beneath the Maginot Line. Supplies and reinforcements could be rapidly moved along this formidable fortification. The tragedy was that it only defended the French border with Germany and Luxemburg and was not extended along the French-Belgian frontier which is the way the Germans eventually came*

Centre left *German panzer units speed towards the Channel*

Map 8 Fall of France 1940

① Belgian Army surrenders 27 May 1940.

② Anglo-French armies cut off. They retreat to Dunkirk from where most were evacuated by 4 June.

③ Remaining French armies surrender by 16 June. Armistice with Germany 22 June 1940.

Above *This artist's impression of the last stages of the Dunkirk evacuation looks south-west. The Germans have the Anglo-French army trapped against the sea and are only a few miles away behind the burning oil dumps. The English coast is some 50 miles away and a tremendous variety of boats, some so small that they can come close into the shore, are helping with the evacuation*

Ardennes, an area so hilly that the French had assumed no effective tank attack could there be made. Careful German planning and inspired leadership by Guderian and by Manstein (whose plan it was) pulled it off successfully. Guderian's armoured columns smashed through the French defences at Sedan and then hurtled eastwards towards the Channel. He got there on May 20—after just ten days fighting—and once he did so, the battle for France was won. The main Allied army was now trapped in Belgium, another French army was cut off in the Maginot area and Paris lay undefended. Remorselessly the Germans pushed the Allied army in Belgium into a smaller and smaller pocket backed by the Channel coast. Eventually, more than half a million men were trapped near the small port of Dunkirk, and appeared to face nothing but total annihilation. On 24 May, however, the German armour halted and, in this most unexpected and welcome breathing-space, a fleet of small boats operating from the south-east of England managed to evacuate about

340,000 mainly British troops before Dunkirk finally fell on 4 June after heroic resistance by the French rearguard. The extraordinary decision of May 24 seems to have been Hitler's. It appears that he was terrified lest the German armour should get bogged down in the muddy Flanders countryside, and he allowed Goering to convince him that the Luftwaffe could finish off the Allied army on its own. The R.A.F. and bad weather combined to prove Goering wrong.

The Germans made no more mistakes. Holland had surrendered on May 15, Belgium on May 28. Mussolini, now sure that France was beyond recovery, invaded the south of France. An undefended Paris was occupied on June 14. With France now completely demoralized, Reynaud's government thought briefly of moving to North Africa and of continuing the struggle from there. It decided against it however and Reynaud handed over to the aged Marshal Pétain to make what peace he could with Hitler. Finally on 22 June the most spectacularly successful campaign in modern history ended when the Führer asked for the French to sign the terms of surrender in the same railway carriage in which the Germans had admitted defeat in 1918. Hitler sat in the chair which Marshal Foch had taken twenty-two years before. An American journalist saw him that day. His face was 'afire with scorn, anger, hate, revenge, triumph'. What Imperial Germany had failed to achieve in the four years of the First World War, his armies had managed in six weeks. The northern half of France including Paris and the whole of the Atlantic coastline down to the Pyrenees passed under direct German rule. The rest was governed from the town of Vichy by a French government headed by Pétain which was ready to collaborate with the victorious enemy. The whole continent of Europe lay at Hitler's feet.

Summer 1940. France has fallen. The swastika flies over Paris

Hitler was now ready for peace. He regarded Britain's position as quite hopeless and expected her to sue for terms. He had no wish to inflict further humiliation on Britain for whom he had a sneaking admiration. If Britain would leave him to rule Europe as he pleased he would leave the British Empire alone. What he failed to realize— and indeed never properly understood—was the mood of the British people and of Churchill's government. By any logical analysis, Hitler was right, their situation was hopeless. Though most of their army had been evacuated successfully from Dunkirk, it had had to leave all its equipment behind. Their air force was small in comparison to the Luftwaffe, their navy could find no answer to the German U-boats. All their allies were conquered so they were completely alone.

Yet no serious thought was ever given to negotiating peace-terms. To his astonishment, all Hitler got was defiance. On June 4, as the Dunkirk evacuation ended, Churchill spoke to the House of Commons and to the world. If Dunkirk was a miracle of deliverance, he said,

This Partridge cartoon, published by Punch *as the British and French armies were cornered at Dunkirk, superbly reflects the attitude of Churchill and the British people in the dark days of 1940 and 1941*

AT BAY

it was also a terrible defeat. 'Wars are not won by evacuation.' But he continued. 'Even though large tracts of Europe and many old and famous states have fallen or may fall into the grip of the Gestapo and all the odious apparatus of Nazi rule, we shall not flag or fail. We shall fight in France, we shall fight in the seas and oceans, we shall fight with growing confidence and growing strength in the air, we shall defend our island whatever the cost may be, we shall fight on the beaches, we shall fight on the landing grounds, we shall fight in the fields and in the streets, we shall fight in the hills, we shall never surrender and even if, which I do not for a moment believe, this island or a large part of it were subjugated and starving, then our Empire beyond the seas, armed and guarded by the British Fleet,

163

would carry on the struggle until, in God's good time, the New World, with all its power and might, steps forth to the rescue and liberation of the Old.' He meant every word he said. It was 'Britain's inflexible resolve to continue the war'. Puzzled and annoyed, the Führer and his advisers turned to planning the conquest of the British Isles.

The Battle of Britain

The plan they produced (which became known as Operation Sea Lion) anticipated 250,000 men being ferried across the Channel in barges and landed along the Kent and Sussex coast. Without air supremacy, however, such an attack would be suicidal. Hitler therefore ordered the Luftwaffe to destroy the R.A.F. and the 'Battle of Britain' began. The odds seemed heavily against the R.A.F. which had only 820 up-to-date fighter-aircraft against 2,600 enemy of which more than 1,000 were fighters, the rest bombers of various sorts. However it did have some advantages. Its fighters were the equal if not better than the enemy's. Experts rate the Spitfire the best, followed by the German Me 109, followed closely by the Hurricane. The other German fighters were outclassed. The British pilots, though trained rapidly, were trained well. They possessed confidence in themselves and their machines and morale was high. Most important of all, the R.A.F. possessed radar, a system of detecting approaching aircraft which was perfected by a team of scientists headed by Watson-Watt between 1935 and 1939. A network of radar stations fed information to the Headquarters of Fighter Command at Stanmore in Middlesex, enabling the incoming enemy formations to be met with a swift and co-ordinated defence.

The battle began on August 8. It was very closely fought. In the first phase the Luftwaffe concentrated on shipping in the Channel, in the second, 1,000 planes struck at R.A.F. airfields and the radar stations. By August 23 the airfields were so badly damaged and the British pilots so exhausted by constant missions that the total destruction of Fighter Command seemed likely. However, on the night of August 23, some German bombers, having flown off-course, dropped their bombs on the centre of London, killing some civilians, instead of on the oil depots on the city's edge. On Churchill's instructions, British bombers raided Berlin the next day in retaliation. For Berliners, the shock was considerable. After months of effortless victory they had never dreamt that they could be in danger. Hitler flew into a rage and ordered the Luftwaffe to switch from the airfields to London. It was a major error. Though 800 Londoners lost their lives, the R.A.F. was given precious time to repair the airfields and radar stations. Moreover, the British Bomber Command was able to damage the barges and supply dumps of the invasion fleet. On 15 September the R.A.F. beat off two

In the late summer and autumn of 1940, London was heavily bombed

Top left Here Churchill visits part of the East End damaged in a raid

Top right Spitfires in formation over South-East England during the Battle of Britain

In one of the most famous photos of the war, St Paul's survives among the flames

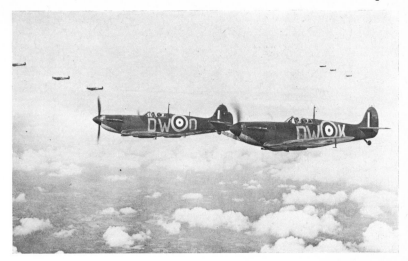

massive attacks on London. Plainly air supremacy over the Channel was far from won. Two days later, Operation Sea Lion was postponed indefinitely.

The Battle of Britain was an event of major importance in the early years of the war. It was Hitler's first serious setback since he came to power in 1933 and the first clear defeat of any part of the German armed forces since war began. Though the British Empire alone stood no chance of breaking the Nazi hold on Europe, so long as the British went on fighting they kept alive the hopes of millions who, though conquered, maintained the will to resist. Many governments in exile were set up in London, of which the Free French, led by de Gaulle, was the most important. In the long term the British Isles provided a base from which in due course a successful onslaught on Hitler's Europe could be launched. If Britain had fallen, the destruction of the Nazi tyranny would have been immensely more difficult. The world's debt to the young pilots of Fighter Command is therefore hard to exaggerate. In the words of Churchill, who was speaking as the Battle of Britain still raged overhead, 'Never in the field of human conflict was so much owed by so many to so few.'

The Invasion of Russia

Britain had other causes for satisfaction in the autumn of 1940. One Italian army ran into trouble in Greece and Albania, while another, invading Egypt from Libya, was heavily defeated by a British army at Sidi Barrani. Hitler's confidence, however, was unaffected. So complete had been the success of his armies that he felt strong enough in the spring of 1941 to plan the invasion of Russia through 'Operation Barbarossa'. The success of Barbarossa would mean his complete domination of Europe and the conquest of the living-space in Eastern Europe which the Nazis believed to be so important for the German people. He had no doubt that success would be quickly won. The Russians were an inferior race. Their army had made a poor showing against the Poles and the Finns. 'You have only to kick in the door', he said to his generals, 'and the whole rotten structure will come crashing down.'

Before Operation Barbarossa could begin, he had to bale out the Italians in the Balkans and North Africa and crush a popular and anti-German movement in Yugoslavia. In March a German army stopped the British advancing into Libya and in April a textbook blitzkrieg campaign dealt with Yugoslavia in twelve days at the cost of only 151 German deaths. Greece was then invaded and before the end of April the swastika was flying on the Acropolis in Athens. The Balkan triumphs ended with a brilliant paratroop attack which won the island of Crete from the British. The Balkan campaign seemed

1941—German successes continue . . .

Above *The Russian plains and prisoners stretch as far as the eye can see*

Right *The British surrender in Crete*

to be an unqualified success. However, it delayed the invasion of Russia by four important weeks. Could the German armies achieve complete victory before the Russian winter set in?

It looked at first as if they could. Despite constant warnings by Churchill and others, the massive attacks launched on 22 June 1941 took Stalin by surprise. 175 divisions drove deep into Russian territory. Whole Russian armies were surrounded. By the end of September Leningrad was cut off from the rest of the country, and by October the German army was less than sixty miles from Moscow. Stalin left the capital, and with Russia's most productive areas in enemy hands, complete German victory seemed certain. On 3 October Hitler broadcast to the German people. 'I declare and declare it without reservation that the enemy in the East has been struck down and will never rise again.' He was counting his chickens. Russian resistance stiffened and fighting with a courage that the Germans had never met before, it managed with the help of autumn mud to slow the enemy advance and hold on to both Leningrad and Moscow. An all-out assault in November took German troops into the Moscow suburbs but then Zhukov, the Russian commander, ordered a Russian counter-attack. Making good use of the wintry conditions, the Russians drove the Germans back as much as 150 miles in some places before the front was stabilized for the duration of the winter.

Outside Moscow late in 1941, Zhukov's troops prepare a counter-attack . . .

World War

So little bothered was Hitler by his failure to win the complete victory in Russia he had planned for 1941 that as the Russians counter-attacked outside Moscow, he declared war on another of the world's most powerful nations, the U.S.A. A war which had been brewing for years between Japan and the U.S.A. began on 7 December 1941 with the surprise Japanese attack on the American naval base at Pearl Harbor. Germany and Japan had been allies for some years and Hitler was extremely annoyed by America's readiness to supply economic and military aid to Britain. It was an astonishing decision astonishingly lightly taken. The Führer had no understanding of the strength of the U.S.A. 'I don't see much future for the Americans,' he said. 'It is a decayed country.'!

Thus the European war became a World War and still the dictators triumphed. Indeed the first half of 1942 saw some of their most spectacular successes. In February, with total naval and air supremacy, the Japanese sank the British battleships *Prince of Wales* and *Repulse* off Malaya and then overran the whole of the peninsula, including the great British port of Singapore. The Dutch East Indies fell in March, the Philippines in May, and by July 1942 they were threatening the north-east frontier of India as well as Australia and the southern Pacific.

While in Leningrad, some inhabitants bombed out of their homes seek shelter elsewhere in the besieged city

169

Pearl Harbour
Above *The deck of the Japanese carrier 'Hiryu', minutes before the attack was launched*
Left *'Battleship Row'—their main target as the attacking pilots saw it*

Anguish in a Singapore side-street after a Japanese air-raid. It was not just the British who suffered during the Japanese attack

Simultaneously, the Germans took the offensive against the British in North Africa. Under the command of Rommel, one of the outstanding generals of World War II, the Afrika Corps captured the important port of Tobruk and in August was preparing to invade Egypt and seize the Suez Canal. Britain's vital sea-link to her Middle Eastern oil supplies and to India were thus in the greatest danger. Moreover in Russia an even greater German offensive began. It was aimed with concentrated force in the southern sector of the front at the Volga basin and the huge Caucasian oilfields. The Volga River was reached on 23 August and the strategic centre of Stalingrad besieged. Two days later German mountaineers fixed the swastika on the top of Mount Elbruz, the highest peak in the Caucasus. Finally, in the stormy waters of the Atlantic, the German U-boats seemed to be winning the Battle of the Atlantic which was as crucial to Britain's survival as the Battle of Britain. Working in 'wolf-packs', the German submarines sunk 1,664 ships in 1942, a tonnage of 17,790,697 which was more than the Allies were able to build in the same period. Not only did this U-boat success make it impossible for the U.S.A. to pose any immediate threat to Nazi Europe but it was bringing Britain, as in 1917, close to economic collapse. In the late summer of 1942 the world-wide victory of the dictators seemed likely.

Chapter 9
The Destruction of Nazism

Four Critical Battles

Map 9 Nazi power at its peak 1942

In June 1942 a Nazi victory looked almost certain. By June 1943 it was difficult to see what could prevent a Nazi defeat. In the four main theatres of war—in Russia, North Africa, the North Atlantic and the Central Pacific, the Allies gained the upper hand, and from early summer 1943 the tide of war began to run strongly in their favour.

Midway The first successes came in the Pacific. Between June 2 and 6 1942 a Japanese carrier fleet threatening Midway Island was severely damaged and forced to withdraw by the Americans. 'Midway', wrote a leading U.S. naval historian, 'was the first really smashing defeat inflicted on the Japanese navy in modern times.' The much larger and more lengthy battle for the Solomon Islands which began in August 1942 further underlined America's growing superiority. When it eventually ended in another U.S. victory, the Japanese had lost some 3,000 aircraft and 6,000 airmen and the striking power of their navy was considerably weakened.

El Alamein While the Americans were triumphing at Midway, a German general, Rommel, known by both friend and foe alike as the Desert Fox, was giving the British a terrible time in the sandy wastes of North Africa. Though starved of supplies by Hitler—who never properly understood the importance of North Africa—Rommel, with his small but excellent Afrika Corps, backed up by Italian divisions, drove through the position of the British and Commonwealth Eighth Army at Mersa Matruh (see Map 10, opposite) at the end of June. At the First Battle of El Alamein (July), General Auchinleck just managed to prevent him breaking through to the Nile Delta. This achievement was more significant than it looked at the time. Rommel's supply position was desperate. Through their control of the island base of Malta, which somehow managed to survive all Axis bombing and blockade attempts to force it to surrender, the British could both stop Rommel's supplies and reinforce their Eighth Army. Consequently when Montgomery (who had replaced Auchinleck) was ready to counter-attack in October, he had a two to one advantage in men and weapons (1,029 tanks against 496) and complete air superiority. The Second Battle of El Alamein lasted from October 23 to November 4 and though Rommel hurriedly returned from a hospital

Map 10 The North African campaign 1942–43

Axis Powers in 1939

Powers co-operating with Axis

Territory occupied by Axis

France — Vichy Governed

Neutrals

Unconquered

0 Miles 300

0 300 Kilometres

0 Miles 600

0 600 Kilometres

Rommel's advance from Tripoli towards Cairo 1942

The 8th. Army meeting and driving Rommel back at El Alamein

Rommel retreats into Tunisia

The 8th. Army advancing on Tunis along North African coast

The Anglo-American Armies advancing from Casablanca towards Tunisia

Anglo-American attack on Sicily 1943

Oilfields in middle-east

173

bed in Austria to take command in person, he could find no answer to Montgomery's continuous and varied attacks. As the Afrika Corps retreated westwards, an Anglo-American army landed behind them in Morocco and Algeria. 'This', wrote Rommel, 'spelt the end of the army in Africa.' In May 1943 240,000 Axis troops, half German, half Italian, surrendered in Tunisia. The North African campaign was over.

Stalingrad Meanwhile, the most decisive battle of the war had begun on the Russian front in the valleys of the Don and Volga (see Map 11, below). At the end of June 1942 the Germans launched another great offensive, this time in the south, its aim being the oilfields near the Caspian Sea which were as vital to Russia's war-effort as the Middle East ones were to Britain's. At first, it made rapid progress, one of the armies advancing 300 miles in three weeks, and by early August it was within 50 miles of Grozny, one of the main oil-producing centres. At this point, the struggle switched to Stalingrad which stands on the Volga where this immense river bends eastwards towards the Caspian Sea. Since it was the communications centre of a wide area and since as long as it remained in enemy hands the German armies attacking the oilfields were threatened by a flank attack, the Sixth Army, under von Paulus, was ordered to capture it. The Russians retreated no further. With remarkable skill and courage they fought for every street and every house, hanging on through August, September, October and November. The conditions

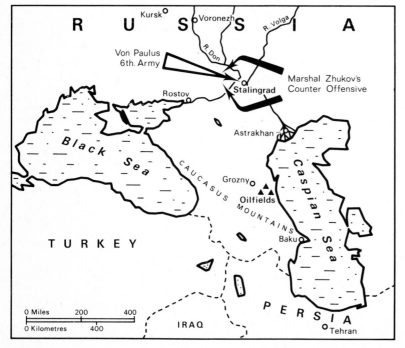

Map 11 Stalingrad 1942–43

were the most terrible of World War II. 'We have fought fifteen days for a single house,' wrote a German officer, 'with mortars, grenades, machine-guns and bayonets. Already, by the third day, fifty-four German corpses lay slain in the cellars, on the landings and the staircases. The front is a corridor between burnt out rooms or the thin ceiling between two floors . . . Stalingrad is no longer a town. By day it is an enormous cloud of burning blinding smoke . . . a vast furnace lit by the reflection of the flames. And when night comes, one of those scorching, howling, bleeding nights, the dogs plunge into the Volga and swim desperately to gain the other bank. The nights of Stalingrad are a terror to them.' Early in September the Germans got through to the river, cutting the defending force in half. Two months later they had reduced the Russian position to a few hundred yards of river bank. But these few hundred yards, dominated by two large factories, never fell. Ordering the defenders to fight on at all costs and sending them a few reinforcements, Marshal Zhukov, the Russian commander, methodically prepared his forces for a counter-attack which he eventually let loose on November 19. Six new armies, including 450 new T.34 tanks and 2,000 guns attacked from both the north and the south of the city, and after four days combat closed pincers of steel around Paulus' army. Such an attack had been obvious to the German High Command for some time but Hitler had refused to allow any retreat from the Volga. 'Where the German soldier sets foot, there he remains', he stormed. Moreover, when it became clear that the Luftwaffe was not, as he had thought, capable of supplying the Sixth Army by air, he refused to allow Paulus to attempt to break out if this meant leaving the city in Russian hands. In December, a relief army was beaten off by the Russians and with all his armies in South Russia now dangerously exposed, Hitler ordered a general retreat. The Sixth Army in Stalingrad was now doomed. The Russians offered surrender terms early in January. These Paulus rejected. 5,000 Russian guns now bombarded the German position. On 24 January 1943 Paulus requested from Hitler permission to surrender. 'Surrender is forbidden' was the reply. 'Sixth army will hold their positions to the last man and last round and by their heroic endurance will make an unforgettable contribution towards the establishment of a defenseive front and the salvation of the Western World.' Since no Field-Marshal in German history had ever been taken prisoner, Hitler made Paulus one on January 30 in the hope of stiffening his resolve to die fighting. It did not. Three days later Paulus and 91,000 men surrendered, all that was left of the army of nearly 300,000 which had entered the city six months before. After this annihilation of the Sixth Army, the Russians not the Germans henceforward called the tune on the Eastern Front. By the summer of 1943 they had 6,442,000 troops in the field against the enemy's 5,325,000 and were building up a marked advantage in

weapons. Having as Churchill put it 'torn the guts out of the German army', they prepared to drive it from their homeland.

The first German Field-Marshal ever to surrender. Von Paulus, front centre, leads the remnants of his Sixth Army from Stalingrad into captivity

The Battle of the Atlantic During 1942 German U-boats sank more than 6 million tons of Allied shipping, mainly in North Atlantic waters, with few losses to themselves. Such losses were higher than could be replaced by Allied shipyards and threatened not only the British and U.S. plans to prepare for an invasion of France but Britain's very ability to continue fighting. 'The only thing that ever really frightened me during the war,' Churchill admitted later, 'was the U-boat peril.' The convoy system which had worked so well in World War I seemed for some time to be ineffective against the new generation of U-boats operating in wolf-packs. New inventions and new tactics, however, began to even the struggle. Allied escort ships were fitted with High Frequency Direction Finders (H.F.D.F.) and ten-centimetre radar. The former enabled them to pick up the short-range radio signals of the U-boats, the latter to detect surfaced U-boats at much greater distances. More long-range aircraft and planes from escort carriers also became available to help in the work of detection. The battle of the Atlantic reached its climax in the spring of 1943. In March the wolf packs sank another 600,000 tons of merchant shipping, but in April began themselves to suffer heavy losses. In May 1943 41 U-boats were destroyed and the Allied losses dropped dramatically. At the end of May Doenitz, commander-in-chief of the U-boat fleet, withdrew it from the North Atlantic. 'The enemy', he noted in his diary, 'holds every trump card, covering all areas with long-range patrols and using location methods against which we have no warning . . . The enemy knows all our secrets and we know none of his.' Though improved U-boats re-appeared in the North Atlantic in the autumn of 1943, they were never able to win back the advantage they had lost.

The Allied Offensive

With the tide now clearly turning, the Allies were quick to take the offensive. In July 1943, two months after the final victory in North Africa, British and U.S. forces successfully invaded Sicily. This coming of war to Italian soil brought about Mussolini's downfall. To his surprise, the Fascist Grand Council demanded the end of his dictatorship on July 24 and the following day the king dismissed him and had him arrested. A new government was set up under Marshal Badoglio which, after six weeks secret and confused negotiations, surrendered to the Allies. In those six weeks, however, the Germans moved fast. Disarming Italian units and taking possession of strong points throughout the peninsula, they were able to provide tough opposition to the Allied forces invading from the south. Moreover a German secret agent rescued Mussolini from his mountain hotel-prison in one of the most daring exploits of the war and re-established the Fascist dictatorship in the north. Mussolini, his spirit broken, was now a puppet of the German army. So the Allied advance through Italy was slow. The countryside was well suited to defensive warfare and the Germans fought every inch of the way. Rome did not fall until June 1944 and the Po Valley was still in enemy hands in the spring of 1945. Differences of opinion between the Allies hampered progress. While Churchill wished to give high priority to the Italian campaign, hoping to break through to Vienna before the Russians advancing from the east, the Americans preferred to give priority to a new front in France. As preparation for the French invasion intensified, the armies in Italy failed to get the supplies they needed if they were to achieve a real breakthrough.

Since 1942, Stalin had been pressing Roosevelt and Churchill to open up a second front in France. So great were the difficulties of a cross-Channel amphibious operation against well-organized coastal defences that the U.S. and Britain did not feel ready to move until 1944. After months of ceaseless planning and preparations which had transformed much of southern England into one great military camp, Operation Overlord was ready in the early summer. Bad weather caused nerve-wracking delays but on June 6 General Eisenhowever, Supreme Commander of the Allied forces, gave the

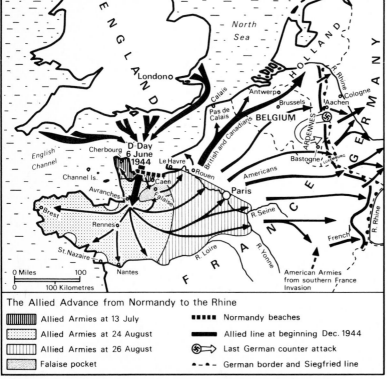

Operation Overlord
Above *Separate concrete caissons were floated across the Channel and then fitted together to make these artificial harbours called 'Mulberries'. Such ingenuity helped considerably towards the success of the D-day invasions*

Map 12 The Allied invasion of France 1944

Allied troops wade ashore from their landing-craft

order to go. On this Deliverance Day (D-Day) more than a thousand bombers blasted the coastal defences, parachute divisions dropped on to key communications centres and more than a thousand ships made for the Normandy beaches between the Seine estuary in the east and the Cherbourg peninsula in the west (see Map 12, opposite). The Germans were taken by surprise. Brilliant Allied intelligence work had convinced them that the main attack would be directed against the Pas de Calais, further to the north. With the aid of tanks specially adapted for dealing with coastal defences, the Allied troops quickly established themselves. By nightfall on D-Day 156,000 troops were ashore along a bridgehead thirty miles long. June and July saw heavy fighting in the Normandy countryside. While the British struggled to take Caen, the Americans managed to advance eastwards and seize the port of Cherbourg. At the end of July, a real breakthrough came. General Patton smashed with his tanks through the German defences near Avranches and swept down into the Loire valley. Disaster followed disaster for the Germans. Part of their army was trapped in the Falaise 'pocket' and 50,000 prisoners taken. By mid-August they were in rapid retreat. As the Americans and Free French neared the capital, the men of the Paris resistance movement rose against the Germans. The underground newspaper *Combat* made its first open appearance. 'Paris', wrote Camus, its editor-in-chief as well as one of Europe's foremost novelists, 'fires all its bullets into the August night. In this vast setting of stone and water, all around this river, heavy with history, the barricades of liberty have risen once more. Once more, justice has to be bought with the blood of men'. By August 25 the city was completely liberated. Still the Allied advance bounded on. Brussels and Antwerp fell at the beginning of September and only as Germany itself came within reach in mid-September did supply problems bring the Allied armies to a halt.

On the eastern front, the Russians were also pressing the Germans back towards their homeland. After the Stalingrad victory a huge area of south Russia between Voronezh and the Black Sea was liberated (see Map 11, page 174). The greatest tank battle of the war with German Tigers pitted against Russian T.34's then took place in the Kursk region (July 1943). It ended in a decisive Soviet victory, and by the winter of 1943–44 the Germans had been driven back beyond the Dnieper River. In the far north, Leningrad was relieved after a siege of 300 days which had cost the lives of a million of her three million inhabitants. Even after the D-Day invasions, the Russians still faced the bulk of the German army. Where 61 German divisions fought on the Western front, 228 remained in the East. Nonetheless, the Russian advance continued. As British and U.S. troops liberated France, the Russians entered Poland and Rumania. By the autumn of 1944 they had reached the River Vistula, where they too were halted by an enemy fighting ever more ferociously the closer he was driven to his own country.

In the Far East, Japanese resistance was also being ground down. The huge naval battles of the Philippine Sea (June 1944) and Leyte Gulf (October 1944, which involved the greatest number of warships ever to take part in a naval battle) destroyed the Japanese navy as an effective fighting force, and the Mariana Islands, from which the Japanese mainland could be directly bombed, were won. On land, the British took the offensive in Burma, and by August 1944 had won back the whole country.

Nazi Europe

'The Slavs are to work for us. Insofar as we don't need them, they may die.'

Official of the Nazi Commission for the East Europe Region.

'Most of you know what it means when 100 (Jewish) corpses are lying side by side, or 500, or 1000. To have stuck it out and at the same time to have remained decent fellows, that is what has made us hard. This is a page of glory in our history which has never been written and is never to be written.'

Himmler (Head of the S.S.) to a conference of S.S. officers.

As the Allied armies liberated parts of Europe which had suffered Nazi occupation, the real nature of Nazism began to become clear to the world. Between 1939 and 1945 Europe experienced a return to barbarism, a new Dark Age, far darker than the so-called Dark Ages of thirteen hundred years earlier. Were the evidence not overwhelming, many Nazi atrocities would defy belief.

The Germans mistreated their prisoners of war worse than anyone else. Sometimes they were shot in cold blood, like the seventy-one Americans in a field near Malmédy in Belgium in December 1944, or they would be forced by the S.S. to dig their own huge grave, be machine-gunned beside it and have their corpses buried by bulldozers, like the 100,000 men, women and children of Kharkov who were captured in the invasion of Russia in 1941 and were an administrative nuisance to the advancing army. Prisoners of war were also used as labour in German factories. In 1944 between 6 and 7 million foreigners were employed. The Krupp armament works and the V-weapon rocket sites for example depended upon foreign labour. Slav prisoners were treated worse than others, dying in their hundreds of thousands from their harsh conditions.

The Nazis also specialized in taking civilian hostages and murdering them in revenge for resistance activities. On 4 June 1942 Heydrich, Himmler's deputy and the terror of Eastern Europe, was assassinated by the Czech resistance. Six days later the village of Lidice near Prague was surrounded by the Gestapo. 172 men over the age of 16 were immediately shot, the women and children taken off to concentration camps, and the village itself entirely destroyed. Exactly two years later, the S.S. descended on Oradour-sur-Glane,

'If any further German troops are assassinated', this street poster in Paris reads, 'Frenchmen held by the German authorities will be treated as hostages and shot'. Such threats were made and carried out by the German authorities all over Europe

a village near Limoges. Here they burnt the women to death in a church, the men in nearby barns. In all, 652 people were killed. But the S.S. had made a mistake—their quarrel was with Oradour-sur-Vayres sixteen miles away. Nor did defeat lessen such cruelty. When the S.S. retreated from Tulle in southern France in August 1944, they left a hostage hanging from each of the lamp-posts in the main street.

The war is over in the Belsen area and Allied troops make former S.S. men carry the bodies of their victims to be placed on a lorry and taken for a decent burial

The worst treatment, however, was reserved for the Jews. In the first two years of the war the Jewish population of the occupied countries was usually moved into ghettoes or concentration camps where, though conditions were bad, survival seemed possible. After the invasion of Russia, however, Nazi leaders began talking about 'the final solution'. 'The final solution', said Heydrich to a small group of Nazi officials, 'will apply to about 11 million people.' They were careful not to put down in writing what exactly was meant by this phrase. In fact it meant the systematic slaughter of the Jewish race in Europe.

In Russia, Special Action Units (Einsatzgruppen) were formed by the Nazis to deal with the Russian Jews. They used two methods—locked vans in which the carbon monoxide fumes of the exhaust were diverted and mass shootings. A German civilian engineer was in Russia in 1942, and on October 5 he came across a unit in action near the town of Dubno. 'My foreman and I', he wrote, 'went directly to the pits. I heard rifle shots in quick succession behind one of the earth mounds. The people who had got off the trucks—men, women and children of all ages—had to undress upon the order of an S.S. man who carried a dog-whip. An old woman with snow-white-hair was holding a one-year old child in her arms and singing to it and tickling it. The child was cooing with delight. The parents were looking on with tears in their eyes . . . I walked round the mound and found myself confronted by a tremendous grave. People were closely wedged together and lying on top of each other so that only their heads were visible. Nearly all had blood running over their shoulders from their heads . . . I looked for the man who did the shooting. He was an S.S. man who sat at the edge of the narrow end of the pit. He had a tommy-gun on his knees and was smoking a cigarette.' Such units were responsible for the deaths of about 1,400,000 Jews.

The rest died in concentration camps like Belsen, Auschwitz and Treblinka. There, more sophisticated methods were used like gas chambers using prussic acid which killed 2,000 at a time. In 1945, the stench of death from Belsen could be smelt in Weimar, five miles away. There were more than twenty large camps and many more smaller ones with room for half a million live bodies. They were run by the S.S. and the inmates, who were usually fed on a diet of turnip soup and a little dry bread, were there to work until they died. These camps saw the most grotesque of the Nazi atrocities— the medical experiments which included freezing male prisoners to death and sterility injections on women and the manufacture of lamp-shades from human skin. By the time the war ended, the 'final solution' had accounted for nearly six million Jews.

The Nazis were by no means the only ones to commit atrocities during the war. What they began, their enemies copied in revenge. Of the 91,000 German prisoners who surrendered at Stalingrad only 5,000 survived their Russian captivity, and for the horrors visited on their people between 1941 and 1944 the Russian armies took a terrible vengeance in Germany in 1945. The British made up for the bombings of London and Coventry early in the war by terror-bombing on a scale far beyond anything they suffered. In 1942 1,000 bombers raided Cologne and in 1945 the beautiful and historic city of Dresden was destroyed with 135,008 casualties. What made the Nazi savagery worse was partly its scale and partly because it was not the excesses of a few isolated sadistic individuals but rather government policy. It was Hitler's racial theories in practice, the National Socialism in action, carried out with typical German efficiency by thousands of Nazi officials in the full knowledge of hundreds of thousands, perhaps millions of their fellow countrymen.

There were some Germans prepared to take the enormous risks involved in opposition to Hitler. They included Christian priests, members of the old aristocracy, generals and intelligence officers. As disaster followed disaster for the German armies, they became

convinced that Hitler must be assassinated and peace made with Britain and the U.S.A. if not with Russia. Hitler was incredibly lucky, surviving at least four assassination attempts between 1941 and 1944. His luckiest escape came in July 1944 when von Stauffenberg, one of a resistance group and as Chief of Staff of the Reserve Army able to attend Hitler's conferences, placed a bomb in a brief-case in the conference room. Just before it exploded, however, one of those present shifted the brief-case and placed it so that a stout table leg stood between it and Hitler. When it blew up, although it killed three people and seriously wounded two others, Hitler himself was only slightly injured. The plotters dithered when they learnt that Hitler was still alive and were soon in S.S. hands. Nazi vengeance was foul. The main plotters were tortured, given a farce of a trial and slowly strangled to death hung by piano wire from meat-hooks. Torture extracted the names of many others in the resistance. In the following purge, 5,000 or so were killed or committed suicide including leading Christians like Pastor Bonhoeffer. The most famous victim of all, however, was Field Marshal Rommel. In order to save his family, Rommel committed suicide. The government announced to the people that he had died of wounds received when in command on the western front and gave him a state funeral.

Dresden, after the Allied bombings in 1945

The End of the Third Reich

In mid-December 1944 Hitler sprang his last military surprise. Against the advice of his generals who were convinced that the war was lost, and to the amazement of his enemies, he ordered a counter-attack in the west through the Ardennes mountains. In this so-called Battle of the Bulge, the Germans were at first very successful, driving the Americans back in disorder. Once their attack slowed, however, through lack of fuel in the Bastogne area (see Map 12, page 178) they had nothing left. A month later, they were back on the line from which they had attacked and lost 120,000 men. By insisting on this last attack, Hitler made it impossible for what was left of his armies to maintain an effective defence of their own frontiers.

In 1945, the nightmare which they had brought down on their European victims descended on the Germans themselves. Thanks above all to the introduction of the Mustang fighter-bomber, the British and U.S. air forces had complete control of the air. The British flew by night, the Americans by day, and so precise had they made their bombing that they began to destroy the main industrial plants and communications centres. Speer, Minister for Armaments, who had trebled German military production by 300% between 1942 and 1944, reported to Hitler on 30 January 1945 that economically the war was lost. Moreover the British were bombing the main cities with the intention of breaking civilian morale. Hitler was by now in a wretched physical condition. 'A stooped figure with a pale and puffy face,' was how one of his generals described him in 1944, 'hunched in his chair, his hands trembling, his left arm subject to violent twitching which he did his best to conceal. A sick man . . . when he walked, he dragged one leg behind him.' His mental instability was becoming more marked. As early as 1942 General Halder noted during the Russian campaign that 'the continual underestimation (by Hitler) of enemy possibilities takes on grotesque forms and is becoming dangerous . . . His decisions are the product of a violent nature following momentary impulses which makes its wish-dreams the father of its acts.' By 1945 his dream-world had largely replaced reality. He therefore remained hopeful. In his opinion, the alliance of Britain, Russia and the U.S.A. was artificial and could not last. 'If we can deliver a few more blows,' he told his generals, 'then at any moment, this artificially bolstered common front may suddenly collapse with a gigantic clap of thunder . . . provided always there is no weakening on Germany's part.' Moreover he convinced himself that his secret weapons—the electric U-boat, the improved Tiger tank, jet aircraft, the V-1 and V-2 rockets—would snatch victory from the jaws of defeat. He was wrong on both counts. The common bond of the Allies, their need to destroy Nazism, was stronger than the growing strains between them, and the precision bombing of their aircraft and the rapid advance of their armies prevented the secret weapons from being produced in any quantity.

In January 1945 the Russians leapt forward again and were only held on the Oder River less than 100 miles from Berlin. In March, British and American troops were across the Rhine. German resistance then collapsed in the west and the Allied armies met on the Elbe south of Berlin into the suburbs of which the Russians entered on April 25. Hitler had decided to stay in Berlin. He continued to direct operations from an underground concrete bunker. Living on drugs, he kept extraordinary hours and could only be contacted through Bormann, his secretary. The other leading Nazis—Himmler, Goering and Goebbels—were now more concerned with their own personal rivalries than in the catastrophe all around them. The lunatic streak in Nazism had taken over.

Orders went out that any deserters from the army should be shot, that any commander who failed to hold a position was punishable by death, that everything of value in the invaders' path including food and clothing shops should be destroyed. This last decree, which meant in fact national suicide, was sabotaged by Speer who had come to realize that he was working for a madman. From April 25 to 28 Hitler played at defending Berlin with an army which no longer existed. The following day he married Eva Braun and dictated his last will. The Jews obsessed him to the end. 'The war,' he wrote, 'was sought and provoked exclusively by international politicians belonging to or working for the Jews . . . In times to come, the ruins of our cities will keep alive hatred of those who bear the real responsibility for our martyrdom: the agents of international Jewry.' On April 30 he committed suicide with his wife. Two days earlier, Mussolini and his mistress had been shot by Italian anti-Fascists and hung up by their feet in a Milanese square. Almost everywhere the fighting was coming to an end and it was left to the unfortunate Admiral Doenitz, whom Hitler had named his successor, to authorize unconditional surrender on May 8.

To stop Nazi ambitions cost Europe about 38 million lives. Eastern Europe had borne the brunt of the struggle and paid the highest price. Russia alone lost 20 million military and civilian dead, Poland 4.3 million, Yugoslavia 1.7 million. In contrast France lost only 600,000, the U.S.A. 406,000, Britain 388,000. Germany herself lost 4.2 million, her allies Italy, 410,000, and Japan, which had fought on till August when the atomic bomb forced her surrender (see page 194), 1.2 million. These figures do not include the 6 million Jews. The destruction of property was so great as to be incalculable. Germany and much of Eastern Europe had been made a waste land.

Russian troops look ahead for signs of resistance as they advance into what is left of Berlin

Germany lost World War II because Hitler's ceaseless and careless aggression brought about the alliance of the three most powerful states in the world, the U.S.A., the U.S.S.R. and the British Empire. As long as each of these stayed fighting and the alliance held, Germany was doomed and the longer the war lasted the more sure was her defeat. Economically she could not begin to match the resources of her enemies. Hitler's hunches, sense of timing and strong nerves contributed much to Germany's remarkable run of victories up to 1942; his arrogance, miscalculations and insistence on no retreat contributed greatly to her speedy defeat between 1943 and 1945. His single greatest miscalculation was the invasion of Russia but there were many others, the bombing of London during the Battle of Britain, the failure to reinforce Rommel in North Africa, Stalingrad. Thereafter his decisions grew ever crazier but the nature of his dictatorship was such that his authority was never seriously challenged until the very end.

Chapter 10
Science and Technology

In this chapter the word 'science' will be used in the sense of *knowledge and understanding of the laws of nature and of the universe reached by hypothesis, observation and experiment*, 'technology' in the sense of *the application of science for industrial, commercial, national or social needs*. Since in recent times scientific and technological ideas have tended, except in times of war, to move quickly across national boundaries, European science will not here be isolated from world developments.

In 1890, many scientists and with them the educated public believed that the basic laws of nature and the universe had been discovered. Nearly two centuries of research and experiment had apparently confirmed the general picture of the universe as sketched by Newton and his successors. There were held to be three basic truths: first, matter consisted of atoms; secondly radiation consisted of waves; and thirdly, both obeyed the principle of conservation of energy. The task of twentieth century scientists would be to fill in the gaps of this logical, mathematical and essentially simple Newtonian structure. Between 1895 and 1915, however, almost the opposite happened. A sequence of major discoveries brought into question many of the most fundamental principles of Newtonian science.

In 1895, Röntgen in Germany discovered X-rays and inspired related research all over Europe. In France, Becquerel discovered radio-activity, while in Britain Thomson and Rutherford, assisted by the young Dane, Niels Bohr, showed that sub-atomic particles must exist and that the atom could no longer be accepted as the fundamental solid and unchanging basis of matter. These and other discoveries also brought into question the wave theory of light and caused Einstein, building on the ideas of Planck, to suggest that light and other forms of radiation must consist of particles, each with its separate amount of energy, rather than of continuous waves. Moreover with first his special (1905) and then his general (1915) theories of relativity, Einstein caused a complete rethinking of Newtonian mechanics and dynamics and, in particular, of the relationships between mass and energy, space and time. The effect of these remarkable twenty years of scientific effort was to impress scientists with how little was really known about the foundations of the natural world and how disorderly and mysterious they seemed in contrast to the order and simplicity of the Newtonian universe.

As the years passed, the pendulum swung back as further research seemed to indicate a more complex yet more complete unity in nature. Quantum theory, as developed by 1927 by Heisenberg, Schrödinger and Dirac from Planck's original work, gave a better understanding not only of sub-atomic phenomena but also of the structure of solids and of the unifying links between physics and chemistry. Advances in the biological sciences, particularly in bio-chemistry and molecular biology, demonstrated new and vital links between the inanimate physico-chemical world and the animate world of cells and species. Biochemical explanations of the origin of life itself no longer seemed far-fetched. At the other end of the scale, Einstein's general theory of relativity and the quantum theories of radiation helped astronomers to reach a better understanding of the evolution and the movements of stars and galaxies. By 1970, a new plan of the universe was emerging, the basic outline of which was provided by sub-atomic particles, protein and nucleic-acid com-binations, quantum and relativity theory. In contrast to the New-tonian plan which assumed that the universe was basically un-changing, the mid-twentieth century plan emphasized continual evolution.

The influence of pure science on technology varied considerably in the 18th and 19th centuries and continued to do so into the 20th century. While the radio and electronic industry sprang up after the successful development of the thermionic valve following J. J. Thomson's discovery of electrons, the aerospace industry was the creation, in the first place, of enthusiastic amateurs like the Wright brothers (bicycle repairers) and Blériot (a manufacturer of lights for automobiles). In general, however, science and technology became interlinked in the twentieth century as never before. Not only were technologists increasingly unable to do without scientists but vice-versa. Much less would have been discovered about the arrangements of atoms without the invention of the electron microscope (1937) or about the galaxies without the radio-telescope (e.g. Jodrell Bank completed 1958).

The Main Phases of Development

In the last hundred years, science and technology have grown faster and affected individual lives more than ever before in history. The thirty years from 1880 to 1910 was a vital phase in this trend. It was as if—so one historian has put it—'the industrial revolution got its second wind'. Electric power became available on a large scale, the telephone, radio, diesel-, petrol- and steam-turbine engines, the aeroplane were all invented. Furthermore, that essential aspect of

189

twentieth century industry, mass production, having first been used in the U.S.A. for the manufacture of rifles, was adapted for auto- mobile production in 1908 by Henry Ford for the introduction of his Model T. Wars often breath new vigour into technology and World War I was no exception. There were of course new weapons— e.g. gas and tanks—but more important for the future were develop- ments in the automobile, radio, aircraft and synthetic chemical industries. World War I caused mass production methods to be introduced in European automobile factories, proved that radio was a practical and comparatively cheap means of communication, revolutionized aircraft design and production and, by cutting off Germany's supplies of natural nitrates, pushed Haber towards his method of making artificial nitrates from nitrogen in the air (see page 69).

Between the wars the pace of technological advance slackened, although major scientific discoveries continued to be made. The depressed economic conditions of these years were certainly one reason for this. The radio, electrical and automobile industries, however, continued to grow and the technological advances necessary for television broadcasting were also made.

World War II caused technology to accelerate even faster than World War I and this time there was no obvious slowdown in the post-war years. Radar and microwave radio, first developed to detect enemy aircraft, transformed the telecommunications industry. The computer, originally created to solve the complex mathematical problems of aiming anti-aircraft guns, was found to have innumer- able other uses in peacetime, the computer industry becoming one of the fastest growing of the 1950's and 1960's. The jet engine and nuclear energy were both developed in order to help win the war as quickly as possible. The former then transformed civil aviation after 1950 while the latter became an important source of power in industrial countries. Similarly, the firing of German V2 rockets on London in 1945 turned out to be a step towards putting Americans on the moon in 1969. The war also caused rapid advances in the chemical industry and in medicine. Wartime shortages drove both the Germans and the Americans to work out how to synthesize rubber. Moreover a natural concern to provide the best possible treatment for their armed forces encouraged the British and Americans to convert penicillin from an experimental laboratory specimen to medicine for mass consumption. In the postwar years synthetic hormones (which had also been developed under laboratory conditions before the war) became available for social purposes (e.g. the contraceptive pill) and even food began to be synthesized (e.g. petroleum wax into animal protein).

One example of the advance of 20th century technology— aviation

December 17th 1903 : the Wright brothers in the U.S.A. make the first ever powered, sustained and controlled flight

German aircraft of World War I : in the background the Gotha, a twin-engined heavy bomber ; in the foreground the Albatross Scout fighter

More than 100 passengers cross the Atlantic at twice the speed of sound and development costs of at least £1,000 million. The Anglo-French Concorde prototype flies over Paris

Science and Technology

During the twentieth century the rapid expansion of science and technology led to new methods of research and organization. Breakthroughs were achieved less often in the 19th century tradition when individual geniuses like Faraday or Pasteur had made their discoveries in their own or university laboratories supported by limited funds. Instead, major advances tended to result from the combined efforts of research teams backed by millions of pounds from governments and industry. For every scientist working in Western Europe in 1900, 20 to 30 would be working in 1970 and whereas in 1900 the wage of a professional glass-blower was often the most substantial expense of a research scientist, the sophisticated research equipment of 1970 might itself be priced in millions of pounds. Realizing that science and technology were essential for national defence and economic progress, the governments of advanced industrial countries were spending 2%–4% of their Gross National Product (i.e. the money value of all goods and services produced in one year) in research and development in 1970. Furthermore they set up organizations to encourage their scientists to pursue projects which were nationally desirable. Ministries responsible for science and technology were established in France in 1958, in Germany in 1962, and in Britain in 1964. Yet so expensive had scientific development become that the nations of Western Europe, conscious that the U.S.A. and U.S.S.R. were racing ahead, joined together on some major projects. For example, the European Centre for Nuclear Research (C.E.R.N.) was set up in 1953 and by 1966 was spending 35 million dollars a year, and Britain and France started co-operating in the production of a supersonic airliner (Concorde) in the 1960's and early 70's which had cost each of them £1,000 million pounds apiece.

Some Important Discoveries

It is worth looking more closely at two of the most important scientific advances of the century to make clear the important part played by war and by government support in the achievement of scientific and technological advance and the problems which this dependence on national and military consideration presented to scientists in the mid-twentieth century.

Penicillin Fleming made the original observations which demonstrated the bacteria-killing powers of penicillin mould in 1928. His discovery was not effectively followed up however until 1938 when Florey and his co-workers in Oxford proved that penicillin was poisonous to bacteria but not to their hosts. Before war broke out in 1939, successful experiments on animals had been completed and a form of penicillin was being prepared for use on humans. Once war had begun, both the British and American governments realized that this antibiotic, perfected for the treatment of human beings and produced in sufficient quantities, could save the lives of thousands of wounded soldiers. Britain lacked the resources to convert the research discoveries into the mass production of penicillin to meet the needs of the Allied armies. So an Anglo-American team involving biologists, chemists, doctors and the American pharmaceutical industry concentrated on this single task with rapid success. By 1943, penicillin was being used in Allied hospitals on the Tunisian and Sicilian fronts and significantly reduced the death-rate from infected wounds. But for the war, it would have taken many years longer to be developed for general human use.

Nuclear Energy While those scientists who helped in the development of penicillin as part of the war effort can never have doubted the value of their work once the war was over, those involved in the Manhattan Project—the code-name for the scientific team working primarily between 1943 and 1945 to manufacture an atomic bomb—were in a very different position.

In 1919, in England, Rutherford had first split the atom, and by 1933 further research, notably by the Frenchman Joliot, had suggested that the splitting of a uranium atom might produce an immense amount of energy through a chain reaction. An uncontrolled chain reaction would produce a violent explosion, a controlled one a powerful source of energy. Before the war, most nuclear research had been done in Europe and this continued until 1942. War forced leading nuclear scientists to concentrate on nuclear weapons. In 1940 and 1941, the combined research of various scientists—Peierls and Frisch at Birmingham, Simon at Oxford, Halban and Kowarski (French exiles in Cambridge)—had shown that an atomic bomb was a possibility and had suggested various methods of making it a reality.

That the Germans might be ahead of them in the nuclear weapons race continuously alarmed Britain and the U.S.A. Einstein voiced this fear to President Roosevelt in 1940 and it was a major reason for the recommendation of the 'Maud' Committee to the British government in 1941 that despite the enormous cost work on nuclear weapons should 'be continued on the highest priority and on the increasing scale necessary to obtain the weapon in the shortest possible time'. The committee also recommended increasing collaboration with the Americans, a wise suggestion since they were in fact requesting the most complex technological assignment ever undertaken in history, for which only the U.S.A. had the resources required.

Consequently in 1942, under the code-name 'the Manhattan Project', a vast team of physicists and engineers set to work under the scientific direction of Oppenheimer and the military direction of

General Groves in various centres scattered across the U.S.A. The two chief ones were Oak Ridge, Tennessee, which produced enriched uranium-235 and Los Alamos, New Mexico, where the weapons were made. In this isolation of Los Alamos scientists from all over the free world worked at continuous high pressure, often eighteen hours a day. 'We were a true community, working for a common purpose,' Oppenheimer recalled. 'We started out by thinking that it might make the difference between defeat and victory . . .' When the British scientist, Tuck, first arrived there, its spirit much impressed him. 'It had all the greatest scientists of the western world: they didn't care who you were or what you were. All they cared about was what you could contribute.'

The project proved far more expensive than expected and more difficult. To produce sufficient U^{235} cost more than 1,000 million dollars. 86,000 tons of pure silver also had to be borrowed from the U.S. Treasury to make a special electro-magnet! By 1944 the whole project was spending 100 million dollars each month. There were many technical snags and a major political quarrel between Britain and the U.S.A. over the extent to which they would share each other's nuclear secrets. Finally however it achieved complete success. When, in July 1945, the first atomic bomb was exploded experimentally in the Alamogardo desert, it was even more powerful than expected. Somewhat ironically, however, the Nazis, the main enemy, had surrendered two months previously. Though in the circumstances the Allied High Command were quite right to attach the highest importance to the destruction of the heavy water plants, their anxiety about the German nuclear research was nonetheless groundless. Only after the war did it become clear how far behind the Germans were.

Among the Los Alamos scientists, doubts were already felt about what they had done. For some, the long journey home to Los Alamos from Alamogardo was not the expected celebration. A bottle of Scotch was drunk in gloomy silence, one of them remembered, with 'each of us groping to understand what we had witnessed'. Some leading scientists like Szilard and Bohr tried to persuade Churchill and Truman against using the new weapon, so much more destructive and more unpredictable was it than any previous weapon. Others pressed for a demonstration of its powers on some uninhabited island before it was loosed on the Japanese mainland. They won little attention. Truman was determined to bring the war to an end as quickly as possible, partly to keep American casualties to a minimum and partly to prevent Russia, who was poised to join the war against Japan now Germany was defeated, from strengthening her position in the Far East. Three weeks after the experimental explosion, a single plane dropped a single bomb on Hiroshima, killing 60,000 people and injuring 100,000 more. Never had the power of science to destroy as well as create been more vividly demonstrated. And never had the question of the social responsibilities of the gifted scientist been more clearly put. 'When you see something that is technically sweet,' said Oppenheimer, 'you go ahead and do it and you argue about what to do about it only after you have had your technical success. That is the way it was with the atomic bomb'. For Oppenheimer, his experiences in those years caused him to rethink his values as a scientist. Speaking on a television programme on the building of the bomb in 1965 he commented 'I am not completely free of a sense of guilt'.

Oppenheimer (on the left) with General Groves in the Alamogardo desert site of the first experimental explosion of the atomic bomb

Science and Technology

Electrons and electronics The electronics industry has from its origin been firmly based on scientific research and is aptly named since it is a direct development of the discoveries of the nature of the electron in the late nineteenth century. The thermionic valve with its properties of amplification and feedback was perfected in 1912 and was the creation of a number of scientists from J. J. Thomson's theory of electrons. Similarly, the first successful radio transmissions by Marconi in 1895 could not have taken place without the earlier theoretical work of Maxwell and of Hertz (1886) nor the first telephone conversations of Bell (1876) without the earlier research of Helmholtz.

Radio communications were rapidly developed in World War I and radio as a source of mass entertainment, news and government propaganda in the 1920's and 1930's. Regular broadcasting began in 1921 in Holland and the U.S.A. with the B.B.C. making its first public broadcasts the following year. Television developed rather slowly. Though the beaming of electrons in cathode ray tubes had been tried successfully in laboratory experiments in the 1890's and Campbell Swinton had established the principles on which modern television broadcasting came to be based as early as 1912, electrical and radio companies were not convinced of its future until the experimental work of Baird in the 1920's. Regular television broadcasting began in Russia and in Britain in the 1930's but not until after the World War II did it become popular. North America led the way in the 1940's, Western Europe and the rest of the world followed in the 1950's and 1960's. In 1953 there were about 31,400,000 television sets in the world—122 per 1,000 persons in North America, 8 per 1,000 in Europe and 1 per 1,000 in Russia. In 1967 the number had risen to 214,000,000 and though the density of ownership was still highest in North America (285 per 1,000), Europe (159 per 1,000) and Russia (96 per 1,000) no longer lagged so far behind.

Solid-state Physics Though men have been using solid materials since the beginning of history, a scientific understanding of the nature of solids was not achieved until after World War I.

The researches of Joffe, Griffith and Taylor into the properties of perfect and imperfect crystals and of the Braggs on the X-ray diffraction of solids led to a better knowledge of the character of metals and of alloys under stress, making possible the production of very tough metals essential for nuclear and space engineering. Further investigation of slightly imperfect crystals and their electromagnetic properties led to the invention (1948) of the transistor by the Americans, Shockley, Bardeen and Brattain. This was a most significant development. Transistors began to replace the more bulky electronic valves in most electronic equipment which became increasingly compact. They also played a vital part in the rapid development of the digital computer which, through its ability to store information and solve problems, seemed likely to transform the century after 1950 as much as the steam engine transformed the century after 1780. The first electronic computer, E.N.I.A.C. was built by Mauchly and Eckert for the U.S. Army in 1945 and much improved versions, E.D.S.A.C. in Cambridge (England) and E.D.V.A.C. in Pennsylvania (U.S.A.), came into operation in 1949. In the mid-fifties, computers began to spread from university laboratories to business and government offices and with the replacement of vacuum tubes by transistors grew in capability while shrinking in size, so becoming commercially more attractive. The so-called 'third generation' of computers, some of which included 500,000 or more separate transistors, appeared in 1964. The 1960's saw an extraordinary growth in the use of computers. In 1961 there were about 4,000 in the world, in 1968 100,000, and the forecast for 1980 was over a million.

Polymers and Plastics A major advance of twentieth century chemistry was the analysis and application of the process of catalysis, one important technological development of which has been the creation of a large number of polymers—including plastics, synthetic rubber and artificial fibres. The first plastics, of which bakelite was the most famous, were developed before World War I but their rapid development came in the 1930's and 1940's after the researches of chemists like Semenov and Melville with the creation of nylon, polythene, P.V.C. and terylene. Further major experimental work by Ziegler and Natta after World War II led to the production of tougher and more uniform polymers and to the continuing growth of the chemical industry, particularly its artificial fibres branch. Some indication of the growth of world demand for plastics was the expansion of Britain's annual production from 27,500 metric tons in 1937 to over 500,000 tons in 1959.

The Biological Sciences There has usually been a close link between advances in medicine with those in the biological sciences and so it continued in the twentieth century. Of the thousands of remarkable medical achievements since 1900, we have already noted penicillin. Another worthy of mention is the discovery and manufacture of vitamins.

For centuries an unbalanced or inadequate diet was suspected of being a cause of many diseases. Captain Cook for example had kept his sailors free from the scurvy—the disease which had brought disaster to so many long voyages of exploration—by providing them with a regular diet of fresh fruit. There was, however, no scientific understanding of the cause of such diseases. Then, in the 1890's, biochemistry—the study of the chemistry of living things—came into its own as a separate discipline and an early outstanding British biochemist, F. G. Hopkins, was convinced by 1906 that in a balanced diet there must be, in minute quantities, certain substances without which disease or stunted growth was likely. Such substances which were later defined neatly as substances 'which make you ill if you don't eat them' and called vitamins were then isolated, analysed and, in a number of cases, manufactured in an artificial form. Vitamins A, B1, B2, C and D2, were discovered and recreated in the 1930's, others in later years. Moreover, in those countries which could afford it, steps were taken to make sure that every child received a vitamin-rich diet. A considerable improvement in health standards usually followed. For example, while in Britain in 1931 80% of English school children showed some evidence of rickets (a bone disease caused by a shortage of Vitamin D), by 1970 it was virtually unknown.

A new branch of biology became possible after the explosion in technological knowledge brought about by the war. New and ever

Above *Watson, on the left, with Crick in the Cavendish laboratory, Cambridge in 1953 when they successfully established the molecular structure of D.N.A. Part of a model of the D.N.A. molecule can be seen in the background*

Left *Hungarian children suffering from malnutrition in the 1920's. Rickets are caused by Vitamin D deficiency*

more-powerful tools of research allowed scientists to unravel some of the molecules of life itself and so the science of molecular biology was born. The investigation of nucleic acids indicated that deoxyribonucleic acid (D.N.A.) might be the key to the handing on of genetic characteristics from one generation to the next in living things. By 1953 Crick and Watson in Cambridge had discovered that D.N.A. possessed a double-helix-shaped molecular structure which carried in it genetic information in a chemical code and by 1956 Jacob and Monod in France had shown how the code worked. These D.N.A. discoveries were among the most significant not only of twentieth century biology but of twentieth century science as a whole. They brought man closer to an understanding of the nature of life itself.

Of innumerable technological developments since 1900, space only remains for two more. These have been chosen because of their tremendous effect on the shape and appearance of our industrial society.

The first is *concrete*. The Romans used a form of concrete in their cities, but the pioneers of modern concrete building were Frenchmen. Reinforced concrete—concrete and steel in combination—was first used by Monier in 1868 though it did not pass into general use until the 1920's. Pre-stressed concrete—concrete with steel under tension giving greater lightness with no loss in strength—was developed by Freysinnet in 1928. These materials made possible building, bridges, dams and so on of new size and form and re-shaped the city centres of the world after World War II.

The second is the *automobile industry*, the effect of which on other branches of engineering, on the organization of modern industry as a whole, on the economics of Europe and North America and on the urban and rural environments of advanced industrial societies, is almost impossible to exaggerate. A steady increase in traffic between the wars accelerated dramatically after 1945 and the motor-car became an essential part of Western European as well as American life. A significant but not immediately obvious consequence of the growth of the motor industry was the spread of mass production methods through industry generally. Linked with computers, mass production methods made possible automation—industrial processes which worked without human labour once they were set in motion. In 1950, a completely automated factory went into operation in the U.S.S.R. producing aluminium pistons, and the 'continuous-flow' chemical plants of the 1960's were essentially automatic with their staffs employed to perform maintenance and 'trouble-shooting' tasks. Furthermore in the ships, aeroplanes and spacecraft of the 1960's automatic direction-finding and control became increasingly used and increasingly sophisticated.

Above left *Mercedes cars being constructed in the Daimler factory in Stuttgart, 1900. Mass production methods did not spread to European car factories until the 1920's and 1930's*

Above *A highly automated petro-chemical plant in the U.S.S.R.*

Science and Society

Scientists and engineers were among the heroes of the 19th century. Their achievements were obvious and considerable. Moreover in 1900 they were optimistic about the future and the general public shared their optimism. The scientist would continue to investigate the truth about nature, the technologist would apply the scientist's discoveries to the benefit of society and so the great progress of the previous hundred years would continue at an ever faster pace. Halfway through the 20th century such optimism, frequently gave way to pessimism. Both science and technology were seen to be two-faced, with remarkable powers to harm as well as heal, to destroy as well as create. Nuclear physics, for example, had helped to create hydrogen bombs as well as atomic power stations. The motor-car offered individuals a new freedom of movement, yet brought city-centres to a standstill and obliterated the countryside under motorways and car parks, while insecticides by temporarily getting rid of harmful insects, managed also to upset the delicate balance of nature.

In the 1950's and especially in the 1960's discontent with the destructive aspects of technology began to express itself in organized ways. For example dismay at the nuclear arms race and at the perversion of science for military purposes led to the so-called Pugwash conferences which began in 1957. They provided a meeting-place for leading scientists from the U.S.A., U.S.S.R., Britain and elsewhere to direct their combined influence towards nuclear disarmament. Moreover, in the sixties, a more broadly-based movement emerged in both the U.S.A. and Western Europe which expressed itself in the publication of numerous books and pamphlets and the formation of many societies devoted to environmental and anti-pollution issues. An influential pioneer was the American Rachel Carson whose eloquent and carefully researched *Silent Spring*, first published in 1962, pointed out to the general public the dangers of the excessive use of agricultural chemicals like DDT. The main aim of this movement was to prevent the powers of modern science and technology from being misused by governments and industry at the expense of ordinary people and their surroundings. The social responsibilities of scientists and the future of technological man were discussed as never before.

Far left *An advertisement for a protest demonstration against 'atomic pollution' including nuclear power-stations*
Left *These barrels, dumped illegally on waste ground in W. Germany in 1973, contained enough highly poisonous chemical waste, including arsenic and cyanide, to poison the whole of the country's water supply*

Chapter 11
Urbanization

The Modern City and its Critics

Urbanization–the rapid growth of towns and cities–has been one of the more striking features of both Europe and the world in the twentieth century. In Europe, it has been closely associated with industrialization and was already marked a century earlier. For example, between 1801 and 1891 the urban population of England and Wales rose from 17% of the whole to 54%, in France, between 1846 and 1891, from 24% to 37%, and in Prussia (the core of modern Germany) between 1816 and 1895 from 26% to 41%. At the turn of the century Europe was the most urbanized continent in the world, yet the move from the countryside into the cities still continued. By 1961 58% of all Frenchmen, 77% of all Germans and 80% of all Britons were townspeople. Moreover the same trend could be seen on the fringe of Europe. Between 1920 and 1966 the proportion of town-dwellers increased in Sweden from 30% to 52% and in Russia over the same period from 18% to 48%. In 1970, therefore, Europe remained the most urbanized of the continents though in recent years her rate of urban growth had been slower than that of some other parts of the world.

Patrick Geddes (1854–1932), an early and influential writer on urbanization, has given an interesting explanation of how and why cities grew. He divided their recent history into two eras. The first era—'palaeotechnic'—was in the first century of the industrial revolution, from about 1770 to 1870. Since the machines of the time were fuelled by coal and made of iron, rapid industrial and urban growth was tied to mining areas, hence the appearance of such conurbations (extensive urban areas) as Birmingham and the West Midlands in Britain, the Ruhr region in Germany and Lille in France. The second, 'neotechnic' era began about 1880 with new sources of energy and invention. Steam gave way to electricity and oil while the telephone, the motor-car and the radio each made communication speedier and cheaper. This neotechnic technology freed industry from the need to be tied to a particular area like the coal or iron fields and, according to Geddes, should have eventually meant the end of the massive industrial city growing from a particular industry and tied to a particular geographical site. In its place industry and cities could be scattered throughout the land on a

Above *A paleotechnic panorama—Le Creusot, a small industrial town of Central France, as it was in the mid-nineteenth century, dominated by the Schneider ironworks*

Right *Too elegant to be a typical office-block but an office-block nonetheless. The headquarters of the Pirelli tyre company, Milan*

smaller more human scale. However, the opposite happened. The neotechnic era continued, yet cities, large or small, continued to grow. What Geddes failed to take into account was the creation of the modern office. Shorthand was perfected in 1837, electric telegraphy in 1840, lifts 1857, typewriters 1867, skyscrapers 1870's, telephones 1876, adding machines 1870's, dictating machines 1887 and carbon paper 1890. In the same period, improvements in photographic and printing methods made modern advertising possible. Furthermore, the development of mass markets made large manufacturing companies as much concerned with the financing and selling of their products as with manufacturing them. Thus the office rather than the factory floor became the centre of decisions. Neotechnic technology might make possible the resiting of the factory but not the office. For major companies, their headquarters had to be near the main banks and in easy reach of the government and civil service. Accountants, advertisers, lawyers and others had to keep close to big business while newspapers, magazines, radio and television all felt bound to stay close to the centre of affairs. Similarly, the big stores felt it necessary to be close to the centres of wealth and fashion. Thus many cities, especially capital cities, acted as a magnet to population and mass transport improvements like buses, underground and electric trains made possible hundreds of thousands of daily movements in and out of the metropolitan centres.

Transport developments made possible another important feature of twentieth century cities – the suburb. Suburbs have a long history. At first they were aristocratic. The nobles of Renaissance Florence built their villas on the surrounding hills, those of Venice along the River Brenta. The coming of the railways in the mid-nineteenth century enabled businessmen to commute and so spread suburban growth along the railway lines. In the early twentieth century first the underground and then the motor bus further encouraged commuting by every social class. Finally mass motor-car ownership spread commuting and suburban growth even further than public transport could reach. Consequently, not only was there a huge increase in the urban population of Europe but the new urban areas tended to sprawl further and further from the centre.

Geddes gave the name 'world cities' to those which by their size and the importance of the business taking place within them dominated the international scene, and in the 1960's this term was applied to twenty-four of the largest urban areas in the world. Of these twenty-four, the largest was New York/North Eastern/New Jersey with a population in the early 1960's of nearly 15 million people, the second Japanese – Tokyo/Yokohama with about 14 million people. Of the remaining twenty-two seven were European:

1. London	12,500,000	No. 3 in the world
2. Rhine-Ruhr	10,500,000	No. 4 in the world
3. Moscow	8,000,000	No. 5 in the world
4. Paris	8,000,000	No. 6 in the world
5. Berlin	4,000,000	No. 19 in the world
6. Randstad Holland	4,000,000	No. 20 in the world
7. Leningrad	3,000,000	No. 22 in the world

Once 'world cities' were viewed with pride by their inhabitants and envied by the provinces. There the talented congregated, fortunes were made and power wielded. The trend towards comprehensive redevelopments within the metropolis, of higher skyscrapers, better communications in the city-centres (preferably of motorway proportions), expanding suburbs again with better communications so that ever more people could enjoy the variety and dynamism of city life, seemed so obviously desirable as to be unworthy of discussion. The twentieth century city, however, has always had powerful critics, and after World War II these critics grew in number and influence. Geddes was one of the earliest. 'Slum, semi-slum and super-slum, to this has come the evolution of our cities,' he wrote in 1915. The huge conurbations of the modern industrial world were too big to survive. Just as Imperial Rome, on the collapse of its Empire, was little more than a necropolis, or city of the dead, so modern industrial cities contained the seeds of their own destruction. This theme was developed by Lewis Mumford (1895-   ) one of the best informed writers about cities past and present. For Mumford, the vast, congested city-centres of the twentieth century were dehumanizing since they limited and perverted human contacts and relationships. 'The metropolitan world,' he wrote in *The City in History* 'is a world where flesh and blood are less real than paper, ink and celluloid. It is a world where the great mass of people, unable to achieve a more full-bodied and satisfying means of living, take life vicariously as readers, spectators, passive observers . . . The poorest stone-age savage never lived in such a destitute and demoralized community.' Nor did the suburb seem to him much better. It was without character or sense of community, 'a proliferating nonentity . . . towards which our present random or misdirected urban growth has been steadily tending. A large-scale pattern of expressways and airfields, sprawling car parks and golf-courses envelops a small-scale shrunken mode of life.'

In the early 1970's, while office blocks in London could stand empty for years and remain profitable to their owners because of soaring land values, the number of that city's homeless grew. In Paris, the banks of the Seine, once the most beautiful and romantic landscapes in the world, were converted into motorways. In Rome, continuous and chronic traffic congestion made ordinary movement a nightmare. At the same time, urban crime rates rose dramatically both in Europe and the U.S.A. The pessimism of Geddes and Mumford, therefore, did not seem far-fetched and the creation of a civilized urban environment became one of the most urgent tasks of the last thirty years of the century.

Five European Cities
London Ever since its foundation in Roman times, London has been the major city of the British Isles, leaving other urban centres well behind. It owes its size and importance to its site on the lowest bridging point on the River Thames and its excellent trading position as the port linking the fertile agricultural heartlands of England with northern Europe. London grew as England's overseas trade expanded, and by the end of the eighteenth century, its population was almost a million. Its growth accelerated during the nineteenth century. By 1900 it had 6½ million inhabitants and by 1961 the London Planning Region which best defines the modern London urban area extended over 4,500 square miles and included 12,500,000 people. It was the largest metropolitan region in Europe and the third largest in the world.

Londoners, like most Englishmen but unlike most Europeans, never much fancied living in apartment blocks. They preferred two- or three-storied houses with a garden if possible. Consequently London was never densely populated and by 1914 it sprawled over an area in a five mile radius of Charing Cross with an additional daily commuting population entering by railway. Between the two world wars the suburbs grew fast. The underground was extended into open countryside making places like Hendon, Wembley, and Harrow ideal for commuting. To the south, the Southern Railway electrified its track and the Surrey suburbs grew. Between 1921 and 1939 the population of London rose by 1,200,000 and, since the new suburbs were planned on a spacious scale (usually 12–14 houses per acre), the built-up area trebled in size. In 1938 the government tried to bring this urban sprawl under control. The Green Belt Act, later confirmed by the Town and Country Planning Act, fixed a belt of open countryside around the city about five miles wide on average between twelve and fifteen miles from the centre. Such, however, were the population pressures on London and the south-east of England, that the Green Belt only partially controlled London's growth. In the city centre, office-building accelerated after World War II. Between 1951 and 1961, office-space increased by more than 30% and the office population by 150,000. By 1961 more than half the 1,400,000 workforce of Central London were office workers and in the years after the war the central skyline was transformed by skyscraper blocks, most of them the headquarters of major companies. Much of the West End, once the residential area of the rich, also became office land, alive by day, dead by night, and the army of commuters further expanded. London leapfrogged the Green Belt and pulled into her sphere of influence old well-established towns like Reading, Luton, Bishops Stortford and Guildford, thirty miles or more from Charing Cross. One consequence of this trend was traffic congestion on a grand scale. Between 1910 and 1962, Central London's traffic was estimated to have increased tenfold without a single major road-building scheme. In 1961, British Rail carried 13% more commuters than ten years earlier. By the mid-60's therefore railways, roads and underground were seriously overloaded.

Before World War II, apart from the Green Belt legislation, London's growth was largely unplanned. In 1944, however, as the war drew to its close, the County of London Plan, usually called the Abercrombie Plan after its chief author, suggested how the London area might be reshaped in the postwar years. Housing was a major consideration of the Plan. German bombing had caused enormous destruction and much of the surviving East End of London consisted of appalling slums. Rehousing at civilized standards was to be achieved partly by the redevelopment of the London built-up area at a density of not more than 136 persons per acre (a low density by European standards) and partly by the creation of overspill towns between twenty and thirty-five miles from central London. So were born the New Towns—of which Harlow in Essex and Crawley in Sussex were among the earliest. They were developments of the garden-city concept pioneered by Howard, a visionary town planner,

The lure of the suburbs. A London Underground poster of the 1920s

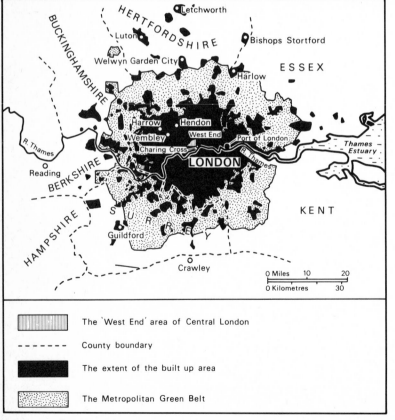

Above *Harlow in Essex, one of Britain's most successful New Towns. The photo is taken from the south-east and looks north-west across the central shopping and entertainments area and the northern half of the town. It is just possible to distinguish, in the top right hand corner, one of the two industrial areas which are carefully set apart from the housing. Another feature is the 'green wedges' of open space which lead right into the heart of the town*

Map 13 London after World War II

at Letchworth and Welwyn Garden City twenty years before. A garden city should be small enough to be genuine community with easy access to open countryside yet large enough to support its own industry and commerce. The London New Towns, which were first built for a population of 60,000 to 80,000 with a possible further expansion to 120,000 to 130,000, proved to be one of the outstanding successes of twentieth century town planning. They attracted industry and impressed their inhabitants as well as international architects and planners. They were copied not only in Britain—e.g. Cumbernauld near Glasgow and Skelmersdale near Liverpool— but also in other parts of Europe like Vallingby in Sweden.

The weakness of the Abercrombie Plan was its under-estimate of future population growth. It assumed that the population of Britain would remain static and that the movement of population into the London area could be reversed. Post-war reality was a steady growth of population and a continuous movement into the London and South-East Region. The South-East Study of 1964 took account of these trends and replanned the future of the region on the assumption that by 1981 there would be another 3,000,000 people living in the South-East, of whom London should absorb about half while the rest should be redirected to cities on the fringe of the South-East region, carefully planned on New Town principles, like Milton Keynes or Peterborough.

Randstad, Holland The Randstad (Ring City) of Holland is the fourth largest metropolitan centre of Western Europe after London, Paris and the Ruhr-Rhine conurbation of West Germany. It is made up of the once separate cities of Amsterdam, Rotterdam, the Hague, Utrecht, Harlem and Leiden which have virtually coalesced to form an urbanized horseshoe opening to the east. All but one of the towns were thriving communities in the late Middle Ages. The exception was the Hague which was a solitary castle before it became the seat of government in the 16th century. Their growth before the nineteenth century was steady rather than spectacular. In 1796 Amsterdam was much the largest with 217,000 inhabitants, the next largest, Rotterdam, had only 53,000. Rapid expansion came in the second half of the nineteenth century with the industrialization of Western Germany and the development of the Rhine as the most important commercial waterway in Europe. Between 1850 and 1913 the population of Rotterdam and of the Hague grew fourfold, Amsterdam more than twofold. This population growth continued in the twentieth century and there was another surge after 1945. A large proportion of the Common Market's trade with the rest of the world passed through the Netherlands and, as the Market prospered so did the Netherlands. A huge new dockland development— Europoort—was constructed to the west of Rotterdam which by 1970 was the busiest port in Europe handling 65 million tons each year. Between 1945 and 1960 the population of the Netherlands increased faster than anywhere else in Europe and the Randstad cities faster than anywhere else in the Netherlands. In 1796, their combined population had been about 400,000, in 1947, about 3 million, in 1970 5.7 million. The estimate for year 2000 was 8 million.

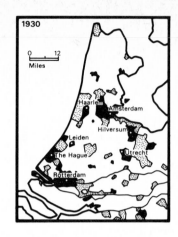

Suburbanization first made its appearance in the Het Gooi region near Amsterdam in the 1870's. Like the English, the Dutch liked living in single family houses surrounded by gardens, so with rising population and prosperity the Randstad cities spread rapidly towards each other. Throughout the twentieth century there was relentless population pressure on these urban areas from two main directions. First there was a continuous movement from the land to the towns; between 1947 and 1960, 278,000 left the land. There was also a continuous movement from the northern and eastern Netherlands to the Randstad cities. By 1960, the Netherlands was the most densely populated country of Europe with 881 people per square mile (the equivalent figure for England and Wales was 772) and the West Netherlands' average was 2,075. The possibility that the open countryside in the centre of the ring might also be swallowed up had become real. Accompanying this population growth were problems familiar to London and Paris, congestion, particularly of traffic, and the centralization of government, education and industry in the Randstad at the expense of the rest of the country.

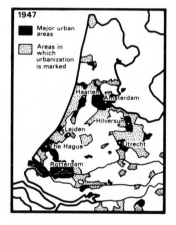

The 1950's and 1960's saw a series of plans to bring the growth of the Randstad under control. The various cities within it were encouraged to maintain a separate identity and where possible to preserve buffer zones of open land between them. The dunes along the North Sea coast and areas inside the horseshoe were to be preserved for recreational and agricultural use and industrial and educational institutions encouraged to move into the Eastern Netherlands. Nonetheless, a 1966 government report commented that 'the urban residential environment will in any case have to be doubled by the year 2000' and further extensive urbanization was expected both along the northern (Amsterdam) rim of the horseshoe and the southern (Rotterdam) one. Much of the modern Netherlands stands on land reclaimed from the sea and further ambitious reclamation schemes in the Zuiderzee and the Scheldt estuary have been planned to cope with future urban growth.

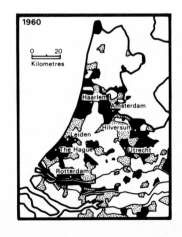

Moscow The recent history of Moscow which overtook Paris in the 1950's to become the third largest conurbation in Europe after London and Rhine-Ruhr indicates that rapid urbanization is not a problem restricted to the capitalist West as some writers have suggested. In 1860 the population of Moscow was about 360,000. By the time of the 1917 Revolution it had risen to 1,700,000. It

doubled again within twenty years, partly as a result of the transfer
of government offices from Leningrad. In 1935, the decision was
made to limit further growth to 5 millions which should be achieved
by the natural growth of the city's population, not by further
immigration from other parts of the country. As in London and Paris,
this decision proved ineffective. The population topped five million
in 1959, reached 6,400,000 in 1963 and, with an annual increase
of about 90,000, the Russian capital continued to act as a powerful
population magnet.

Again as in London and Paris, an increasing proportion of
Moscow's work force were white-collar or office workers, though in
Moscow they manned the various state-owned agencies rather than
the head offices of business corporations. Like the Randstad, Moscow
contained a high proportion of the country's educational institutions
as well as being the main centre of culture and communications
and again like the Randstad, a policy of decentralization was pursued
from the late 1950's which had some success with industry and
commerce though less in other fields. Suburban growth was also
marked but more strictly controlled than in the West. In the 1930's
a basic unit for suburban expansion—the mikrorayon—was laid
down. This mikrorayon or microdistrict was 75 to 125 acres in
extent, housing between 5,000 and 15,000 people and providing
essential daily services like shops and public amenities. In Stalin's
time, these districts were usually rectangular blocks facing onto a
central courtyard, constructed by mass-produced industrialized
building methods with little or no attempt to blend with the locality.
Their oppressive monotony was strongly criticized in the Soviet
press and in the 1960's more variety and imagination was encouraged.

As in London, the planners tried to limit Moscow's suburban
growth by a Green Belt. A ring motorway, on average about 11
miles from the city centre, was completed in 1962 beyond which the
Green Belt ran for a further six miles. The building of 'dachas'
(private houses chiefly for summer use) was severely restricted as
was any expansion of the towns bordering the Green Belt. The
pressure of population continued however and, like London again,
satellite towns beyond the Green Belt like Krykovo, eighteen miles
to the north-west of the city-centre were begun in the 1950's.
Consequently, Moscow developed its own huge population of daily
commuters—about a million by 1960—but without the traffic congestion which became so characteristic of the cities of Western
Europe in the 1960's. There were a number of reasons for this.
The Russian capital created one of the most efficient public transport
systems in the world. The Moscow Metro which opened in the
1930's had sixty miles of track in 1960 with further extensions
planned. It was cheap, clean and efficient and linked up with main-

line commuter services of a similar standard. In contrast, the road system was poorly engineered for traffic movement, some of the major developments like the Sadovoye Ring having been built by Stalin more as a monument to his government than as an improvement to ease the flow of vehicles. However, the comparatively low standard of living and the absence of a large motor industry in Russia meant that private car-ownership and thus vehicle densities lagged well behind Western levels. In 1961 there were only 730,000 cars in Russia (1 to every 300 people). A rapid increase was expected in the 1970's and 80's but car-hire and taxis were to be encouraged rather than widespread private ownership. By these means and by the maintenance of their cheap and efficient public transport, the Moscow authorities hoped to keep congestion at bay.

Paris The city of Paris which was founded in Roman times was already a thriving metropolis of 200,000 inhabitants in the Middle Ages. At the time of the 1789 Revolution its population had grown to half a million. By 1850, it had passed a million and had burst its city walls, the last of which were built in 1841. In the second half of the nineteenth century, suburban tentacles extended outwards along the main railway lines as Paris developed into a major industrial city. The industrial suburbs of St. Denis and Aubervilliers date from this period, and by 1900 the Paris 'agglomeration' as the French planners aptly name it contained four million inhabitants. In the early years of the twentieth century, the Parisian underground (the Metro) was begun which improved transport movement in the city centre and railway and bus services encouraged further suburban growth. Between the wars, uncontrolled property speculation on a huge scale caused further suburban growth, usually of low density housing with few local amenities and dependent on central Paris for employment. By 1935 the population of the agglomeration was more than six million. During World War II and immediately afterwards, Paris only grew slowly but from 1954 another period of rapid expansion began. Commercial and industrial activity in the central area throve and gaps in the suburbs were filled with high-density housing estates again dependent on the city centre for employment and with limited facilities. Sarcelles, for example, a northern suburb, grew from 8 to 35 thousand inhabitants between 1954 and 1962. By 1972, the population of the agglomeration had reached 8,500,000 and the ever-lengthening suburban tentacles had swallowed up old-established towns like Poissy and Pontoise.

For centuries, the domination of Paris over the rest of France was more marked than that of any other European capital. Both before and after the 1789 Revolution, governments tended to follow a policy of centralization so that all the major decisions affecting politics and society were made in Paris. Industry, commerce and education were similarly centralized. In 1947 a young geographer, Gravier, published a book, *Paris et le Désert français* (Paris and the French Desert). He argued that the domination of Paris was excessive and that provincial France had become a desert of ideas, culture, education and decision-making. His book won much attention. From 1949, successive governments did their best to encourage decentralization policies. Their success, however, was limited. There was some movement of industry but commerce, education and culture stayed put. Nor did industry move very far, usually to nearby towns like Mantes and Melun which became just further additions to the central agglomeration. In 1968, though Paris employed only 21% of the French labour force, 80% of car production, 75% of radio and television engineering and 68% of precision engineering was Paris-based. Moreover 40% of managers, 48% of engineers and 72% of researchers were Parisians. And large-scale office building in the 1960's had further concentrated big business in Paris. In 1968, nearly two-thirds of France's major companies were directed from Paris.

In the early sixties, intensive thought was given to the future development of Paris. One landmark was the proposal of 1960, usually known by its initials as P.A.D.O.G. which suggested major transport improvements and the encouragement of new commercial and administrative centres (suburban nodes), near but not part of the old historic centre of the city. However like the Abercrombie Plan for London, P.A.D.O.G. assumed that the population growth of Paris could be limited and that no developments outside the 1960 urban area were necessary. Between 1962 and 1964, however, the number of Parisians rose by 165,000 per year and a total population of twelve millions was expected in 1985 and fourteen millions in 2000. Consequently a Master Plan (Schéma Directeur) was drawn up in 1965 which allowed for such population growth. This Master Plan suggested a policy up to the year 2000. It rejected the London approach with its ring of New Towns on the grounds that such a concentric plan must eventually lead to the strangulation of the centre. In contrast, the Parisian planners suggested two axes, the one running about 75 kilometres on the north side of the agglomeration, the other about 90 kilometres on the south side, both in a south-east–north-west direction. Along these axes old towns would be extended and new towns built, one of which Cergy-Pontoise was planned to have 130,000 inhabitants in 1985 and 700,000 in 2000. The Master Plan continued the P.A.D.O.G. policy of suburban nodes, and by 1973 one of these, La Défense, had become a dramatic and highly controversial addition to the Parisian skyline.

In fact in the early 1970's, there was massive concern both inside and outside France that the character of central Paris, one of the best-loved cities in the world, was being destroyed by thoughtless

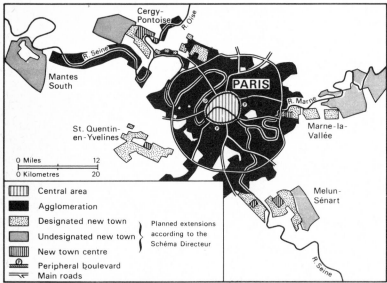

Above *Where once young lovers wandered and the Impressionists painted! The banks of the Seine downstream from the city centre with the skyscrapers of La Défense in the background*

Map 15 Expanding Paris

commercial expansion. The towers of La Défense which interrupted a famous view past the Arc de Triomphe, the destruction of the fascinating old market of Les Halles and the proposal to build another expressway along the banks of the Seine were all criticized as examples of destructive rather than beneficial urban growth.

Venice An even more dramatic example of the threat that twentieth century industrial and technological advance could offer to beautiful and historic cities was given by Venice. Venice stands on islands in a lagoon on the north-east coast of Italy. It was founded in the ninth century, became a major commercial and political power in the late Middle Ages and, in the fifteenth century, ruled much of the Eastern Mediterranean. Between 1400 and 1700 the city was a centre of

Above *Booming Mestre—Marghera*

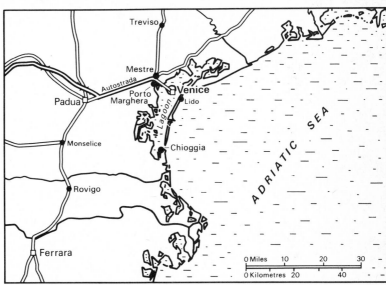

Map *16 Venice and its environs*

extraordinary artistic creativity which found expression in the design of the city as a whole as well as in particular paintings and palaces. Few would deny that it is among the most beautiful cities in the world and in 1970, it was attracting nearly two million tourists each summer. On the mainland banks of the same lagoon stand the towns of Mestre and Porto Marghera, both of which grew rapidly as industrial centres, especially after World War II. By 1970, Marghera was the largest chemical refining centre in Italy and the combined Mestre Marghera population numbered more than 250,000. On 4 November 1966 Venice suffered the most catastrophic flooding in her history. The waters of the lagoon rose 6½ feet above their normal level and driven by exceptional winds overwhelmed every part of the city. Damage in the region of £27,000,000 was

215

Drowning Venice

done in that one night. The inquiries which followed this catastrophe brought a number of facts to light. Venice was sinking into the lagoon at an average rate of about an inch every ten years. A major cause of this sinking was the pumping of water from the artesian wells beneath the city. Most of this pumping was done by Mestre-Marghera industry. Moreover, the 1966 flooding was only the worst of a series and the increasing frequency of these floods was due at least in part to alterations to the shape of the lagoon in the interests of industry. New channels had been dredged to give larger ships access to Mestre-Marghera and parts of the lagoon reclaimed for industrial developments. In addition many of Venice's most famous buildings were shown to be seriously damaged by the air of the lagoon, chemically polluted by the Mestre-Marghera refineries. In the opinion of many experts, the survival of the whole city was now in question.

> "Oh Venice! Venice! When thy marble walls
> Are level with the waters, there shall be
> A cry of nations o'er thy sunken halls,
> A loud lament along the sweeping sea!"
> Lord Byron

Chapter 12
Religion and Psychology

'When churches fall completely out of use
What shall we turn them into?...'

Philip Larkin, 1955.

Religious belief has always been an intensely personal and private matter. Consequently, it is not easy for historians to make any firm statements about trends in religious belief.

Nonetheless, there can be no doubt that twentieth century Europe has seen a sharp decline in both Christian belief and practice. In 1851, the regular church-going population of England and Wales amounted to about half the whole population; in 1901 it had fallen to 26%, in 1966 to 21%. The decline in Britain was probably as great as anywhere in Western Europe, since the trend appears to have been more marked in the Protestant North than in the Catholic South. In Russia and Eastern Europe, communist governments were bitterly hostile to the Christian religion. They considered it to be part of the centuries-old exploitation of working people and persecuted Christians into a small if stoutly resisting minority.

The decline and the reasons for it date back deep into the nineteenth century. Advances in science, especially the biological theories of Charles Darwin, and in historical methods, caused the Bible, which had previously been accepted by almost all Europeans as the word of God and therefore literally true, to be regarded more as a collection of myths the truth and value of which were no longer obvious. There was therefore a marked drift away from Christianity among educated Europeans. At the same time, the organized Christian churches which had grown up slowly over centuries in a mainly rural and aristocratic society found it most difficult to reach into the new industrial cities and more democratic society of the later nineteenth century. 'The greatest defeat of modern Christianity', wrote a Christian scholar in 1969, 'was its failure in the industrial revolution'. In 1880 Berlin, a city of more than a million inhabitants, had only seating in its churches for 25,000. In the 1940's, French priests coming into close contact with French working men for the first time in Nazi forced labour camps discovered that their countrymen were not so much anti-religious as without any religious experience whatever.

217

Developments in Protestant Theology

The terrible human disaster of World War I inspired a new theme in Christian thinking. In the previous two centuries, Christian writers had tried to look at Christian ideas and practice from the viewpoint of human reason, and where reason and belief seemed in conflict to adapt belief to reason. In his commentary on St. Paul's Epistle to the Romans, Barth, a Swiss theologian, took a very different view. Man, he emphasized, is an imperfect, corrupt creature whose powers of reasoning are as imperfect as his will. What we know of God we know only through divine grace, and the certainty of faith, which is something beyond explanation by human reason alone, must be the foundation of Christanity. German Protestantism also produced another powerful thinker in Bonhoeffer, whose courageous opposition to Nazism led to his death in a concentration camp at the age of 39. He believed that the religion of the old traditional church of Europe was virtually dead because the churches by their organization, cares of the present world, ceremonies and attitudes had tended to become obstacles rather than links between man and God. This point of view was vividly expressed by Simone Weil, a French writer who was converted to Catholicism. 'I love God, Christ and Catholic faith', she wrote 'but I have not the slightest love for the Church . . . What frightens me is the Church as a social structure.' Christanity could nonetheless live, Bonhoeffer argued, since its essence was a personal encounter with God through the secret discipline of prayer.

The disconnecting of real Christianity from church-centred religion was continued by a number of German, British and American theologians after the war. Christianity became a matter more of inner facts and of concern with the evil and injustices of the present world than with some 'beyond' or life after death. There is no evidence however that these changing ideas won it greater support among Europeans.

The Roman Catholic Church

In contrast with the Protestant Churches, the Roman Catholic Church could look back over a continuous history nearly 2,000 years long and possessed a comprehensive set of beliefs and rules of behaviour based on the Bible and on the decrees of the Councils of the Church tried and tested in the course of this long history. In 1900, it was an extensive international organization strictly controlled by its hierarchy (i.e. the chain of command from the Pope and Cardinals in Rome down to the bishops and local clergy) which expected obedience from those below them. Using its parish priests, its schools and its political influence (especially in Italy, Spain and Ireland), it kept more direct control of its members than did most other churches. While from the late 19th-century many Catholics became increasingly concerned about social problems and sympathetic towards democracy and socialism, the senior clergy remained sternly conservative. Its attitude to new approaches to religious belief was similar, the Modernist movement which aimed to rethink basic teachings in the light of recent historical and scientific research being firmly curbed in 1910. It regarded Communism as a more serious threat to Christianity than Fascism and consequently managed to compromise first with Mussolini and then with Hitler. The failure of Pope Pius XI to make any major public protest against Hitler's treatment of the Jews caused much criticism both during and after the war.

After 1945, the Catholic Church aimed to maintain its influence in democracies through Christian Democratic parties (as in Italy and West Germany) and in social matters stood out against such 'liberal' and 'permissive' trends as easier divorce, abortion and artificial contraception. By its own system of censorship it warned its members away from books and films which it considered dangerous. Though it managed to maintain its numbers better than the Protestant churches, it too experienced a serious decline. In 1943 Catholic France was shocked by a book by Godin entitled *France, pays de mission* (later translated into English as *Pagan France*). Godin's main theme was that since only one in ten Frenchmen took their religion seriously, France was in as great a need of Christian missionary activity as the more obviously pagan territories overseas. Such activity came in the form of the bold experiment of the worker-priests. In 1944 Cardinal Suhard founded the 'Mission de Paris', the aim of which was to give Christianity meaning to the industrial workers by priests working alongside them in their world, not, as in the past, standing outside it at the porch of their parish church. The experiment failed, crushed between the traditional anti-clericalism of the workers and the hostility of the more traditionally-minded Catholic leaders. Its failure further underlined the enormous problems of those who aimed to add a religious dimension to the twentieth century industrial city.

In the late fifties and throughout the sixties, however, there were unprecedented stirrings within the Catholic Church. Much of this was due to Pope John XXIII (1958–63) who called together the Second Vatican Ecumenical Council. The Council lasted from 1962 to 1965 and was attended by 2,500 delegates, including not only Catholics from all over the world but representatives of most of the other major Christian churches which previously had been regarded as among Catholicism's most dangerous enemies. Pope John's aim was to achieve an 'aggiornamento' (renewal) or in other words to bring his Church in touch with the present. Both John and his successor, Paul VI (1963–   ) devoted much of their energies to

St Peter's Square, Rome, October 1962. Pope John XXIII being carried to open the Second Vatican Council

building greater trust between the various Christian sects. At the Vatican Council, discussions were held which would have been unthinkable even ten years earlier. Topics included the extent of error in the Bible, the Christian attitude to the Jews, the right of individuals to question the decisions of bishops and of the Pope himself, the desirability of Catholic priests remaining unmarried, contraception and Freud's psycho-analytic theories. The spirit of the Vatican Council revitalized Catholics and Christians everywhere.

Its consequences, however, are hard to measure. There can be little doubt that it seriously divided the Catholic Church. Many delegates on returning home hoped to turn the suggestions of the Council into reality. Others, the more conservative, believed the open questioning of well-tried traditions to be potentially disastrous and wished to return to the old ways as quickly as possible. A third group, the most radical, felt that the Council was at best the 'beginning of the beginning' and that the task of modernization had barely begun.

The question of artificial contraception brought these divisions out into the open. To the reformers, artificial contraception was essential since over-population was the most serious threat to the world's future; constant childbearing was a severe strain on both women and their families and the traditional Catholic ban on artificial contraception was an outdated, unhealthy rule laid down by celibate priests with no experience and consequently little understanding of sexuality and marriage. The conservatives, however, stuck to the traditional view that contraception was an unnatural interference with the life-creating process which must remain subject solely to God's will. After long consideration, Pope Paul VI finally committed himself against the reform. In the encyclical (or letter) 'Humanæ Vitæ' he forbade the use of the contraceptive pill. It was a most controversial decision. As one Catholic reformer put it, the Pope had taken his decision 'against the opinion of important bishops and cardinals . . . contrary to the resolutions drawn up by the lay advisory body, against the views of leading moral theologians, at least in Europe. At no point in his letter does the Pope admit the limitations of knowledge concerning individual human experience, he never asks, he always knows.' Perhaps the most significant feature of the contraception debate was not so much the Pope's decision but the fact that educated Catholics put forward their views to their priests, bishops and the Pope and expected them to be taken into account. The traditional authority of the Church was being questioned and criticized with unpredictable consequences for the future of Catholicism.

Christianity in Russia

The Christian Churches of Eastern Europe had been separated from the Roman Catholic Church for centuries, the Greek Orthodox Church headed by the Patriarch of Constantinople having broken away in 1054. The Russian Orthodox Church had in turn won its independence from Constantinople under the leadership of the Patriarch of Moscow in 1589. By the late nineteenth century the Russian church was closely linked to the Tsar's government. Directed by Pobedonostsev from 1880 to 1905, it was a powerful enemy of all reformers and critics of the Tsar's rule. It was also corrupt.

Once the Bolsheviks had seized power, they immediately took action against the Church. Its priests, said Lenin, were 'the defenders of serfdom in cassocks'. In December 1917 all Church property was confiscated, all seminaries for the training of priests closed and church marriage replaced by civil ceremonies. Tikhon, the recently

elected Patriarch of Moscow, retaliated by excommunicating (excluding from all the privileges of Church membership) the Bolshevik leaders. The new government then launched a violent anti-Christian campaign in both the cities and the countryside. One priest in Georgia reported how young schoolchildren, hungry after playing energetic games, were asked by Communist leaders whether or not they were hungry. When they said they were, they were told to ask God for their daily bread. When nothing arrived in answer to their prayer, they were told to ask Lenin for their daily bread. As soon as they did, a truck arrived carrying bread, cheese and fruit. 'You see now', they were told, 'it is not God that provides bread but Lenin'. Nor were the Communists content with propaganda. In 1922 the Metropolitan (Archbishop) of Petrograd was shot with three of his colleagues and the same year Patriarch Tikhon was imprisoned. He was released the following year but only after he had publicly promised complete loyalty to the new government. On his death the government refused to allow the appointment of a successor. For the next fifteen years what was left of the Church was in a sorry state. The Metropolitan Sergius acted as head and, like Tikhon, after a spell in prison, declared his loyalty to the government. By doing so he provoked a bitter split among his followers. Then

Russian Orthodox priests pictured as drunkards and exploiters of the poor in this Bolshevik anti-religious painting on the wall of a Moscow monastery

came another wave of persecution, headed by the League of Militant Atheists whose aim was the total destruction of religion in Russia. By a decree of 1929 Stalin limited religious activities to the holding of church services, and this right was further reduced by the confiscation of thousands of church buildings. By 1939, of the 54,000 churches which had existed in 1917, only a few hundred were still open. Of many thousands of monks, nuns and priests, only a few hundred remained, of 163 bishops, only 17.

Hitler's attack on Russia in 1941 changed the Church's fortunes. Without hesitation, Sergius instructed all Russian Christians to rally behind Stalin in that hour of national disaster and raised money to equip a new tank division. Stalin responded to this gesture. In 1943 he allowed a new Patriarch to be elected—Sergius was chosen unanimously—and permitted him and his successor Alexis to make real progress in rebuilding the Church's organization. By 1947 the number of bishops had risen to 74; of priests to about 30,000; 67 monasteries and 8 seminaries had been reopened and contacts with Eastern European Christians renewed. But such success appeared to alarm the government and in 1959 persecution began again. Churches like Riga and Perm cathedrals were closed, senior churchmen like the Archbishops of Chernigov and Kazan imprisoned and priests and their congregations constantly harassed. In 1970 the Russian Orthodox Church still survived and millions of Russians were still Christians, but they were a persecuted minority with an uncertain future.

The Ecumenical Movement
The most positive response by Christians to the growing secularization of the twentieth century was the Ecumenical (literally 'worldwide') Movement. Its aim was to heal the centuries-old divisions between the numerous Christian churches so that, more united, they might better withstand the forces which were threatening to destroy Christianity as an active influence on human affairs. In the opinion of Temple, archbishop of Canterbury (1942–44), it was 'the great new fact of our era'.

Its beginnings can be dated from the International Missionary Conference at Edinburgh in 1910. Between the wars, it grew in vigour. In 1921 the International Missionary Council was established, and in 1929 the Universal Christian Conference on Life and Work met at Stockholm in Sweden. Its guiding spirit was Söderblom, archbishop of Uppsala, who was determined to bring about greater unity among Christians and a greater concern among the churches for the social problems of the time. Two years later another important conference was held at Lausanne in Switzerland—the first World Conference on Faith and Order. This was the inspiration of an

American, Brent, then Bishop of the Philippines. Though the Catholics were absent, 400 delegates attended, including representatives of the Eastern Orthodox churches. The ecumenical spirit encouraged mergers between churches. In 1925, four Canadian churches merged as the United Church of Canada, and in 1947, the Presbyterians, Congregationalists, Methodists and Anglicans combined as the United Church of South India.

The creation of the World Council of Churches at Amsterdam in 1948 was another step forward. Representing 147 churches from 44 countries, its intention was to increase co-operation and understanding between its members and encourage greater unity among Christians everywhere. Still, however, the Roman Catholics stayed outside. When therefore Pope John XXIII, the head of the Roman Catholic Church, called representatives of the world's churches to his Vatican Council in 1962 and both he and his successor gave active encouragement to the Ecumenical Movement, Christians the world over rejoiced.

Alternatives to Religion

If Christianity appeared to be dying, no alternative philosophical or spiritual system of belief successfully took its place in the first seventy years of the twentieth century. Between the wars, political movements like Communism and Fascism looked as if they might. Both had mass appeal and demanded self-sacrifice by their followers in the service of what each claimed to be a noble cause. Both had a vision of a better future, a clear definition of good and evil and the certainty that each possessed a monopoly of the truth. World War II however destroyed Fascism and did little to increase the appeal of Communism, immense though the expansion of Russian power proved to be. After the war, Communism flourished in Asia but stagnated in Europe. West of the iron curtain, some writers and intellectuals, especially in Italy and France, remained passionate supporters of Marxist-Leninism but their numbers increased hardly at all between 1945 and 1970. Communism was imposed on Eastern Europe by force of Russian arms and there seems no good reason to suppose that it ever won the hearts and minds of ordinary people outside the official ruling party. A joke frequently heard in Hungary in the sixties went like this:

Q. What is philosophy?

A. Searching in a dark room for a black bed.

Q. What is marxist philosophy?

A. Searching in a dark room for a bed which isn't there.

Q. What is marxist-leninist (i.e. official communist) philosophy?

A. Searching in a dark room for a bed which isn't there and shouting 'I've found it'!

Otherwise Europeans developed a considerable suspicion of traditional beliefs and attitudes—whether based on religion or custom. Authority, whether of the church or of the state or of the father within the family or of the husband within marriage or of the teacher in university or school, was increasingly questioned and the right of the individual to make up his own mind and follow his conscience whatever the issue increasingly emphasized. At the same time, as European nations tended to measure national success or failure in terms of economic growth, so individuals tended to treat the acquisition of material goods as one of life's main aims. 'This is the modern morality', commented the American economist J. K. Galbraith in 1967, 'St. Peter is assumed to ask applicants only what they have done for the G.N.P.'. But just as the late sixties saw a fundamental questioning of the desirability of economic growth for its own sake, so more and more individuals began to wonder whether greater material prosperity was really their life's chief goal. 'The creed of the English is that there is no God and that it is wise to pray to him from time to time,' commented Alasdair MacIntyre, writer and broadcaster, in 1963. Much of urban industrial Europe shared this English muddle.

Psychology

Religion is much concerned with men's souls and minds, so too is psychology. As the twentieth century progressed, psychiatrists came to rival priests as guides to modern urban man in his moments of difficulty or depression.

Psychology, the science of mental life, emerged as a scientific discipline in the second half of the nineteenth century. It grew out of physiology, the scientific study of functions and processes of living things which in Germany was dominated by four eminent professors—Helmholtz, Brücke, Ludwig and Du Bois Reymond. Together they were determined to destroy vitalism—the belief that life itself could not be explained solely in scientific terms. As one of their colleagues put it, they were convinced that 'no other forces than the common physical-chemical are active within the organism'. Psychology grew out of physiology and two of the greatest modern psychologists, Pavlov and Freud, studied under members of this group, Pavlov under Ludwig, Freud under Brücke.

Pavlov and Behaviourism Pavlov (1849–1936) was born in Ryazan in Central Russia. Though his grandfather was a peasant, his father was a priest which enabled him to be educated at the local priest's seminary and then, having showed a lively interest in science, to continue his studies at the University of St. Petersburg. Darwin, whom he first read when he was still a boy, deeply influenced him for the rest of his life. Biology and physiology were his chief interest and having been well taught in physiology at university he decided to become an experimental physiologist.

Pavlov was highly eccentric with a single-minded devotion to his work which kept him indifferent to most other things including poverty while he was young and fame and administrative responsibility when he was older. He was fortunate to marry a wife who was prepared to accept him for what he was and take care of him even though this meant tolerating a husband who turned up at the marriage ceremony without any money for either the ceremony or the honeymoon and who took her for such strenuous walks during her first pregnancy that she had a miscarriage! As a scientific researcher however, he was clear, thorough and uncompromisingly honest, choosing a few problems and pursuing them with unceasing skill and vigour to bring them nearer to solution. In fact, during his long career, he worked on three major experimental subjects. The first, the nervous system of the heart, was the subject of his thesis for a Doctorate in Medicine which he gained in 1883. The second, the main digestive glands, on which he worked until 1902, won him international fame. The third, the working of the higher nervous system of the brain, lasted the next thirty-four years until his death and strengthened his reputation both inside and outside Russia.

He did not approve of the Bolshevik Revolution. Indeed he believed that the events of 1917 were one of the greatest disasters in Russian history and said so publicly. When asked to comment on the great social experiment the Bolsheviks were conducting, 'for such an experiment', he said, 'I would not give a frog's hind leg'. The Communists, however, were proud of him and his scientific ideas tended to reinforce rather than contradict Marxist theory. In 1922 he asked Lenin to allow him to go abroad. Lenin refused, offering instead to increase his food rations, an offer which Pavlov turned down. When in 1927 sons of priests were expelled from the medical schools, Pavlov resigned in protest from his position as Professor of Physiology and protested further to Stalin when scientists began to be admitted to the Russian Academy of Sciences because they were good communists rather than scientists.

In 1933 however, he made his peace with the government. It was spending heavily on scientific research and he seemed to be persuaded that a new society inspired by respect of scientific truth was in the making. After his death, the experimental work which he had begun was continued and extended in Soviet research centres.

His most important experimental work was done after 1902. In order to eat and digest food, dogs need to salivate and do so naturally by a reflex action when food is placed in their mouths.

Pavlov (3rd from the left) with four other eminent European scientists, outside Buckingham Palace in 1936. He had just had a meeting with King George V

Pavlov noticed that some dogs began to salivate as soon as they saw the person who usually fed them and he was able to construct an experimental situation where a bell was repeatedly sounded just before food was placed in the mouth of the dog. The eventual result was that the sound of the bell caused salivation without food being presented at all. The bell he named the 'conditional stimulus', salivation as a result of the bell sounding the 'conditional reflex' and this process of changing animal behaviour became known as 'conditioning'. The implications of this apparently simple experiment were enormous. Pavlov had shown that animal and probably human behaviour and learning were to an important degree instinctive reactions to repeated outside stimuli and would alter according to the nature of the stimuli. He also discovered that the dogs which he forced to make difficult choices between two stimuli often became disturbed, and thus he began to explain neurotic behaviour.

As significant as his discoveries was his method of reaching them. Pavlov was only interested in what he could observe and would only draw conclusions from the behaviour which he observed in his strictly controlled experiments. As a scientist, he believed that there was no other valid method of working. The school of psychology which is inspired by his discoveries and his methods is usually known as Behaviourism.

Freud and Psycho-analysis Sigmund Freud (1856–1939) was born in Freiberg, Bavaria, then part of the Austrian Empire. His father was an unsuccessful Jewish wool merchant and moved with his family to Vienna in 1859. Freud was a clever young man and entered

Vienna University when he was seventeen. Biology and physiology won his interest and after lengthy research on the spinal system of fish, he gained an M.D. degree in 1881. There was growing anti-semitism in Vienna in this period so he had no chance of promotion within the university. He therefore went into private practice as a specialist in nervous disorders, and after some years of appalling poverty was able to make a living.

He became interested in hysteria, in hypnosis as a method of treating it and in the nature of the mind as revealed through his study of nervous disorders as well as in the disorders themselves. He discovered that things usually ignored or misunderstood like dreams, slips of the pen and tongue and early childhood memories seemed to offer important clues to the nature of emotional difficulties and of the individual personality. He pioneered a new method of treatment which he called 'free association'. In 1897 his father died and in the months that followed he subjected himself to his own methods, making use of his dreams and childhood memories to achieve self-analysis at this emotionally critical time of his life. Three years later he published what came to be considered his most important book— *The Interpretation of Dreams*—which not only discussed the nature of dreams but outlined a new theory of the mind. To begin with, it was barely noticed. The first edition of 600 copies took eight years to sell. Freud, however, kept writing. In his *Psychopathology of Everyday Life* he attempted an explanation of slips of the tongue and pen and in 1905, in *Three Essays on Sexuality*, he gave an account of the importance of sexual desires in the development of individual emotions and personality. He now began to attract gifted followers first in Vienna and then, after a visit to the U.S.A. in 1909, internationally. He worked vigorously for the rest of his life to build up a 'psycho-analytic movement' which would spread his ideas throughout the world.

Freud was a most difficult man to work with. He found it almost impossible to tolerate disagreement with his new ideas and those who did disagree were treated as rivals. He broke with his two most gifted followers, Adler and Jung, for this reason. 'Psycho-analysis is my creation', he wrote in 1914. 'For ten years I was the only one occupied with it ... Nobody knows better than I what psycho-analysis is'.

As he grew older, he grew increasingly interested in the application of his psycho-analytic ideas to past and present society, the progress of which caused him growing despair. When the Nazis took over Austria in 1938, he fled to London where he died the next year.

Freud's greatest contribution to our understanding of human nature was his idea of the 'unconscious', a vital area affecting human behaviour but generally unknown to individuals and only to be reached through dreams, slips of the tongue and so on. He also emphasized that a basic force affecting human behaviour and development was the 'libido' or sexual instinct. As he grew older, he also emphasized the power of the instinct of aggression. Human mental and emotional characteristics take the form they do as a result of the channelling of these instinctive forces into socially acceptable forms. The vital time for this channelling is in early childhood and much depends on the child's early relationships with his or her parents which even in infancy are powerfully sexual in character. The events of early childhood largely determine adult personality.

These ideas were of great significance. By discovering the 'unconscious' mind he destroyed the old confidence in man as an essentially reasonable creature who pursues his own happiness in an orderly and civilized way. By his emphasis on sexuality he defied religious and respectable prejudice and compelled the world to recognize the importance of this elementarly biological force. By stressing the significance of early childhood and family relationships in the development of personality, he brought about a revolution in child rearing and in the education of the young in both Europe and the U.S.A. His general theory of human nature provided immediate insights into aspects of human experience both in the past and present which other psychologists, historians, artists and writers have since developed.

Freud was bitterly criticized during his lifetime. Most contemporaries found his insistence on the importance of the sexual drive, even in infancy, profoundly shocking. 'Freud went down deeper, stayed down longer and came up dirtier than anyone else' commented one humorous critic. His own followers were inclined to agree that he overstressed sex. Alfred Adler, for example, insisted that in most people the drive for power, success and esteem is at least as powerful as the sexual drive, while Jung argued that Freud thought too much in terms of individuals, failing to realize that in any given society the whole is larger than the sum of its parts and that in any civilization there is a great 'collective unconscious' which includes religion, traditions, legends, and folk customs and which subtly influences individual behaviour. Other psychologists were and are strongly critical of the Freudians for their lack of scientific method. They depend, their critics argue, too much on hunch and insights based too often on the rambling thoughts of unbalanced individuals rather than on observed human behaviour. Seldom did Freud and his followers attempt experiments to confirm his hunches. Nonetheless, few hunches have influenced twentieth century society more deeply.

Chapter 13
The Arts in the Modern World

Never in human history have there been so many experiments and new movements in the arts as in the twentieth century. Never has there been so strong a rejection of the artistic ideal of the past and a determination to work out new methods and rules. Never have artists been so violently persecuted as in Nazi Germany and Soviet Russia; and never, despite great official encouragement in Western Europe since the war, have they been so unable to bring the general public to an understanding and appreciation of their work.

Painting
In 1900, Paris was the artistic capital of the world at a time when the art of painting was particularly flourishing. In the previous thirty years first the Impressionists and then the Post-Impressionists had broken away from the tradition of centuries that an important duty of a painter was the accurate imitation of the material world. Cézanne, one of the leading Post-Impressionists, had been particularly concerned to reach the simple geometrical shapes which he was convinced were the essence of all natural forms. From 1907 to 1912 Picasso and Braque, working as Braque put it 'rather like mountaineers roped together', pioneered the Cubist style (opposite) which logically extended Cézanne's studies of geometrical shapes by representing the object seen from a number of viewpoints on the same

Left *Pointing the way towards the Cubism of Braque and Picasso, this strongly geometric 'Landscape with a Mill' was painted by Cézanne in 1900–1906*

Right *Braque: 'Violin and Jug'. A characteristic Cubist work painted in 1910*

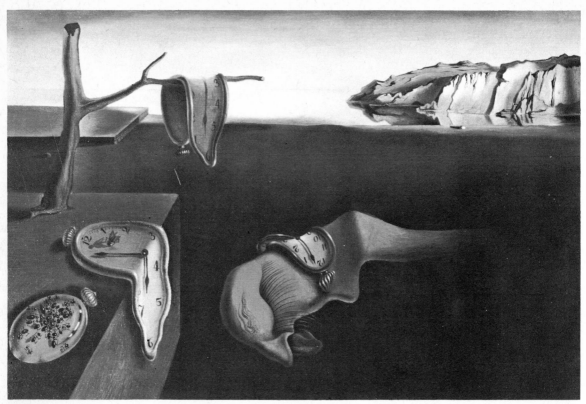

canvas. From these Cubist paintings, in which the object is often barely recognizable, to complete abstraction, when no object is represented at all, was only a small step.

The fantastic dreamlike quality of Dali's 'Persistence of Memory' painted in 1931 makes it a good example of Surrealism

Between the wars, numerous completely abstract paintings were produced, of which the most important were those of Mondrian and Nicholson. Another interesting movement was Surrealism, which was powerfully influenced by Freud's ideas on the 'unconscious' level of the human mind and the importance of dreams for its under-standing; important surrealists were Magritte and Dali. After World War II, Paris could no longer maintain her unchallenged position, first New York and then London emerging as major artistic centres in their own right. Experimentation in style and technique became more frantic. The borders between painting, sculpture, architecture and science became less clear. It is most difficult to evaluate such recent work. Some art critics take the view that the fifty years since 1925 do not really compare with the period from 1900 to 1925 which they rate as one of the great ages in the history of painting.

Picasso (1881–1973) Pablo Picasso was born near Barcelona in Spain. His father was an art teacher. From infancy, he showed a precocious

Picasso : 'Les Demoiselles
d'Avignon', 1907

talent for drawing and at the age of fourteen, though considerably younger than the other students, he passed in a day an examination for which a month was allowed and joined an advanced class at the Barcelona School of Fine Arts. A brief visit to Paris in 1900 intoxicated him and he moved there to work in 1901 when he was still only 19. His talent soon came to the notice of the distinguished dealer Vollard. His paintings began to sell and he became a respected figure among the artists of Montmartre. Then he began to make the rapid and dramatic changes in style which were to become characteristic of him. His first years in Paris saw his 'Blue Period' which in 1905 gave way to his 'Rose Period'. In 1907, however, he shocked even his closest associates with his 'Demoiselles d' Avignon' (Young Ladies of Avignon) which, influenced by African sculpture and heavily distorted, was quite unlike anything he had previously done before. His friendship with Braque had developed into the working partnership of Cubism. So closely did they work together that 'there was a time' Braque noted, 'when we had difficulty in recognizing our own paintings'. Cubism electrified and divided the artistic world and made Picasso internationally famous. By 1912, however, he was experimenting further. By sticking shapes of oilcloth or paper to his canvas, he developed the 'collage'

233

Picasso : 'Guernica', 1937

technique and when on the outset of World War I Braque joined the French army, their close association and with it Picasso's involvement with Cubism ended. Between the wars he further extended his artistic range, turning his hand to sculpture, book illustrations and ceramics as well as painting. Previously he had shown little concern for politics but he found that he was unable to stand aloof from the events of the 1930's. He was bitterly opposed to Franco's rise to power in Spain. In 1937, he produced firstly 'The Dream and Lie of Franco', a savage criticism of the Spanish general in comic-strip form and then, when German bombers in support of Franco's force bombed women and children in the streets of Guernica (see page 145), weeks of enraged activity produced a huge masterpiece, 'Guernica', eleven feet high and twenty-five feet long. In it, he vividly expressed his horror both of this tragedy and of warfare in general.

During World War II, Picasso stayed in Paris, where despite his known hatred of Nazism and the Nazi hatred of his paintings the occupying German forces left him alone. For the rest of his long life, he continued working as energetically as ever. His output was prodigious and his inventiveness hardly paralleled in the history

of art. 'The essential in this time of visual poverty', he once wrote 'is to create enthusiasm' and enthusiasm and vitality pulse through all his work. His influence on other artists, poets, sculptors and dancers, as much as painters, was profound. As a close friend, the poet Apollinaire, put it, 'Picasso is among those of whom Michelangelo said that they deserved the name of eagles because they surpass all others and break through the clouds to the light of the sun'.

Sculpture

In 1900, in Paris, a great exhibition was held of the sculptures of Auguste Rodin (1840–1917). It is often said of Rodin that he recreated a lost art, such vitality did he give to twentieth century sculpture after the comparative emptiness of the previous three centuries. Like the Post-Impressionists in painting, Rodin was less interested in the illustration of reality than in conveying impressions and often used distortions to achieve the desired effect (e.g. his monument to Balzac). His fame was international as was his influence, which deeply affected young sculptors like Epstein from New York, Brancusi from Roumania and Lipchitz from Russia.

In the same ways as Negro masters influenced early twentieth century paintings, so 'primitive' art, especially of Ancient Greece, Egypt and Mexico, affected sculpture. Another influence was modern machinery, the shape and power of which fascinated Italian and Russian sculptors and which placed a variety of new material at the sculptors' disposal. In the 'Manifesto for Futurist Sculpture' (1910) Umberto Boccioni demanded that artists should use 'glass, wood, cardboard, cement, horsehair, leather, cloth, mirrors and electric light bulbs'!

Sculptors, like painters, began producing completely abstract works. The major figure between the wars was the Roumanian Brancusi (1876–1957) who, like Picasso, made Paris his home. For him, Rodin's Balzac was the beginning of modern sculpture, and following Rodin's lead and believing that 'what is ideal is not the external form but the essence of things' he concentrated on reaching what he conceived to be the essence of the object he was shaping. Sometimes the result was totally abstract, sometimes not. Strongly influenced by Brancusi were two English sculptors, Henry Moore and Barbara Hepworth. Like him, they approached their material—mainly stone for Moore and wood for Hepworth—as carvers rather than modellers cutting away to reach the essence for which they were searching.

Since World War II, Boccioni's hopes have been realized and almost every material has been put to sculptural use. Moving sculpture (mobiles), a good example of which is the work of the American Calder, has been another significant development.

235

Henry Moore (1898–   ) Moore was the son of a miner of Castleford in Yorkshire. When he was still a boy, he became passionately interested in Michelangelo who was first introduced to him as 'the greatest sculptor who ever lived' by his local Sunday School Superintendent in Yorkshire. His obvious artistic talent was encouraged at Castleford Grammar School but, following his parents' wishes, he first trained to be a teacher. Only in 1919 after his return from some months of trench warfare did he join the Leeds College of Art. Two years later, he won a scholarship to the Royal College of Art in London where he quickly won the respect of its director, Sir William Rothenstein. His teachers at the College, however, did little to encourage and more to discourage him. One professor having viewed a painting of Moore's commented 'this man has been feeding on garbage'. He learnt most from visiting the British Museum, from primitive Egyptian, American and African sculpture and from the work of a brilliant Frenchman, Gaudier-Brzeska who had worked in London before his death at the age of 23 on the Western Front. During the 1920's and 1930's, Moore taught sculpture at the Royal and Chelsea College of Art, unappreciated by most of his colleagues and pupils but with plenty of opportunity for creative work of his own. Until 1940 he lived in North London with a holiday cottage in the Kent countryside. When his London studio was bombed in 1940, he moved with his family out of London, making his home and workplace in a Hertfordshire village. His English reputation was established in the 1930's, and by 1950 he was internationally

A Moore bronze figure stands against a background of historic Florence during the 1972 Exhibition

famous. There were exhibitions of his work all over the world, one of the finest being on a hillside overlooking Florence, Michelangelo's birthplace, in 1972.

When working in stone, Moore carved directly into it, seeking, like Brancusi, the form. A sense of vitality rather than beauty was his aim. 'For me', he wrote in 1934, 'a work must first have a vitality of its own. I do not mean a reflection of the vitality of life, of movement, physical action, freshly dancing figures and so on but that a work can have in its pent-up energy, an intense life of its own independent of the object it may represent'. His achievement in communicating this vitality has caused many critics to place him in the mainstream of European sculpture, the equal of Michelangelo or Rodin.

Architecture

The modern movement in architecture was launched before World War I. Young architects were in revolt against the nineteenth century tradition of clothing new buildings in old styles (e.g. making a railway terminus look like a mediaeval town hall) and, impressed by the creations of engineers like the Eiffel Tower, determined to build buildings which openly declared, rather than concealed, their actual function. In Germany, in 1907, Peter Behrens was appointed by A.E.G., a large electrical firm, its design consultant with responsibility for everything from factory buildings to notepaper. Behrens was not himself a great architect but he attracted to his office three young men, Gropius (1883–1969), Mies van der Rohe (1886–1969) and Le Corbusier (1887–1965). These three, along with the American Frank Lloyd Wright (1869–1959) came to dominate the modern movement in architecture after World War I.

After the war, Gropius designed and established the Bauhaus (Home of Building) in Dessau in Germany. The courses there emphasized the unity between all branches of art and design and their application to industrial production. At the same time, individuality was strongly encouraged with the result that some of the most lively and talented artists of Europe joined its teaching staff. By the late twenties a recognizable architectural style was emerging which Gropius himself described as 'International Architecture'. The Bauhaus itself is a good example.

The spread of Fascism in Central Europe drove the best architects from Europe and ensured that 'International Architecture' would become really international. Le Corbusier worked in Brazil, North Africa, Tokyo and India; Gropius became a Professor in Harvard; Mies van der Rohe in Chicago, each with tremendous influence on local architects.

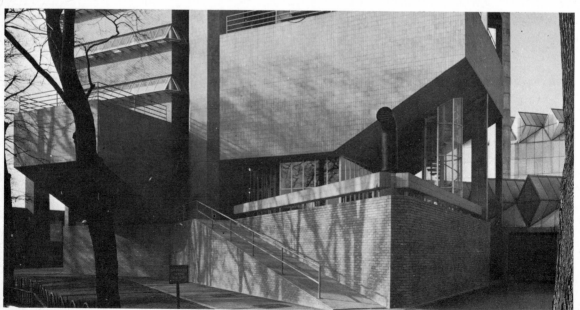

Gropius : part of the Dessau Bauhaus, 1925

After World War II, a number of architects like Nervi in Italy and Aalto in Finland gave the international style a strong personal, even regional character and in Britain in the 1950's and 1960's young architects like Alison and Peter Smithson and James Stirling, developed the 'Brutalist' style which allowed the function of the building to dictate in a harsher more brutal way its appearance.

Ludwig Mies van der Rohe (1886–1969) Van der Rohe was the son of a stonemason. His main training as an architect came through practical experience working first for Paul, a second-rate architect but gifted furniture designer, and then from 1908 to 1911 for Behrens. In the 1920's he remained in Germany, most of his work being experimental designs for skyscrapers or housing schemes. In 1929, however, he was responsible for the German Pavilion at the International Exhibition at Barcelona. His simple but elegant building with its carefully designed interior and furnishings, including the famous 'Barcelona' chair, made his international reputation. This he followed the next year with the Tugendhat Home in Brno, Czechoslovakia, which, again basically simple in design but meticulously finished in every detail, was soon classed among the most beautiful houses in the world. In 1930, Mies succeeded Gropius as Director of the Bauhaus but, three years later, the Nazis closed the school down. In 1937, after uneasily trying to work within Nazi Germany, he moved to the U.S.A. where at the Illinois Institute of Technology (I.I.T.), in Chicago, he obtained the post of Director of Architecture with the responsibility of designing the new campus. Now in his maturity, he was much more productive than in Europe. As well as the I.I.T. campus, he was responsible for many major buildings of which the Lake Shore Apartments, in Chicago, the Seagram Building, New York, and the Gallery of the Twentieth Century, Berlin, are among the most important.

Mies' buildings have an exceptional simplicity and purity of structure. Their beauty comes from their proportions and the relationship of glass and steel. His influence on post-war American and international city-centre architecture was immense, far greater in fact than that of Gropius or Le Corbusier. He was not, however, without his critics. Lewis Mumford condemned his buildings as 'elegant monuments of nothingness'. Others argued that his apparently practical buildings were, for those who had to work in them, impractical in the extreme.

Music

As in painting and architecture the years immediately before World War I were an important period of change in the history of music. The late nineteenth century had been dominated by composers of the romantic style. Chief among these were Brahms (d. 1897),

Stirling : the Engineering Faculty, University of Leicester, 1959–63

Wagner (d. 1883), Tchaikovsky (d. 1893), Verdi (d. 1901), Dvorak (d. 1904) and Mahler (d. 1911). Extending a musical tradition more than two centuries old, and expressing their most important musical ideas in the form of symphonies, concertos and operas, their style was characterized by its length, by its depth and variety of orchestration and by its emphasis on feeling rather than form.

While this tradition was further developed by Richard Strauss in Germany, by Prokofiev and Shostakovitch in Russia, other composers charted new courses. Folk music was a strong influence on the Russian Stravinsky and on the Hungarian Bartok (1881–1945). They broke sharply with the Romantic style producing condensed, rhythmic and often discordant compositions. After World War I Schönberg (1874–1951) went in a radically new direction. In 1924, in his 'Piano Suite', he abandoned all previous methods of composition and harmony which were founded on the eight-note scale in favour of his own methods based on a twelve-tone scale. His compositions greatly interested his fellow composers but their apparent discord and lack of melody prevented them gaining favour with the concert-going public. Even more radical experiments with electronic sounds were tried after the war by composers like Stockhausen and Cage. These, even more than Schönberg's works, proved incomprehensible to most of the musical public.

Stravinsky (1882–1971) Igor Stravinsky was born in Russia, the son of an opera singer. He began by training as a lawyer, but on the advice of the composer Rimsky-Korsakov, began his musical career in 1902. He first became internationally known through his work for Diaghilev, whose Russian Ballet made a tremendous impact on Western Europe, especially in Paris in the years immediately before World War I. For this Ballet he wrote a number of works, notably *The Firebird*, *Petrouchka* and *The Rite of Spring*. This last caused a riot when it was first performed in Paris, so unusual were its irregular and syncopated rhythms, discords and lack of continuous melody. World War I and the Russian Revolution of 1917 cut him off from his former home. Until 1939 he lived for the most part in France, after 1939 in the U.S.A. Between the wars, when he had to earn a living as a conductor and soloist as well as a composer, he produced a number of concertos and symphonies which, while highly individual in character, were comparatively traditional in shape. Much of what he composed in his so-called neo-classical period, has become like his music for the Russian Ballet, standard works in the modern concert programme.

After World War II, partly as a result of his friendship with Robert Craft a young American musician, he became interested in Schönberg's twelve tone system which strongly influenced his compositions in the 1950's and 1960's.

Stravinsky's music aroused continuous controversy throughout his life, both his sympathizers and critics finding it hard to adjust to the marked changes of style which he made during his musical career. There can be little doubt, however, that he was one of the most original, fertile and influential figures in twentieth century music whose compositions while advanced won a wide and appreciative audience.

Literature

'The world is suffocating' wrote Rolland (1866–1944), a major French writer, 'let us open the windows and bring in the fresh air, let us breathe like heroes'. Many contemporary writers—D.H. Lawrence for example—shared his views. They rejected the values of what they considered to be the old, stuffy middle-class civilization of the nineteenth century in favour of the new, the vital, and the heroic. Others like Kafka (1883–1924) in *The Trial* and Mann (1875–1955) in *Dr. Faustus* and T. S. Eliot (1888–1965) in *The Waste Land*—probably the most influential single poem of the century—deeply questioned past and present values without much hope for a more vital, heroic future. The despairing theme that life was basically absurd or meaningless was increasingly heard, culminating in the poetry and plays of Samuel Beckett.

One alternative to despair came from French Existentialism as developed by writers like Sartre and Camus who insisted on the reality of the conscious mind and of the individual's ability to choose. Another alternative was left-wing political commitment. The anti-Franco cause in the Spanish Civil War not only inspired Picasso's 'Guernica' but in Malraux's novels and the poetry of Auden some of the best writing of the 1930's. Political changes also made possible some of the most inspired writing of post-war Europe, the Russian novels of the 'thaw' after Stalin's death. Pasternak's *Dr Zhivago* and Solzhenitsyn's *The First Circle*, both epic novels inspired by the astonishing history of twentieth century Russia, tower above any other works produced in this period.

Albert Camus (1913–60) Camus was born in French Algeria. He never knew his father who was killed in World War I. Gifted in many directions, he took a degree in philosophy, kept goal for the Algerian football team, worked as a journalist and directed many amateur theatricals before going to Paris in 1939. After the defeat of France, he became an active member of the Resistance, editing the underground newspaper *Combat*. Before the war ended, he had also published two of his most important works—*the Outsider*, a novel, and *The Myth of Sisyphus*, a book of essays. For a few years after the war, he stayed in journalism and politics but eventually devoted himself entirely to writing. Another major novel, (*The Plague*), was

The Robber Chief's Banquet at Olympia

Drawn by J. Duncan

published in 1947 and in 1951, an analysis of the role of the writer in the modern world, *The Rebel*. In 1957 he was awarded the Nobel Prize for Literature only to die at the tragically early age of 47 as the result of a car crash in 1960.

Camus was a friend of Sartre and much influenced by existentialist philosophy. Convinced that religious belief was impossible, he was fascinated by the problem of how good and evil could be defined in a godless world. Meursault, the 'hero' of *The Outsider* commits the motiveless murder of an Arab on an Algerian beach but at his trial 'is condemned' Camus himself wrote 'because he doesn't play the

Experimental staging on a grand scale: Reinhardt's design for a production of The Miracle *in 1911*

game ... He refuses to disguise his feelings and immediately society feels threatened.' In *The Plague*, set in the plague-ridden city of Oran a central question is how best to combat evil both internal and external, to be a saint without God. Like many fellow-existentialists, Camus felt strongly the absurdity of much of life, and like Sartre, his political views were firmly left-wing. In the 1950's however, he parted company with Sartre, whose extreme revolutionary and pro-Stalinist views he found unacceptable. The combination of his skill as a novelist, the perceptiveness of his thinking about modern writing and his political involvement made him, at the time of his death, one of the most influential writers of the post-war world.

Theatre

As in the other art-forms, the period 1900 to 1914 was one of intense experimentation in the theatre; particularly in methods of production. Both in Berlin where Reinhardt took over the Deutsches Theatre and in Moscow, where Meyerhold began his connection with the Arts Theatre, new techniques like spot-lighting, cycloramas and revolving stages were married to increasingly imaginative productions. The twentieth century proved to be one of the most fertile in the history of drama. An Irish cultural revival at the beginning of the century was headed by playwrights such as O'Casey (1880–1964). Many dramatists attempted to combine poetry and drama, the most successful being the Spaniard Lorca. The sense of the absurdity of life which affected so many twentieth century writers took dramatic shape in the so-called 'Theatre of the Absurd' which included Pirandello (1867–1936), Ionesco (1912–) and Beckett (1906–). Probably the most influential figure in twentieth century theatre was Bertolt Brecht.

Bertolt Brecht (1898–1956) Brecht was born in Augsburg, Germany, the son of the manager of a paper mill. After a war service of hospital duties and then university studies in science and philosophy, he decided on a theatrical career. His first plays *Baal* and *Drums in the Night* were written in 1922 and international fame came six years later with a musical, *The Threepenny Opera*, written in collaboration with the composer Kurt Weill. By this time, he had become a convinced Marxist and the next year, working again with Weill, he wrote and produced *The Rise and Fall of the City of Mahagonny*, a Marxist attack on the barbarism of the modern city. In 1932 came *St. Joan of the Stockyards* which, set in the stockyards of Chicago, was a condemnation of capitalist exploitation of the working classes. Nazi Germany was no place for any advanced artists, least of all for Marxist ones. In 1937 he left Germany and after stays in Denmark and Finland ended up in the U.S.A. During the war he wrote some

of his finest plays, among them *Galileo* and *Mother Courage and her Children*. His communist sympathies brought him in front of the Un-American Activities Committee in 1946 and the following year he returned to Germany, to the eastern, communist zone, to direct the Berlin Ensemble. Here his genius as a producer became clear with considerable impact on both sides of the iron curtain, though, ironically, more on the non-communist side than the communist.

A production in 1927 in Baden-Baden of Brecht's City of Mahagonny. The placards read, 'for the mortality of the soul,' 'for earthly reward,' 'against law and order,' 'for natural indiscipline'

Brecht believed in 'epic theatre' in which major historical and political ideas should be put over to audiences of all classes. It was not the business of the producer to present an illusion of reality. The audience must always remain aware that the stage was merely a stage and the actors merely actors. What mattered was the message and is presentation in as vivid a way as possible. For this purpose he used a variety of aids including music, loudspeakers, back projection, film excerpts and so on. Despite his chequered career— the Nazis burnt his books, the West distrusted his communism, the communists his advanced theatrical methods—he is a major figure in twentieth century art.

A scene from an early French film—Journey to the moon, 1902

The Cinema

The basic technique of making moving pictures was perfected by Edison in the U.S.A. and in 1895 the first European commercial film show was given to an audience of thirty three in a Parisian cafe. The following year a London audience saw a film of the Derby the same day as the race was held and in 1905 the first European cinema was opened. Up to World War I, the European cinema held its own with the American but during the war the American film industry, centred in Hollywood in California, established a lead in the field of mass entertainment which it never lost. The popular success of cinema meant that politicians and others gradually began to make use of it for propaganda, advertising and education. The Russian communist government encouraged their film industry and in the 1920's allowed their producers, which included men of genius such as Eisenstein, artistic freedom enough for some superb films to result. The Nazis followed a similar policy and for them Leni Riefenstahl produced propaganda films of horrifying beauty and power (e.g. films of the Nuremberg rallies and the 1936 Olympic Games). In the freer climate of post-war Western Europe, there was

245

Jesse Owen, from Riefenstahl's film of the 1936 Olympics

Part of the celebrated 'Odessa Steps' sequence from Eisenstein's Battleship Potemkin, *showing the citizens of Odessa, who have gathered at the top of the Odessa Steps to watch the mutiny on the battleship* Potemkin, *being shot down by the Tzar's White Guards*

something of a film renaissance. Both in Italy and in Sweden many excellent films were made but the most exciting work was done in France by the so-called 'New Wave' producers such as Godard and Truffaut.

Eisenstein (1898–1948) Eisenstein came from a Jewish family of Riga. His father was a prosperous shipbuilder. Educated in the Institute of Civil Engineers, he was swept up by the 1917 Revolution, of which he was an ardent supporter, into the Red Army and then, because of his interest in theatrical design into the Proletcult, the first 'workers' theatre'. He learnt much from Meyerhold who had already an international reputation. Not until 1923 did he have any experience of film-making but within a year he had produced his first full-length film, *Strike*. In 1925 he made one of the world's great films *The Battleship Potemkin*, inspired by an episode in the unsuccessful revolution of 1905, and in 1928, *October*, a celebration of the successful revolution of 1917. These were acclaimed by film critics the world over and, from 1929 to 1932, he travelled widely especially in the U.S.A. and Mexico, but various projects to make further films with American backing bore little fruit. He was now at the height of his powers. As he worked on his next Russian film—*The General Line*— he wrote 'we felt young and surging with the creative energy of a new Renaissance and we saw a perspective of boundless new possibilities in the future'. Such optimism turned out to be misplaced. Stalin's Russia became an increasingly difficult place for artists with minds of their own. Convinced communist though he was, Eisenstein found it hard to satisfy the rigid and inconsistent government censorship. He only produced three more full-length films—*Bezhin Meadow* (1934–36), *Alexander Nevsky* (1938) and *Ivan the Terrible* (1943–46)—and the last in particular won much official disapproval. It was not generally released until after Stalin's (and Eisenstein's) death. He therefore devoted much of his creative energy to working and teaching at the State Institute of Cinematography.

Eisenstein not only made films of remarkable power and humanity, he also pioneered techniques which were widely copied by other film makers. In particular he made frequent use of non-professional actors and of 'montage'—the editing, selecting and repeating of certain shots in order to make clear in visual terms complex ideas vital to an understanding of the film.

Mass Pleasures

None of the art-forms or the individual artists described above can be described as popular. A steady improvement in education and the more rapid spread of ideas and information through television, paperback books and newspapers have made them more popular than ever before but still the overwhelming majority of Europeans would not go to a Picasso exhibition, a Brecht play or an Eisenstein film, nor would they buy a Stravinsky record nor a Camus novel.

Enjoyment of these art forms is still restricted to a small intellectual and social elite. In some cases—e.g. modern music and branches of modern literature—recent work has become so difficult and at first sight so ridiculous that the gulf between this élite and everyone else is more wide than ever before.

We know very little about the art-forms and entertainments which were widely popular in this or in earlier centuries. The necessary systematic historical research has barely begun. All that can be offered here is a bare outline.

Before 1950, two of the most popular forms of entertainment in Europe were the cinema and radio. The popular films tended to be Hollywood Westerns, Romance or Biblical spectaculars. After 1950 however, the cinema went into rapid decline, unable to meet the challenge of television. Britain, which had begun television broadcasts just before the war, began in 1946 with 1,750 subscribers. The French T.V. service began in 1944 soon after the Liberation, the German in 1952 and the Italian in 1953. By 1968, there were more than 15 million sets in Britain, 14 million in W. Germany, 9 million in France and 8 million in Italy. Simultaneously, the number of people going to the cinema fell rapidly. Where in 1955 Britons went to the cinema a total of 1,182 million times, in 1968 the corresponding figure was 237 million. In Germany the number fell in the same period from 767 million to 192 million.

Cinemas were closed or converted into dance or bingo halls and the long-term future of the film industry became doubtful. The growth of television also affected radio which between 1920 and 1960 offered a great range of entertainment and educational programmes. Increasingly it concentrated on music and news, particularly useful for whiling away the time in a traffic jam. Nor did newspapers fare well against the challenge of television especially in those countries where commercial television could attract advertising away from them. The British press suffered badly in the 1950's and 1960's, a number of popular papers like the *Daily Herald* going out of circulation and even famous quality newspapers like *The Times* struggling to survive. Quality newspapers on the continent however, like *Le Monde* and the *Frankfurter Allgemeine Zeitung* managed to increase their circulation and their profits.

In Russia and Eastern Europe television spread less rapidly. For this reason and because all the mass media—radio, newspapers, cinemas as well as television—remained under the firm control of the government, the effect of television was more gradual than in Western Europe.

One form of entertainment which was equally popular on both sides of the iron curtain was organized sport. The rules of most sports had been defined in Britain and then exported to the rest of the world at the end of the nineteenth century. They attracted more and more participants and spectators and were encouraged by dictators and democratic governments alike. Most sports made excellent entertainment on television and by the 1970's, the potential audience for the Olympic Games or for the World and European Cup football matches could be counted in hundreds of millions.

'Popular' music whether deriving mainly from the U.S.A. between the wars or from Britain and the U.S.A. in the 1960's, also won huge support in Western Europe, especially among the young, and also penetrated the iron curtain. The invention of long-playing records, of transistor radio, of electrical instruments and amplifiers combined with the rapid growth in the spending power of young people, allowed 'pop music' to become internationally big business and made its stars like the American Elvis Presley and the British Beatles and Rolling Stones, world famous.

There was a school of thought, expressed in Britain by George Orwell and Richard Hoggart, that the developments in mass culture this century whether directly controlled by the state as in Eastern Europe, or manipulated by business interests as in Western Europe, are for the most part extremely worrying. The mass media are easily available to everyone and easily controlled from a central source. They tend to kill regional and local culture and to replace the varied and vital popular culture of the past with a mindless uniformity, either dictated by the government or by what the controlling companies believe to be what the general public wants most of the time. 'They are in the end,' Hoggart concluded in *The Uses of Literacy* (1957), 'what D. H. Lawrence described as "anti-life". They are full of a corrupt brightness, of improper appeals and moral evasions: ... they tend towards a view of the world in which progress is conceived as a seeking of material possessions, equality as a moral levelling and freedom as the ground for endless irresponsible pleasure.' Time will tell whether so pessimistic a view is justified.

Chapter 14
Women and Society

Feminism before World War II

A feminist movement is one dedicated to winning greater rights for women. In 1900, such a movement had already been in existence for many years in many parts of Europe and had already a number of achievements to its credit. In particular it had won better property and divorce rights for married women. In the first years of the twentieth century, since universal male suffrage had now become general, its chief preoccupation was votes for women. This and other political rights were won, but only gradually. The franchise was gained in Norway and Sweden in 1907, Russia in 1918, Weimar Germany in 1919, Britain in 1928, France in 1945 and Italy in 1946. Not until 1950 was female suffrage general throughout Europe and even then Swiss women remained voteless. There were some direct consequences of this political development. Some, though not many, became politicians, some, though very few, became government ministers. In addition governments, conscious of the female electorate, became more concerned than once they had been with welfare services, especially those affecting children and the home.

Other changes occurred in the female way of life, status and attitudes which were more profound and far-reaching than winning the right to vote in elections. 'It would seem', wrote an eminent sociologist, R. M. Titmuss, in 1958, 'that the typical working-class mother of the 1890's married in her teens or early twenties, and experiencing ten pregnancies, spent about fifteen years in a state of pregnancy and in nursing a child for the first year of its life. She was tied for this period of time to the wheel of childbearing. Today, for the typical mother, the time so spent would be about four years. A reduction of such magnitude in only two generations in the time devoted to childbearing represents nothing less than a revolutionary enlargement of freedom for women, brought about by the power to control their own fertility.' Alongside this greater freedom from childbearing were better educational and career opportunities, labour-saving equipment in the home and a longer expectation of life. For most European women, therefore, the twentieth century placed more time at their disposal, more scope for independence, yet also a growing impatience with what seemed to many of them their unequal position in society.

Different governments reacted in different ways to the potentially revolutionary trend of female emancipation. Liberal democracies like Britain and the Scandinavian nations tended to move slowly and fitfully towards greater female equality, but considered many aspects of female emancipation to be no business of government. In countries where the Roman Catholic church was powerful as in Italy or Spain, the trend was regretted and demands for easier divorce, abortion and the encouragement of contraceptives were bitterly opposed. In many ways, the most interesting reactions came from totalitarian governments like the Nazis in Germany and the Communists in Russia. Both regarded themselves as modern, forward-looking movements which had shaken off the inadequate values of the past and both believed that there was no aspect of society, not even marriage and the family, which was not the concern of the state.

The transformation of the way of life of many women by World War I as seen by a French magazine of the time

Women in Nazi Germany

The feminist movement began comparatively late and had achieved little in Germany before World War I except in education. During the Weimar Republic, however, it made rapid strides. Women won the vote, a number became members of the Reichstag and one of its leaders, Gertrud Baumer, became a senior official in the Ministry of the Interior. National Socialism however, was a self-consciously masculine movement. 'Only by marriage', Hitler wrote in *Mein Kampf*, 'should a woman qualify to be a State Citizen. Otherwise she is merely a State Subject.' Before the Nazis came to power, they had campaigned for the abolition of female suffrage and once in power they made the removal of women from public positions official policy. While in 1933 there had been 30 female members of the Reichstag, by 1938 there were none. In Hitler's opinion, female emancipation was yet another branch of the international Jewish conspiracy to destroy Western civilization. 'The message of woman's emancipation,' he wrote, 'is a message discovered solely by the Jewish intellect and its content is stamped by the same spirit.' One of its dangers lay in the encouragement it gave to women to move out of their proper sphere. 'A woman's world', in his opinion 'is her husband, her family, her children, her

Hitler and Goebbels, both look unusually ill at ease, in the company of young women in the Rhineland

home ... We do not find it right when the woman presses into the world of the man. Rather we feel it natural when these two worlds remain separate.' Women therefore should not be encouraged to enter employment, particularly once they were married. Dr. Frick, Minister of the Interior, insisted that 'the mother should be able to devote herself to her children and her family, the wife to her husband. The unmarried girl should be dependent only upon such occupations as correspond to the feminine type of being. As for the rest, employment should remain given over to the man.' As for education, 'the aim of feminine education,' Hitler wrote in *Mein Kampf*, 'is invariably to be the future mother.' In the Nazi world, the woman had one simple but vital function, to breed well in the service of the Fatherland, which in effect meant providing the cannon-fodder for Hitler's armies of conquest. 'All I want to do,' the Führer declared in 1936, 'is to create to the greatest possible extent the possibility of founding a family and having children, because our people need them above all things.'

Many steps were taken between 1935 and 1939 to give reality to these views. The women's organizations of the Weimar Republic were either nazified or disbanded. The leader of the Nazi Women's Movement, Frau Scholtz-Kluck, defined the only honourable occupation for German women as 'to minister in the home ... to care for the man, soul, body and mind ... from the first to the last moment of man's existence.' Though one in three high school students were women, the proportion of places in higher education was restricted to one in ten. Efforts were made to drive them from the professions and from employment generally. In 1933, they were no longer able to act as judges, in 1936 to officiate in any capacity in the law-courts (though after 1936, labour shortages put paid to this policy). Marriage and childbearing were encouraged by various measures. Bachelors and spinsters were heavily taxed. Interest and tax rebates were made available for every child born to a married couple. Divorce was made more difficult, abortion and the spreading of contraceptive information forbidden. Books like those of W. Reich which recommended free sexual activities and radical changes in marriage were banned.

The Nazi attitude to sexual relations was weird. Hitler who only married Eva Braun on the day before they both committed suicide appears to have had limited and abnormal affairs with women. He was obsessed with the dangers of syphilis which in his opinion was the result of Jewish or Communist sexual excess! He was something of a puritan and official policies reflected his distaste for any breach of conventional morality. Homosexuality, prostitution and pornography were all declared illegal. In some parts of Germany police stopped women smoking in public and the excessive use of make-up by women was officially criticized. However, a double standard of morality was allowed to operate, one for ordinary Germans, the other for Nazi party members and the military. Nazism was a strikingly masculine movement and homosexuality could not but flourish in its ranks, while prostitutes and pornography became the privilege of favoured party groups, especially the S.S.

The Nazi policy towards women was generally successful. The birth-rate rose fast, women were deployed either into marriage or into subordinate employment. Indeed so great was the reluctance of the Nazis to distract their women from the responsibilities of 'sacred motherhood' that, even at the crisis of the war, German women contributed less to their war-effort than their British counterparts. Intolerable though such success must have been to the intelligent, independent women who had begun to make their mark on German life before 1933, there can be no doubt that Hitler could count on the enthusiastic support of millions of German women throughout the twelve years of his dictatorship.

Women in Soviet Russia

'A wife isn't a jug—she won't crack if you hit her a few times'.
(Russian proverb).
'A wife's duty is to obey her husband as the head of the family, to be loving and respectful, to be submissive in every way and show him every compliance and affection.' (Tsarist law).

In much of Russia before the 1917 Revolution, women were little better off than slaves. They had to follow their husband at his bidding, and could not apply for a passport without his permission. Their property became his on marriage. In the east, they still had to wear veils and sometimes tolerate polygamy. In some areas, they could still be sold off to the highest bidder. In the factories, they had to work long hours and were seldom paid as well as men. In the countryside, they had to be agricultural labourers as well as keep the home. Contraceptives were virtually unknown, abortions the work of the local 'wise' woman who possessed neither medical knowledge nor instruments. Female education was limited. Even in the upper classes, only 'disreputable' women would consider joining higher education courses.

Communist theory however was as revolutionary about the family and the social position of women as it was about politics and economics and the key writer was not so much Marx as his closest collaborator Engels, whose book *The Origin of the Family, Private Property and the State* was published in 1884. Engels argued that 'the modern monogamous family is founded on the open or disguised domestic slavery of women and modern society is composed of molecules in the form of monogamous families.' In modern industrial society, the bourgeoisie exploit the proletariat, similarly in the modern monogamous marriage, the man exploits the woman. Genuine social equality would only come, Engels believed, when an economic and social transformation brought about 'the reintroduction of the female sex into public industries.' For this to happen, 'the care and education of children must become a public matter', and the monogamous family, with the wife caring for home and children and economically dependent on her husband, the breadwinner, must cease to exist. Just as Marx believed that the Communist economic and political revolution was inevitable and that the state would wither away, so Engels believed that the sexual revolution was inevitable and that the monogamous family would wither away.

Immediately after their successful seizure of power in the autumn of 1917, the Bolsheviks passed a number of laws which dramatically improved the position of Russian women. Most of the husband's legal authority over his wife was ended and women were granted complete economic, social and sexual self-determination. They could choose where they wished to live and how they wished to be addressed. Divorce, contraception and abortion were to be much more freely available. Homosexuality, incest and adultery were no longer to be treated as criminal offences. Plans were made for nurseries and crèches to be built, housekeeping collectivized and maternity leave so organized that women could, as Engels had prophesied, enter employment on a genuinely equal footing with men. In 1919 the Working and Peasant Women's Department of the Communist Party, usually known as the Genotdel, was formed. Its purpose was to educate women and to involve them in political activity. During the Civil War it organized women for active service in support of the Red Army. They nursed, dug trenches, erected barbed wire and spread Communist propaganda in the provinces. Some actually took part in the fighting. Vera Alexeyeva, for example, before the Revolution a worker in a cigarette factory, became the captain of a Bolshevik cavalry unit hunting the Whites in the Ukraine. When the war ended, the Genotdel organized women's congresses and local discussion groups in order that women might gain confidence enough to take part in local and national politics. Much discussion revolved round the question of how women could best be relieved of the drudgery of their own private housekeeping so that they could play a more public and productive role in the building of the new society. Alexandra Kollontai, active politician, novelist and one of the ablest of the female Bolsheviks, believed that communal living must replace the family. Trotsky agreed. People must, he wrote 'group themselves even now into collective housekeeping units'. It was only 'by the creation of model communities' that women could easily be liberated. A number of communes were formed in the early twenties.

Military training for young Bolshevik women in 1919

In the late twenties, however, there was a change in the official attitude. Keeping the new Communist society going politically and economically was proving a desperately hard struggle and experiments with free love, marriage and the family seemed luxuries that the country could ill afford. Many party leaders took the view that compared with their immediate task of creating the first genuinely socialist state in history, relations between the sexes was a trivial distraction. They also linked a considerable increase in juvenile deliquency and an atmosphere of social breakdown with excessively rapid female emancipation. Opportunities for women proved fewer than had been expected since in the depressed economic conditions of the twenties new jobs remained limited. With the government's emphasis on the expansion of heavy industry at all costs, money for nursery schools and crèches was in short supply. And centuries of

traditional behaviour could not be altered overnight. In many parts of Russia, especially in the more rural and primitive areas, there was a violent backlash against feminism which was regarded as wicked and disgusting. In 1928, for example, Zarial Haliliva, a keen feminist of twenty who dared to go to the theatre unveiled and wore a bathing costume on the beach, was 'tried', 'found guilty' and 'executed' by being cut up alive by her father and brothers. In the same year in Uzbekistan, an area of Southern Russia to the east of the Caspian Sea, there were 203 cases of anti-feminist murders.

With Stalin in power, the pendulum swung back to a more traditional approach to marriage, the family and sexual relations. In 1929 the Genotdel was abolished, in 1934 homosexuality again made illegal, in 1935 parents made responsible by law for their children's education and behaviour, in 1936 divorce was made more difficult, abortion outlawed and financial bonuses given to women with six or more children. In fact in the thirties there were striking similarities between Hitler's and Stalin's policies towards women. Like Hitler, Stalin was keen to encourage motherhood within marriage and for much the same reason. He feared that a falling birth-rate would weaken the Russian race and with reduced manpower the Red Army would not be able to hold its own against Russia's dangerous neighbours. By the mid-thirties, therefore, *Pravda*, the official government newspaper, was expressing views which must have made Engels turn in his grave. 'Only a good family man can be a good Soviet citizen ... marriage is the most serious affair of life ... one of the basic rules of Communist morals is that of strengthening the family.' Free love and a disorderly sex life were, in *Pravda*'s opinion, 'bourgeois', while two central ideas of Engels—the lessening of the importance of marriage and the withering away of traditional mono-gamous family life—were described as 'foul' and 'poisonous'. Stalin himself, who treated his family appallingly, took care to project the image of 'Uncle Joe', the kind, sympathetic, solid and wise family man. On one famous much-publicized occasion, he took his children to visit his old mother in Tiflis. The government newspapers covered the event in depth, even reporting what the children thought of their grandmother's jam. In Stalin's Russia, there was no doubt that women were subordinate. 'He wanted us to work hard and fulfil the Plans', one Moscow woman commented, 'but he kept us in our places, never appointing women to high political office.' In the forties, co-education was cut back and the inheritance laws adjusted in favour of men.

However, throughout the Stalinist period, Soviet women main-tained their right to work and to education. As their numbers passing through higher education increased, so did their numbers holding positions of responsibility in medicine, science, education and the civil service. After Stalin's death, the pendulum swung back towards greater independence for women. In 1955 abortion was legalized once again, contraceptive advice became available and welfare services supporting the mother in work were expanded. In 1964 divorce was made easier, and despite memories of the experiments in communal living of the 1920's which had ended in failure, discussions began again about the creation of new types of communes. By the late sixties, Russian women were probably as emancipated as anywhere in Europe. Their opportunities in education and employment were, comparatively speaking, excellent, as were their legal and political rights. But they were not obviously more emancipated. In many jobs they were still paid less for equal work, and within the home, though their relations with their husbands were far more equal than those of their grandparents, housekeeping remained the woman's responsibility even when both partners were in full-time employment. A survey of 160 working couples in Leningrad showed that in 69 cases the woman alone did the housework, in 26 the grandmother, in 17 the wife and children while in 48 the husband lent a hand. As far as official attitudes were concerned, Soviet marriage and family life appeared to have stabilized and the party line seemed to accept it as it was, avoiding both the Engelian revolutionary and Stalinist counter-revolutionary extremes.

Recent Developments in Western Europe

A new phase of American and Western European feminism began in the sixties. In the opinion of a number of clever, eloquent women with a flair for publicity, the winning of political and legal rights at the beginning of the century had proved a hollow victory. The world was still overwhelmingly dominated by men. Neither in proportion to their numbers nor their abilities were women gaining a fair share of top jobs nor were they often getting equal pay for equal work. Both inside and outside marriage they tended to be treated either as sexual objects or as housewives whose chief function was either to provide pleasure or care for the man and his children. Such treatment 'has succeeded,' wrote the American Betty Friedan in 1963, 'in burying millions of American women alive.' The suburbs were little better than 'comfortable concentration camps' from which it was almost impossible for women to escape. Centuries of social conditioning, these Women's Liberation writers further argued, had caused most Americans and Europeans, whether male or female, to accept woman's second-class status as natural. Only by the ending of this social conditioning would women become genuinely liberated and this, as Engels had argued a century earlier (see page 254) was only possible if monogamous marriage and family life as experienced in modern industrial society withered away. The comparison between the oppressed industrial worker of the nineteenth century and the oppressed housebound housewife of the twentieth was often made. Wrote Germaine Greer in 1970: 'Women represent the most oppressed class of life-contracted unpaid workers, for whom slaves is not too melodramatic a description.' In her opinion, the time was ripe for revolution, the first step in which should be the well-tried method of the industrial worker, 'the valid withdrawal of labour.'

The Woman's Liberation Movement gained wide press and television coverage and its ideas caused considerable interest. Its influence, however, is hard to calculate. Certainly in the sixties and early seventies European women were able to gain greater rights. In Norway, for example, women were able to be ordained priests while in Britain women were at last allowed to work in the London Stock Exchange, and in France, Mme Giroud was appointed Minister of the Female Condition. Simultaneously, similar action was taken within the E.E.C. Article 119 of the Treaty of Rome which stated that member nations would 'ensure and subsequently maintain the application of the principle of equal remuneration for equal work as between men and women workers' had been ignored for fifteen years. In 1973, however, the European Commission decided to prosecute the Dutch whom it considered to have done least to bring about equal pay. Whether such developments came because of or despite the Women's Liberation Movement is hard to say.

Chapter 15
The End of European Empires

Introduction

> 'I contend that we are the first race in the world and the more of the world we inhabit, the better it is for the human race. The objects we should work for are—the furtherance of the British Empire and the bringing of the whole uncivilized world under British rule'.
>
> Cecil Rhodes, Prime Minister of Cape Colony (1890–95).
> 'White people are only overcome by death'.
> A West African child watching a Swiss missionary playing the organ.

Both Cecil Rhodes and the West African child were commenting in their own way on one of the most notable features of the world in the first years of the twentieth century—the complete, confident and barely challenged domination of the white European races. In 1914, the British Empire extended over 10,500,000 square miles with 400,000,000 inhabitants, the French over 4,500,000 square miles with 52,000,000 inhabitants, the German over 1,000,000 square miles with 14,000,000 inhabitants. Europeans controlled all Australasia and virtually all of Africa. With the exception of Japan, Asia was under either their direct political control or their indirect economic influence. The ramshackle Ottoman Empire of Turkey and the Middle East survived at least in part because the European nations could not agree how to divide it up. Only the Americans maintained a clear independence, yet the 'master races' of both North and South America were originally of European stock. The rapid expansion of European rule had taken place very recently, for the most part since 1875, but the European empires were founded on such superiority of weapons, wealth, technology and government that their future seemed likely to be measured in centuries. Within half a century, however, they were, to all intents and purposes, destroyed.

The Beginnings of Nationalism

When the First World War ended, European rule spread further. The German colonies in Africa and the Pacific were not given their independence, but were divided up amongst the victorious Allies; so too were the Ottoman possessions in the Middle East. Syria for example became a French responsibility, Palestine a British one. Moreover in Persia where British influence was already strong before the war, it grew stronger still. Nonetheless the First World

War was very damaging to European empires. The First White Civil War, as some writers called it, not only caused tremendous destruction within Europe but a serious loss of confidence amongst Europeans as to the real value and strength of the 'civilization' of which they had been so proud. Thoughtful non-Europeans also began to question whether the European rule imposed upon them was as powerful and as permanent as they had previously thought. Furthermore, as the war drew to its end, two major nations—the U.S.A. and U.S.S.R.—were openly critical of the European empires. As the fifth of his Fourteen Points (see page 50), President Wilson of the U.S.A. demanded 'a free, open-minded and absolutely impartial adjustment of all colonial claims, based upon the strict observance of the principle that . . . the interests of the population concerned must have equal weight with the claims of the government (i.e. the European one) whose title is to be determined'. His plan was first to establish the League of Nations and then to give the League the responsibility of administering the German and Turkish Empires. Britain and France, however, demanded their complete takeover by the victorious Allies. The result, which was included in the Versailles Settlement, was a compromise, suggested by Smuts, the South African Prime Minister. The old German and Turkish Empires became 'mandates', to be administered by the victors under the supervision of the League on the principle 'that the well-being and development of (their) peoples form a sacred trust of civilization'. There were to be three types of mandate. Territories in Category A, which included most of the Middle East, were expected to become independent before long; Categories B (most of the German colonies) and C (colonies with tiny population) were thought to need a 'mandatory' administrations for many years to come.

'Well, he (Wilson) has saved his precious principles, but we got our colonies', said Lloyd George when the bargaining was over. A vital principle, however had been established: independence from colonial rule was desirable. Moreover, in some areas of the world at any rate, the rule of non-Europeans by Europeans was under regular international supervision. Wilson's efforts gave considerable encouragement to those who all over the world were beginning to unite to overthrow European rule.

Further and more active encouragement came from Communist Russia. Lenin had developed a persuasive theory explaining recent European colonization, which he termed 'imperialism'. In his book, *Imperialism, the Highest Stage of Capitalism*, published in 1916, he argued that the European capitalism had had to expand economically across the world in order to survive and that European empires had come into existence in order to preserve the capitalist system.

Workers of the world united in their struggle against imperialism by the 'International'. A Russian poster, artist unknown, 1919

'Capitalism', he wrote, 'has given to the world a system of colonial oppression and of the financial strangulation of the overwhelming majority of the people of the world by a handful of 'advanced' countries . . . The booty is shared between two or three powerful marauders, armed to the teeth (America, Great Britain, Japan).' Communists therefore were very critical of imperialism. They believed that just as, within Europe, working people were exploited by the middle class so, in a world setting, the rest of the world was exploited by Europe, by the U.S.A. and by Japan. On 8 November 1917, Lenin's government, almost as soon as it had seized power, issued the Decree on Peace which called for the liberation of all African and Asian colonies. Opponents of European (as long as it was not Russian!) rule could henceforward count on sympathy and support from Communist Russia as well as from European socialists generally, who found much accuracy in Lenin's description of imperialism as a form of economic exploitation.

Even before 1914 there had been resistance to European rule. In 1882, for example, there was a revolt in Egypt against British influence, in 1885 in Indochina, against the French, and between 1905 and 1909 a campaign of terrorism against the British in India. All were easily suppressed but they indicated a resentment which was lurking not far from the surface. In 1905, the defeat of Russia by the Japanese sent a ripple of surprise and excitement through

Asia. The European was not invincible after all. 'It almost seems that the East is waking from its slumber', wrote an Englishman in Persia in 1906, 'and we are about to witness the rising of these patient millions against the exploitation of an unscrupulous West'.

The awakening came between the wars. Nationalist movements, the aim of which was to get rid of the Europeans and establish self-governing nations in the former colonies, grew in vigour. In Turkey, Kemal, an able soldier, rallied his defeated countrymen, and in 1923 successfully defied the victorious Allies to win better terms than had originally been agreed. He then proceeded to modernize the country and ended centuries of European bullying and interference. In India, Gandhi led a movement which inspired such a desire for national independence among so many Indians that by 1930 senior British officials knew, that although they might imprison thousands of nationalist demonstrators, the end of their rule was only a matter of time (see pages 268–272). In Indonesia, Dutch rule was challenged by a number of nationalist organizations of which Sukarno's Communist-backed P.K.I was the most violent and effective. Similar events took place in French Indochina. The French crushed a serious uprising in 1930 but were unable to prevent nationalism from growing in secret (see pages 273–274). On the African continent nationalism developed faster in the north and south than in the centre. In Egypt, Zaghloul's Wafd party gave the British many difficulties between 1919 and 1925, though it failed in its aim to win complete independence. The most serious anti-colonial rising of all between the wars was, however, the Riff Rebellion in Morocco. Here Abd-al-Karim organized guerilla forces strong enough to defeat a Spanish army of 50,000 men, and, for a short time in 1921, to set up an independent Riff Government dedicated to 'modern ideas, laws and civilization'. It was overthrown by combined French and Spanish forces in 1926 and Abd-al-Karim was sent to an island exile.

In South Africa, there was a unique situation because the most powerful nationalist movement was a white one, representing the Afrikaner (Dutch) population of South Africa against British domination. The Afrikaners were as determined to reduce the political rights of the black population as they were to weaken British power. The main task of black nationalist movements, like the African National Congress, and Kadalie's Industrial and Commercial Workers Union, was to preserve rather than to extend what few rights non-Europeans already possessed. They were not successful (see pages 277–281).

Another people whose position deteriorated between the wars were the Abyssinians. Apart from Liberia, Abyssinia was the only African country to have maintained its independence, having defeated an invading Italian army in 1896. Mussolini avenged this defeat in 1935 by attacking and conquering the whole country. It was a sign of how the mood of the world had changed that when he did so he had to defy the League of Nations and ignore a storm of protest not only from Africa and Asia but from many parts of Europe too. The age of confident imperialism was over. As an Indian historian put it, 'With the solitary exception of Churchill, there was not one major figure in any of the British parties who confessed a faith in the White Man's mission . . . The French were more brave in their words but the faith had gone out of them also'. The question now was how to hang on to what they had got.

World War II

Then came World War II. For the European position in Asia, it was catastrophic. Between December 1941 and May 1942 the Japanese forced the surrender of the Americans in the Philippines, the British in Malaya, Singapore and Burma, and the Dutch in Indonesia. Not content with military triumph they proclaimed a policy of 'Asia for the Asians' and publicly humiliated their white prisoners with the aim of destroying the myth of white superiority. 'Wherever the victorious Nippon (Japanese) armies have brought the New Order', a Singapore newspaper noted in February 1942, 'Europeans may be seen naked to the waist, doing tasks that Asians only were made to do before. (They) cut ludicrous figures slouching their way through work which even Asian women would tackle better'. The Japanese then established the Greater East Asia Co-Prosperity Sphere, the intention of which was to re-organize the economy of South-East Asia to benefit Japan but was presented to the Asian peoples within it as a co-operative economic venture for the benefit of all, in which their independence would be respected. The attitude of the Japanese to the various Asian nationalist groups varied from area to area. Broadly speaking, however, so long as they were triumphant they were no more anxious to encourage genuine independence than the Europeans had been, but as the tide of war turned against them after 1943, they encouraged Asian nationalism in the desperate hope that it might provide effective support against the advancing Allies.

The Scramble for Freedom

Overwhelmingly defeated though they eventually were, the Japanese had fatally wounded the European empires in Asia. Two days after the Japanese surrendered, the Indonesians proclaimed their independence which, after five years of fighting, the Dutch finally recognized in 1949. In 1945, the new Labour government in Britain decided to give independence to India, which came about two years later (see pages 270–272). Other British possessions followed India's example, Ceylon and Burma in 1948, Malaya in 1957. The French, however, were not going to allow their empire to slip away so easily. They quickly re-established themselves in Indochina, suppressed a major uprising in 1946 and were only driven out in 1954 after eight years of war (see pages 274–275).

In the spring of 1955, a conference was held at Bandung in Indonesia. It was, as an American journalist put it at the time, 'the most formidable and ambitious move made in this generation to apply the principle of Asia for the Asians'. Twenty-nine states were represented, twenty-three for Asia and six for Africa, while representatives of nationalist movements in areas still under European control were also there. President Sukarno of Indonesia gave the opening address, taking as his theme 'let a New Asia and Africa be born'. Though the Bandung conference achieved few concrete results, the mere fact that it could take place at all showed that the world, and Europe's place in it, were changing fast.

'No-one knows when the hour of African redemption cometh. It is in the wind. It is coming. One day, like a storm, it will be here'. Such was the prophecy of Marcus Garvey, an American Negro, who led a 'Back to Africa' crusade in the 1920's. The storm broke in 1956. If 1945 to 1955 was the decade of Asian nationalism, 1956 to 1966 was Africa's. At the beginning of 1956, only Egypt, Sudan, Ethiopia, Libya and Liberia were free from European control. Ten years later, the whole continent was independent except Southern Africa (South Africa, Rhodesia, Angola, Mozambique and South-West Africa), and a number of tiny, coastal territories further to the north. Tunisia and Morocco gained their independence in 1956, and so did Algeria after a vicious war with the French from 1954 to 1962, but the event which won the greatest attention and set in motion the process of decolonization in Africa south of Sahara was Britain's readiness to grant independence to the Gold Coast (Ghana) in 1957. In 1958 the first of France's sub-Saharan colonies, Guinea, won its independence. The dam then burst. In 1960 sixteen African states were granted independence. They were chiefly ex-British and French colonies but also included the former Belgian Congo (see pages 275–277). In December 1960, the United Nations issued a Declaration on the Granting of Independence to Colonial Countries and Peoples. 'The subjection of peoples to alien subjugation, domination and exploitation constitutes a denial of fundamental human rights, is contrary to the character of the United Nations and is an impediment to world peace and co-operation'. Except in Southern Africa, European imperialism was a thing of the past.

The rapid collapse of the European Empires may look surprising but it is less surprising than the fact that they were created at all.

Europe and the World

Even in the years of the 'scramble for Africa' the British and French governments, which ended up with the two largest empires, paid little attention to empire-building. Indeed it was said with some justification of the British Empire that it was acquired 'in a fit of absence of mind.' Throughout their existence they were something of a confidence trick since their stability depended largely on the complete assurance by Europeans of their right to rule and the acceptance by the indigenous inhabitants that this right was somehow natural. For example, Lugard, the maker of British Nigeria, took control of 250,000 square miles of Northern Nigeria containing 10,000,000 inhabitants with an army of 3,000 soldiers of whom 150 were European. He then governed it with one European to every 50,000 inhabitants. Both France and Britain tended to run their

Black Africa about to burst free from its chains. A cartoon from Punch *by Ronald Searle*

colonies on a shoestring. Before the war, Britain managed to spend less than a million pounds a year on the Gold Coast while between the wars the whole of French West Africa, 1,800,000 square miles containing 17 million people cost the French less than £20 million pounds per year. Once Europeans began to doubt their right to rule and the indigenous population to resist, the days of European colonisation were numbered.

Moreover the empires by their very nature ran counter to some of the most passionately held political beliefs of late nineteenth and twentieth century Europe. The desirability of imperialism was soon questioned by Europeans and not just by Communists (see pages 262–263). If one was a British liberal, it was difficult to justify the imprisonment of Gandhi or the Nehrus in India in the 1920's and 1930's. If one was a French nationalist one could hardly claim that the nationalism of the Vietminh in Indo-china was unreasonable. If one was a Belgian democrat, the Congolese demands for 'one man, one vote' in 1959 were difficult to resist. European political ideals thus undermined European empires and with increasing effectiveness with the passing of time. Perhaps most important of all, Europeans lost faith in the notion that the White races were superior. In 1900 most had assumed that races were not equal and therefore argued that the less advanced coloured races would benefit from European rule. By 1950, most Europeans, shocked by the appalling results of Hitler's lunatic racial theories, were much less ready to tolerate any form of racial discrimination or the rule of one race by another.

Where Europeans did not feel this sense of doubt e.g. the Portuguese in Angola and Mozambique or the Afrikaners in South Africa—or if they did so and could not afford to admit it since individually they had so much to lose—e.g. the French settler population of Algeria of the British settler population in Southern Rhodesia—they were ready to hold fast to their privileged position, come what may. It was in these areas that the end of European empires was or seemed likely to be the most bitter.

The fashionable verdict on European imperialism is harsh. Europeans did ravage the rest of the world for economic gain; they murdered, killed and maimed non-Europeans in order to conquer and to hold land to which they had not the remotest right. They needlessly subjected their colonial subjects to decades of racial humiliation and contributed to further chaos and bloodshed by obstinacy or haste in the difficult period of decolonization.

Nonetheless there was a positive side to European imperialism. It brought new and higher standards of medicine, science and education to many parts of the world. If it meant economic exploitation, it also encouraged industrial and agricultural developments which eventually brought real benefit to the indigenous inhabitants. The survival of the British Commonwealth and of close ties between France and her former colonies into the seventies indicated that, besides hatred and humiliation, imperialism also encouraged sympathy and understanding between the races. While the empires lasted, the colonial officials also maintained a degree of peace and order in the areas under their control, the importance of which is easy to underestimate. The troubles of the Indian sub-continent, for example, of Nigeria and of the Congo in the years after independence were so terrible that it must remain an open question whether for millions of Indians and Pakistanis, Nigerians and Congolese, the end of European rule really was a blessing.

Four Examples
The British in India 'If anyone imagines that England would let India go without staking her last drop of blood, it is only a sorry sign of absolute failure to learn from the World War . . . and ignorance of Anglo-Saxon determination.' Hitler.

India, a vast sub-continent with a population of nearly 300 million in 1900, a fascinating history and an exotic and colourful civilization, was Britain's proudest imperial possession and gripped the British imagination in a way no other colony did. It was moreover a profitable captive market for British manufactured goods and the main source of manpower for the large army which defended British interests in the East. Before World War II it was as unthinkable to many Britons as it was to Hitler that Britain should give to Indians their independence. British governments, however, thought differently. Under continuous pressure from a powerful and skilfully led nationalist movement, they concluded, first reluctantly but then with conviction, that Indian independence was desirable. The chief source of dispute between Britain and the nationalist leaders was not so much independence itself as the form it should take and the speed with which it should come.

Indian nationalism, a European journalist joked in the 1920's, 'has its Father, Son and Holy Ghost.' The 'Father' and 'Son' were Motilal and Jawaharlal Nehru, leading members of the Indian National Congress which from its foundation in 1885 was the most respectable and eventually the most effective of the Indian nationalist organizations. Motilal was a rich and successful lawyer who had a great respect for the British, keen though he was on Indian self-government. He had his son educated at Harrow School and Trinity College, Cambridge and made sure that he trained as a lawyer at the Inner Temple in London. After Motilal's death in 1931 the younger Nehru was accepted as the leader of Congress and became the first

Prime Minister of independent India in 1947. The 'Holy Ghost' was Mohandas Gandhi a deeply religious, indeed saintly, Hindu whose political actions, often inspired by inner voices, frequently baffled both the British and his Indian colleagues. Gandhi was quite unlike any other twentieth century political leader yet was the earliest and most successful in making nationalism a mass movement outside Europe.

Gandhi was born in 1869, the son of the chief minister of one of the Indian princes. He too trained as a lawyer in London and in 1894 began practising law among the Indian communities of South Africa. Personal experience of the South African colour bar took him into politics. He founded the South African Indian Congress and in 1906 started to use, against the South African authorities, the tactics of non-violent resistance which were later to make him world-famous. In India, where his activities in South Africa had aroused much interest, he became known as 'Mahatma' or 'Great-souled'. He returned to India in 1915 and settled in Ahmedabad where he founded an 'ashram' or commune which, contrary to traditional Hindu custom, was open to the lowest caste in Indian society, the so-called 'untouchables'. Furthermore he insisted that his followers should wear only handwoven cloth and should encourage the use of the hand-loom and the spinning-wheel. He considered modern industrial society to be evil and strongly opposed the growth of Indian industry along European lines. He also began to use his passive resistance tactics against the British authorities. By 1917, he had won better contracts for Indians going to work in South Africa and for indigo workers in the Bihar province.

Two years later he was campaigning on a national scale, this time against the repressive measures then being taken by the British government against the nationalists. The British were in fact blowing hot and cold on Indian nationalism. In 1917 they had declared that they were in favour of 'the gradual development of self-governing institutions with a view to the progressive realization of responsible government as an integral part of the British Empire', and, by the Montagu-Chelmsford reforms on 1918, they gave greater political responsibilities to Indians while keeping key posts in British hands. However, when the nationalists, especially those returning from Europe at the end of the war, demanded much greater political advance and there had been outbreaks of violence in Calcutta the British decided to take powers to arrest nationalist leaders and to hold them in prison without trial. In protest, Gandhi organized a national fast and there were massive demonstrations all over the country. In Amritsar, five Europeans were killed in rioting. Three days later, on 13 April 1919 in defiance of a government order against meetings, a crowd of about 20,000 met in a public

Gandhi near the end of his life, in Calcutta, 1946

269

park. British troops under the command of General Dyer blocked the only exit and machine-gunned the demonstrators, killing 379 and wounding 1,200. The Amritsar Massacre was both a tragedy and a turning-point. Indian opinion was united as never before against the continuation of British rule and British confidence in their ability to maintain it was severely weakened.

Gandhi and Congress began to work together more closely, and Gandhi launched another campaign against the British. Amritsar showed, he said, that there could be 'no co-operation with this satanic government.' In 1922 he was arrested and further embarassed the British by pleading guilty at his trial. 'I am here,' he said to the judge, 'to invite and to submit to the heaviest penalty which can be inflicted upon me for what according to the law is a deliberate crime but what appears to me the highest duty.' He was sentenced to six years imprisonment but released after two.

During the 1920's, the British allowed a bit more self-government here, a bit there. Indians began to be trained (in England) as officers for the Indian Army and the Royal Indian Navy was created. Indian industry was protected from British competition by tariffs. For the nationalists however the pace was too slow. They demanded that by the end of 1929 India should be given the same Dominion status within the British Empire as Canada and Australia enjoyed (i.e. complete self-government). When this demand was rejected, they began on 26 January 1930—a date now celebrated as India's Independence Day—a campaign for *Purna Swaraj* (complete independence). Gandhi's contribution was the March to the Sea, a 200 mile walk accompanied by huge crowds which ended on the seashore. There he made some salt out of sea-water an illegal act which symbolically attacked the government's hated salt monopoly. In desperation, the British arrested 100,000 people including Gandhi and the two Nehrus. The civil disobedience campaign continued until the following year when the salt monopoly was abolished and the political prisoners released. In 1935 Britain brought complete self-government nearer when she accepted the Government of India Act which gave much greater powers to the Indian provincial governments. By 1937 Congress had won control of eight of the eleven provinces. However, its success was spoilt by a widening split between the mainly Hindu Congress and the Moslems, the second largest religious group in India, who were represented by the Moslem League.

The outbreak of war in 1939 showed the Indians that they still had some way to go before they really controlled their own destinies, since Britain declared war on their behalf without bothering to consult their leaders. Gandhi, who hated every form of violence, swung Congress against supporting Britain. As the Japanese advanced towards India's eastern frontier, he coined the slogan 'Quit India' and prepared for another civil disobedience campaign. In this crisis of the war, the British moved fast. They arrested Gandhi and many other Congress leaders. The worst rioting since the Indian Mutiny had no effect on their policy, as long as the war lasted; India was sternly ruled.

Once the war was over and a Labour government had replaced the Conservatives in Britain, things became very different. The Labour Party favoured independence for India, and a mutiny in the Royal Indian Navy followed by Hindu and Moslem rioting in Calcutta which caused more than 4,000 deaths convinced Attlee that the time had come for Britain to withdraw. In February 1947 he appointed Earl Mountbatten Viceroy of India. His task was to transfer power to an Indian government by June 1948.

The already serious divisions between Hindus and Moslems then got out of hand. There were ferocious riots and in the Punjab Nehru reported 'ghastly sights and behaviour that would degrade brutes.' Civil war seemed close. Jinnah, leader of the Moslem League, demanded the partition of India by the creation of an independent Muslim state, Pakistan. Despite the strong opposition of Gandhi, most of the Congress leaders realized that partition was the only answer. Mountbatten agreed with them and acting swiftly in the hope of preventing any further slide into chaos won the agreement of both the Moslem and Hindu leaders to a partition plan which fixed independence day for both India and Pakistan as 15 August 1947. Independence did come on that date but it was accompanied by frenzied religious violence. Worst affected was the Punjab where millions of refugees were moving in search of new homes. Refugee trains were stopped, attacked and sometimes with their passengers all dead or dying sent on to their destination. At least 600,000 people lost their lives and about 15,000,000 their homes. 'Partition', Gandhi declared, 'is a spiritual tragedy. I do not agree with what my closest friends are doing; thirty-two years of work have come to an inglorious end.' Nonetheless, he devoted himself to stopping the violence. In Mountbatten's opinion, 'single-handed he did more in the Bengal area than 50,000 troops had been able to do in the Punjab.' It cost him his life. In January 1948 a young Hindu extremist assassinated him because of the tolerance and kindness he had shown towards the Moslems. On the evening of his murder Nehru broadcast to the nation. 'The light that has illuminated this country for these many many years will illumine this country for many more years and a thousand years later . . . it will give solace to innumerable hearts. For that light represented something more than the immediate present; it represented the living, the eternal truths, reminding us of the right path, drawing us from error, taking

this ancient country to freedom.' However high the price, India had won her freedom. 15 August 1947 was moreover the beginning of the end of the British Empire.

Lord Mountbatten, Viceroy no longer, addresses the Constituent Assembly of newly independent India, August 1947

II: The French in Indo-China If Indian independence was the beginning of the end of the British Empire, Indo-Chinese independence was the beginning of the end of the French Empire. The main colony was Cochin China with the capital at Saigon and the neighbouring areas of Annam, Cambodia, Laos and Tonkin were also under French control. The Vietnamese were the most numer·ous population group in Indo-China and had not taken kindly to the French occupation. A nationalist movement, already strong before 1914, gathered strength during the war when the French forcibly recruited 100,000 Vietnamese to fight in Europe. Between the wars the French pursued a more rigid policy towards the nationalists than the British in India. They refused to consider even moderate plans for reform and for representative self-government. They preferred to rule through native kings whom they felt that they could rely on, like the French-educated Bao Dai who succeeded to the throne of Annam and Tonkin in 1925. Such a policy failed to win over the nationalists. There was a major revolt in 1939 which was quickly and severely suppressed. French policy continued unchanged and, in secret, the most extreme of the nationalist groups, the Communists, grew in strength.

The outstanding Vietnamese Communist was Ho Chi Minh, much of whose life during these years up till 1945 when he emerged as the leader of independent Vietnam is hidden in mystery. He was born around 1890. He left Indo-China in 1911, and, having worked in a French cargo boat, eventually settled in Paris, where he carved an uncertain living as a photographer. He became deeply involved in French politics, and was a founder-member of the French Communist Party. It seems that he attended the Versailles conference in the hope of putting the case for Indo-Chinese independence to President Wilson but failed to get a meeting with him. Lenin's ideas about the nature of imperialism greatly influenced his thinking about the nature of French rule in Indo-china, as was shown in a speech he made at an important Socialist Congress at Tours in 1920. 'French imperialism', he declared 'came to Indo-China half a century ago; it conquered us at bayonet-point and in the name of capitalism. Since then, in addition to being shamefully persecuted and exploited we have been hideously martyred and poisoned ... by opium, alcohol, etc. . . . . It is impossible to reveal to you, in the space of a few minutes, all the atrocities which have been perpetrated by the capitalist bandits.' He seems to have moved from Paris to Moscow in 1924 and was then constantly travelling to and from Russia, China and Siam. He was a political organizer of rare talent. In Paris he helped to found the Intercolonial Union, in Moscow the Peasants' International. In China he trained Vietnamese nationalists in underground political activity and in guerrilla warfare, while in Siam, disguised as a Buddhist monk, he is said to have organized Communist cells in Buddhist monasteries! When he returned to Vietnam in 1941 after an absence of thirty years, he had a near-legendary reputation among Vietnamese nationalists.

During World War II, the French government in Saigon remained loyal to the Vichy government which collaborated with Hitler in Europe and therefore collaborated with the Japanese. For their part, the Japanese allowed the French to keep some authority. Meanwhile in a mountain cave in Northern Indo-China close to the Chinese border the Indo-Chinese Communist party led by Ho Chi Minh prepared themselves to fight for independence. They reorganized themselves as the League for Vietnamese Independence or Vietminh and issued this call to arms. 'Men of wealth and position, soldiers, workers, peasants, intellectuals, civil servants, traders, young men, women, you who are full of patriotism! At this time, national liberation must come before all else. We must unite in the task of overthrowing the Japanese and French fascists and their jackals. Fighters of liberation, the hour has struck. Rise high the banner of insurrection. The sacred call of the fatherland rings in your ears.' Small guerilla units concentrating on propaganda rather than violence began to spread Vietminh influence southwards. When Japan was defeated by the Allies, Ho Chi Minh declared Vietnam independent in the name of the Committee of National Liberation and for a brief period the whole country was under Vietminh control.

The French however had every intention of reviving their Indo-Chinese Empire and were soon back with forces far greater than the Vietminh could muster. There was little trust between the two sides. While Ho Chi Minh was in Paris negotiating with the French government, Admiral D'Argenlieu, the French representative in Indo-China, took action against the Vietminh. Before long there was open violence between the French and Vietminh, the worst being in the northern port of Haiphong where more than thirty French were killed. The French then demanded that the city be evacuated but, as people began to leave the city, the officers on board a French cruiser in the harbour mistook them for Vietminh reinforcements, opened fire upon them and killed at least 6,000. If there ever had been the chance of a negotiated agreement, it had now passed. On 20 December 1946 Ho Chi Minh called for a general uprising against the French. 'It would be better to sacrifice everything', he declared 'than sink back into slavery. Let him who has a rifle, use his rifle, let him who has a sword use his sword! And let those who have no sword take up pickaxes and sticks!'

The French were confident of victory. The Vietminh was not so much an army as a bunch of guerilla units and the French army had already dealt successfully with guerilla groups in Algeria. Under the brilliant leadership of General Giap, however, the Vietminh proved too much for them. Supplied from a sympathetic China and supported by most of the Vietnamese peasantry, Giap soon had the French on the defensive. The struggle then became, as Ho Chi Minh explained to an American journalist, like one between an elephant and a tiger. 'The tiger does not stand still . . . He will leap upon the back of the elephant tearing huge chunks from his hide, and then he will leap back into the dark jungle. And slowly the elephant will bleed to death.' In 1954 Giap felt strong enough to end his guerrilla methods, and he launched four divisions armed with the most modern artillery against an isolated French garrison at Dien Bien Phu. After furious fighting described as worse than Verdun, Dien Bien Phu was taken. The French were finished.

The same year, an international peace conference met at Geneva to decide the future of Indo-China. Cambodia, Laos and Vietnam were all to become independent. Vietnam was to remain temporarily divided along the ceasefire line with North Vietnam under Vietminh control and South Vietnam under the control of the government of Bao Dai who had remained an ally of the French. Elections were to be held in 1956 to reunify the country. These elections never took place. If they had it seems likely that the party of Ho Chi Minh would have won 80% of the votes and unified the whole of Vietnam. Instead the U.S.A., disturbed by the spread of Communism throughout the world, moved in to fill the vacuum left by the French and to prop up the South Vietnamese government against guerrilla forces supplied from and reinforced by North Vietnam. When Ho Chi Minh died in 1969, Vietnam was still divided and a war of appalling atrocity still raged.

There were many reasons why the French failed in Indo-China. For one thing they underestimated their opponents and the extent to which the peasant population genuinely supported the Vietminh. For another not only was Ho Chi Minh an extremely skilful politician and Giap a guerrilla general of genius but Communist China was able to provide them with a good supply of modern weapons at a critical point in the war.

The Belgians in the Congo Leopold II, King of the Belgians, had led the European 'scramble for Africa' with his conquest of the Congo. His main aim was the profits to be made from rubber and ivory. From these beginnings and for the first half of the twentieth century, the Belgian Congo was governed with economic gain being the chief interest of the government. Belgian and foreign companies were granted large areas of land for their operations and were also made responsible for keeping order in that area. While industry was advanced by international standards, agriculture, which was essential to the well-being of most of the native Congolese population, remained backward. Until the fifties the Belgians assumed that they could go on running the Congo indefinitely and made no attempt to provide the Congolese with the experience of responsible government or industrial posts. Higher education and senior posts in every walk of life were almost entirely actively restricted to Belgians. Political activity was banned. In the late fifties however news of the achievements of nationalists in other parts of Africa began to cause unrest among the Congolese. January 1959 saw considerable rioting in Leopoldville, the capital. The Belgians panicked and completely reversed the policy which they had followed for nearly eighty years. Not only did they now allow political activity but caused elections to be held and complete independence given, all in the space of eighteen months.

*Map 17 Southern Africa
in the 1960s*

The Congo 1961—Congolese soldiers force their Belgian captives to pose in a humiliating fashion for the photographer

The result was anarchy. The elections were won by the Congolese National Party whose leader, Lumumba, a former postal clerk, became Prime Minister. Since no Africans had been trained as army officers, Belgians stayed on in senior army positions but their troops mutinied. Widespread racial violence erupted and Europeans began to flee the country. Meanwhile the south-eastern and copper-rich province of Katanga had declared itself independent under the leadership of Tshombe, who was supported by powerful European business interests including the most important mining company, Union Minière. The Belgian government sent in troops to protect its citizens. Lumumba, convinced that they were coming to restore Belgian rule, appealed to the United Nations for aid. A United Nations force made up mainly of Tunisian and Ghanaian troops arrived in the Congo. The situation worsened, since the USA and USSR decided to treat the crisis as another Cold War struggle. In September 1960 Lumumba, who was planning a military expedition against Katanga, was dismissed from his position as Prime Minister and then kidnapped and murdered in Katanga in mysterious circumstances. A year later, Hammarskjold, the Secretary General of the United Nations, also died mysteriously, this time as a result of an air crash on his way to Katanga to negotiate with Tshombe. Anarchy continued unchecked. With no disciplined army or police force left, rival guerrilla groups, sometimes led by European mercenary soldiers and paid for by foreign organizations, fought it out amongst themselves while the UN army, without clear direction, tried desperately to bring back peace. Tshombe was driven into exile in 1962, returned in 1964 to become briefly Prime Minister of the whole country, only to be driven into exile again in 1965. A firm peace only came with the establishment of a military dictatorship under General Mobutu.

Southern Africa; a Break in the Pattern

'I want to state here unequivocally now that South Africa is a white man's country and he must remain master here' Dr. Verwoerd, having just joined the Senate as a Nationalist in 1948.

South African history in the twentieth century has two main themes. The first is the success of the Afrikaners (Europeans of Dutch origin) in gaining political control at the expense of the English-speaking Europeans. The second is the steady deterioration of the political position of the non-whites, about 80% of the population.

In 1902, a war (the Anglo-Boer War) between the British and the Afrikaners had ended in a British victory. The previously independent Afrikaner republics of the Transvaal and the Orange Free State became part of the British Empire. However by the South

Africa Act of 1910 all white South Africans were granted full voting rights as well as their own parliament and Dominion status within the British Empire. This meant in effect self-government. Afrikaners outnumbered English-speaking South Africans by a ratio of 3 to 2, so if a self-consciously Afrikaner political party could win the support of most Afrikaners, it would be sure of political power. Such a party came into existence in the form of Dr Malan's Nationalist Party. It first achieved success in the elections of 1948 and then maintained itself in power, steadily increasing its majority between 1948 and 1974 as more and more of the English-speaking population voted for it as well as Afrikaners. In the same period, South Africa withdrew from the British Commonwealth. In 1960, the white electorate voted by a small majority in favour of South Africa becoming a republic, which meant that they would have their own President rather than the Queen as head of state. The following year, Dr Verwoerd, South Africa's Prime Minister, led his country out of the Commonwealth after bitter criticism of his racial policies at the 1961 Commonwealth Conference in London.

In 1900, the total population of South Africa was about six million, with about four non-Europeans to every European. By 1960 the total had more than trebled but by then there were five non-Europeans to every European. In 1900, non-Europeans did not have many political rights, but they had some. In Cape Province for example they could vote if they had certain educational and financial qualifications. They could also stand for parliament, though none of them did. The deterioration of the non-white political position began with the South Africa Act which at the same time as it gave such opportunities to the Afrikaner put an end to the right of non-whites to stand for parliament. Disgusted by Britain's lack of concern for the non-white population a number of African chiefs were brought together in 1912 by Seme, a young Zulu lawyer. 'We have discovered', said Seme, 'that in the land of their birth Africans are treated as hewers of wood and drawers of water . . . We have called you together so that we can together find ways and means of forming our own national union for the purpose of creating national unity and defending our rights and privileges.' So the African National Congress was born. In 1931, the Native Land Act allowed just 7% of the country to the non-white population and turned a million non-whites into homeless refugees. Pass laws were introduced which ended freedom of movement for non-whites and the Masters and Servants Act made it a crime for an African to disobey his employer's orders. All these measures the ANC strongly but unsuccessfully opposed. Nor was its delegation successful in its attempts to win international support at the Versailles Peace Conference. Despite some temporary success in the organization of a trade-union movement non-white political leaders were unable to reverse this trend between the wars and in 1936 Parliament abolished non-white voting rights.

The victory of the Nationalists in 1948 was a further blow to the non-whites. Before 1948, though the Europeans had increased their racial dominance at the expense of the non-whites, they had done so in a haphazard and half-hearted manner. The new Nationalist government was neither haphazard nor half-hearted. It believed firmly in European racial superiority and had a systematic political programme to maintain it. 'Either we must follow the course of racial equality', declared the Nationalist election manifesto of 1949, 'which must eventually mean national suicide for the white race, or we must take the course of separation'. Separation would be achieved through the policy of 'apartheid' or separate development. Non-whites would be allowed political rights but only in homelands to be created for them.

The land the government made available as 'homelands' turned out to be entirely rural and amounted to only 14% of the country. The Nationalists' message to those millions of urban Africans whose labour in the mines and factories played so large a part in keeping the dynamic South African economy booming was simple—no political rights ever. A series of laws made 'apartheid' a physical as well as a political reality. Every citizen was registered by the state and classified according to race. Marriage and sexual intercourse between races was forbidden, so too were multiracial meetings entertainments and education. Even multiracial social gatherings were discouraged. Local authorities were encouraged to set aside particular areas for particular racial groups and thousands of families, usually non-white, were uprooted from their homes so that separation was achieved. Almost without exception the most attractive areas were allotted to the whites so were the best seats on public transport, the best beaches and even the best park benches.

The ANC strongly opposed these laws and, following Gandhi's example, organized massive campaigns of civil disobedience. In 1952 there was a campaign against the segregation laws which ended with 8,000 people of all races being gaoled. Despite the ANC's intention to avoid violence, serious racial riots with loss of life occurred in Port Elizabeth, East London, Johannesburg and Kimberley. The government refused to alter course. Armed with the Suppression of Communist Act which gave it wide powers to take action against anyone it considered a threat to the country, and with further powers of banishment, house arrest and imprisonment without trial for periods up to 180 days, it ruthlessly harassed the ANC and other opposition groups. In 1956, 156 people were arrested and charged with high treason. After five years of legal

argument, the charges could not be made to stick and were with-drawn. Non-white protests grew more violent. Between 1956 and 1958 African women led a series of demonstrations against the hated pass laws and disturbances in the Durban area led to the death of nine policemen. The ANC then split. A group led by Sobukwe, despairing of ever bringing reform to South Africa by peaceful means, broke away from the ANC to form the Pan-African Congress and set to work organizing another round of mass demonstrations against the pass laws. Confronted by one of these at Sharpeville in 1960, the police lost control and opened fire, killing 69 Africans and wounding another 178. For a few days South Africa seemed on the edge of revolution. But again the government acted with ruthless decisiveness. It arrested nearly 12,000 Africans and banned the ANC and PAC. There was no revolution.

The opposition then went underground, turning to sabotage and secret conspiracy. It soon suffered a further crippling blow when most of its leaders were arrested following a police raid on a house in the Rivonia suburb of Johannesburg. Eleven of them were brought to trial and seven, including Nelson Mandela, the most charismatic African leader, were sentenced to life imprisonment.

Black South Africans burn their hated pass books

For the next decade, African nationalism seemed broken. The South African government stayed firmly in control and energetically hunted down any reviving opposition. Economically, the country went from strength to strength and though the gulf between the living standards of white and non-white grew wider, with thousands of African workers still being paid wages too low to prevent them and their families from starving, many other non-whites became undeniably better off. In the early seventies, opposition to the government took on two new forms. The first was strike action for higher wages which, particularly in the Durban area, won much publicity and some success. The second was the emergence of self-confident black leaders in the 'apartheid' homelands like Matanzima in the Transkei and Buthelezi in Kwa Zulu. They demanded that the government should keep its word and give them real independence rather than treat them as 'puppet states'. Nonetheless, European rule in South Africa looked as secure as ever. Furthermore the South African government openly defied all the efforts of the United Nations to end its control of South-West Africa and gave quiet but vital support of troops and money to the neighbouring European governments of Rhodesia, Angola and Mozambique in their efforts to crush the African nationalist guerrilla forces active in their territories.

Up to 1974 Southern Africa was unique in that it was the only area of the world where Europeans were able to maintain their control over a considerably larger non-European and hostile majority despite continual harsh criticism from almost every other nation in the world. There are a number of reasons why this was so. The first was the unusual character of the Afrikaners. They first arrived in the Cape area of South Africa before the Africans moved down from the north and they came as farmers to settle not as businessmen to make quick profits and go away again. They were a deeply religious and old-fashioned people who believe that it was God's will that South Africa should be their country and that the races simply should live separately with the black subordinate to the white. As Dr Verwoerd put it in 1960, 'We believe we hold the fort here, not only for the sake of white civilization but also for . . . Christianity'. They were therefore unusually determined to hold on to 'their' land and unusually convinced of their own righteousness. Secondly, South Africa was very wealthy. Even before 1900 it had become the world's largest producer of gold and diamonds and, during the twentieth century, it was found to possess almost every mineral required by a mineral-hungry world. By 1970 foreign, mainly British and American, investment could be measured in thousands of millions of pounds, and however much the British and American governments might disapprove of 'apartheid' they were

281

unlikely to lead or indeed support any international action to force the South African government to change its ways. As a consequence of its wealth, the South African government could arm itself with the most modern weapons. By 1970, its armed forces were stronger than those of all the African nations south of the Sahara put together. Military experts also believed that it had the resources to manufacture nuclear weapons if need arose. Moreover, until 1974 when Portugal decided to grant independence to Mozambique, South Africa had a useful cushion of white-controlled states between her and the hostile world.

Europe in the Middle East

Events in the Near East after World War II vividly illustrated the marked decline in the international power of the nations of Western Europe.

Zionism, the movement which aimed to create a national home for the Jewish people who for the best part of two thousand years had been scattered all over the world, was already an active force at the end of the nineteenth century. The British were sympathetic and in 1904, without bothering to consult the Ugandans, offered the Zionist leaders Uganda as a national home. This offer being turned down, Balfour, the British Foreign Secretary, let it be known to the Zionist leaders in 1917 that his government 'viewed with favour the establishment in Palestine of a national home for the Jewish people'. Between the wars about 326,000 Jews settled in Palestine, which was a responsibility of Britain under a League of Nations mandate. The Arabs who populated Palestine were never consulted about a possible Jewish settlement and there was acute religious hostility between the Jews and the Moslem Arabs, especially over Jerusalem which each religion considered to be its 'Holy City'. There was continual fighting between the two peoples, and in 1939, anxious to gain Arab support in the coming struggle against Germany, Britain went against the Balfour Declaration and forbade Jewish settlement in Palestine. When World War II ended, Britain was in a dilemma. The frightful suffering of the Jews of Europe at the hands of the Nazis suggested a policy of encouraging further Jewish settlement yet this was bound to lead to Arab-Jew conflict. Britain wavered and both Jews and Arabs turned to terrorism. The postwar British government had neither the will nor the resources to cope and suddenly withdrew, leaving the United Nations to try to find a solution. Before the UN could take any effective action, the Jews proclaimed the state of Israel which they had at once to defend against the combined attack of five Arab states. This war lasted a year and ended in an Israeli victory. Israel joined the United Nations, her right to exist recognized by most of the world except the Arabs.

European involvement with Palestine and Israel was however far from over. In 1956, both the British and French were very suspicious of the activities of Nasser, who had recently become President of Egypt. The French knew that he was sending arms to the rebels in French Algeria while the British believed him to be anti-British and a danger to their influence in the Arab world. In the July of that year he further upset them by nationalizing the Suez Canal Company, a mainly French and British company which controlled the Suez Canal. Eden, the British Prime Minister, decided that Nasser must be taught a lesson, and so with the French and the Israelis, who regarded Nasser as the most dangerous of their Arab enemies, he concocted a plan. Israel was to provoke a war with Egypt. Britain and France would then move in on the Suez Canal area claiming to act as peacemakers. Nasser would be made to look a little man with a loud mouth and the Suez Canal would be saved.

The plan was put into operation on 29 October 1956. It proved a fiasco. The Israelis played their part well enough but Britain and France moved in too slowly to prevent the Egyptians from blocking the canal. Then it became clear that they had misjudged world opinion which was almost completely hostile. The Russians informed them that they could expect to be crushed 'with every kind of modern destructive weapon' if they did not at once end their attack on Egypt. More immediately serious was the refusal of the Americans (who were furious that they had not been consulted) to provide the necessary financial support to allow Britain to head off a nasty financial crisis. 'How splendid it is to hear the British lion roar once more', wrote a patriotic Briton to *The Times* in the first days of the crisis. He did not roar for long. After barely a week the advance was halted and a humiliating withdrawal followed. There were clearly new kings in the international jungle.

In the autumn of 1973, another Arab-Israeli war brought another crisis to Europe. Though the nations of Western Europe, with the exception of Holland, were careful not to take sides, the Arab oil-producing states reduced their oil supplies to Europe. This action caused consternation, so dependent had the Western European economy become on a steady supply of Middle Eastern oil. That winter, when the Shah of Persia took his usual ski-ing holiday in Switzerland the economics ministers of the major Western European nations queued to arrange meetings with him, in the hope of getting supplies back to normal. The world had changed somewhat since 1907 when Britain and Russia could divide Persia into 'zones of influence' and move their troops in and out of that country as they pleased.

Chapter 16
The Cold War

Origins

Big European wars had in the past been followed by big peace treaties (e.g. Vienna 1815, Versailles 1919) which had settled the shape of Europe in the years to come. No such clear-cut settlement was made in 1945 because the victors, on the one hand the U.S.A., Britain and France, on the other, Russia, fell out with each other so fast. The shape of post-war Europe did not fully emerge until 1949 and was as much due to the position of the various armies at the end of the war as to properly negotiated agreements of the traditional sort.

This is not to say that the victors failed to agree on anything. First at Teheran in 1943, then at Yalta in 1945, Stalin, Roosevelt and Churchill managed to agree on the broad outlines of a world settlement once Germany was defeated. They also co-operated to plan and then make real a new international peace-keeping organization, the United Nations Organization, which came into existence on 24 October 1945. They agreed that the Nazi leaders should be tried as war criminals and from November 1945 to October 1946 their leading legal experts formed a court at Nuremberg for this purpose. At these Nuremberg Trials twenty-one leading Nazis were tried, twelve were sentenced to death, six to long terms of imprisonment and three acquitted. The victors also co-operated in the drawing up of peace treaties with Germany's former allies—Italy, Hungary, Bulgaria and Rumania—and with Finland, with whom Russia had fought a separate war from 1939 to 1940. These treaties were signed in February 1947. On three major European issues, however, there was no agreement—on the future of Germany, on the boundaries of Poland and on the types of government in the countries of Eastern Europe. As the suspicions grew between the major powers in the years immediately after the war, chances of negotiated agreements on these issues faded. The balance of power rather than negotiations settled them.

These suspicions went deep. Communication between Russian-dominated and Communist Eastern Europe and American-dominated and capitalist Western Europe became increasingly limited and hostile. Consequently the term 'Cold War' was soon used to describe the international politics of the twenty years or so after 1945 which

The Big Three at Yalta, looking friendly enough. Churchill is on the left, Roosevelt, who was in fact dying, is in the centre, and Stalin is on the right

The 'star' of the Nuremberg Trials. Goering in the witness box. April 16th, 1946

284

were neither real peace nor real 'hot' war. Even before the fighting against the Nazis was over, Roosevelt, the American President, felt the suspicions growing. A week before his death in April 1945, he wrote to Stalin: 'It would be one of the greatest tragedies in history if, at the very moment of victory now within our grasp, such distrust and lack of faith should prejudice the entire undertaking after the colossal losses of life and material and treasure involved.' Not long afterwards, Churchill too wrote to Stalin. A quarrel between Communist nations and English-speaking men would, he said, 'tear the world to pieces and all us leading men on either side who had anything to do with that would be shamed before history.' These warnings had no effect. When the victorious allies met at Potsdam, near Berlin, in the late summer of 1945, they achieved little. There were disputes about whether Germany should be partitioned and should pay reparation, about the boundaries of Poland, about the extent of Russian power in Eastern Europe and the Eastern Mediterranean. No solutions could be agreed and both sides began to go their own ways with less attempt to seek the other's co-operation. In the spring of 1946 the French, British and Americans stopped while the Russians continued collecting reparations from their occupation zones of Germany. Moreover the Russians placed more and more obstacles between their zone and the West, thus cutting it both economically and politically from the rest of Germany. Similar policies affected all of Eastern Europe where their armies were still stationed. In March 1946 Churchill was visiting the U.S.A. and made a much reported speech at Fulton, Missouri, in the presence of President Truman. 'A shadow has fallen upon the scenes so lately lighted up by Allied victory', he declared. 'From Stettin on the Baltic to Trieste on the Adriatic, an iron curtain has descended across Europe. Behind that line lie all the capitals of the ancient states of central and Eastern Europe ... This is certainly not the liberated Europe we fought to build up.' Referring to the Russians, he commented 'There is nothing that they admire so much as strength and nothing for which they have less respect than military weakness.' The Russians, convinced that this speech represented the American as much as Churchill's views, reacted sharply to it. It was, Stalin said in an interview with the official government newspaper, *Pravda*, nothing less than 'a call to arms against the U.S.S.R.'

Stages

America and Britain had agreed at Yalta that Eastern Europe should be in Russia's 'sphere of influence', but, behind the 'Iron Curtain', the Russians were hard at work establishing not just their influence but their political and economic control (see pages 292 ff). America therefore reacted strongly to developments in Greece in the winter of

Churchill, with Truman, Roosevelt's successor, in Fulton, Missouri

1946–47. Also at Yalta the Big Three had agreed that when the war ended Greece should be a British sphere of influence. Though Stalin seems to have kept his word, the Yugoslav and Bulgarian Communists backed the Greek Communists in the civil war which broke out in 1944 while Britain backed the anti-Communists. By February 1947 the Communists held the upper hand and Britain, no longer able to afford to continue her support, informed the Americans that she was going to withdraw her troops. This led to a decisive development —effectively the declaration of war—of the Cold War. President Truman and his advisers, certain that Russia was bent on world domination, that Communist action in Greece was, as in other parts of Eastern Europe, part of a carefully organized Moscow design and that Europe as a whole was in such a bad way that a Communist takeover of the whole continent was a real possibility, announced on 12 March 1947 the so-called 'Truman Doctrine'. 'At the present moment', Truman declared, 'nearly every nation must choose between alternative ways of life. One way of life is based on the will of the majority and is distinguished by free institutions ... The second is based on the will of a minority forcibly imposed on the majority. It relies upon terror and oppression ... I believe it must be the policy of the U.S. to support free peoples who are resisting attempted subjugation by armed minorities or by outside pressures.' He asked the American Congress to vote 400 million dollars for the defence of Greece and promised military support for both Greece and Turkey.

287

Since the Russians had in fact given no significant help to the Greek Communists, they regarded Truman's speech a belligerent move to increase American power in Europe. Not surprisingly, therefore, they interpreted the Marshall Plan (see pages 310–11) for the economic recovery of Europe as nothing more than the Truman doctrine in economic disguise, a skilful plot to lure Eastern Europe away from their influence. In October 1947 Stalin set up Cominform, an organization to co-ordinate and expand the work of European Communist parties throughout the world, and in January 1949 Comecon, an organization to balance the Marshall Plan in Western Europe by unifying Eastern Europe economically.

In 1948 there was nearly a hot war, the cause Berlin. The old German capital had an odd position in post-war Germany. Though 176km. inside the Russian zone, it too was divided into four occupation zones so that West Berlin was controlled by American, British and French troops who had road, rail, canal and air links with the rest of Germany but along corridors easily blocked by the Communists. A crisis blew up over German currency reform. No agreement could be reached in the Allied Control Commission (the joint Russian-American-British-French military government of Germany set up in 1945), so in June America, Britain and France went ahead with the reforms in their own zones. The Russians walked out of the Commission in disgust and, acting probably more from sheer anger and bloody-mindedness than from carefully considered policy, blockaded West Berlin by cutting all its land links with West Germany. America and Britain, however, had little choice but to see the blockade as a challenge to their determination to halt the spread of Russian power, and it quickly became a trial of strength. The West won. In a remarkable combined and sustained operation, the American and British airforces, flying all manner of aircraft round the clock, airlifted the supplies necessary to keep West Berlin going. Simultaneously inside the city energetic measures were taken to prevent a communist takeover of the city government. In May 1949, after months of secret negotiations, the Russians ended the blockade.

The Berlin crisis further hardened the division of Berlin, of Germany and of Europe. In April 1949 the North Atlantic Treaty Organization (N.A.T.O.) was formed. It was a military alliance made up of Belgium, Britain, Canada, Denmark, France, Iceland, Italy, Luxembourg, the Netherlands, Norway, Portugal and the United States, intended to prevent any further Russian expansion in Europe. The Russians for their part saw it as 'a weapon aimed at the establishment of Anglo-American world domination'.

That Europe should have been so divided so soon after such suffering in World War II was regrettable but it was not very

Scenes during the Berlin Airlift:
Above *Loading one of the planes*
Right *Unloading the railway wagons in a siding of an airfield in West Germany*

surprising. The Cold War had in fact been in existence since 1917. The revolution of 1917 which brought the Bolsheviks to power created a Communist society which believed that capitalism with its encouragement of private property was fundamentally evil and preached its destruction through world revolution. Britain was the founder of modern capitalism, the U.S.A. the most powerful capitalist state in the world, and most Britons and Americans believed that the preservation of private property was essential if a genuinely free society was to exist. Throughout the 1920's and 1930's Communist Russia had existed isolated and uncomprehended by the rest of the world. The Grand Alliance of 1941–45 of Russia, the U.S.A. and Britain was an extraordinary and unnatural alliance brought into existence by Hitler's ceaseless aggression, and it always had its strains. Stalin distrusted everyone. Churchill and Roosevelt were no exceptions. 'Churchill is the type', he once said, 'who, if you don't watch him, will slip a kopeck out of your pocket. Roosevelt is not like that. He dips his hands in only for the bigger coins.' He was convinced that they delayed the opening of the Second Front in Western Europe in order to weaken Russia, and bitterly complained of their failure to keep him informed about the North African and Italian campaigns. On their side, the British and Americans found it hard to forget the Nazi-Soviet Pact of 1939. Hatred of Nazism was the cement of the alliance. Once Nazism was destroyed, it was bound to crumble.

The Cold War after 1945 was the result of fear and misunderstanding arising from the fundamental differences between capitalism and communism and the immense military power of the U.S.A. and the Soviet Union. An important misunderstanding by the Western nations was their failure to realize how much more atrociously Russia had suffered at Germany's hands than they had. Russia's insistence on the creation of 'friendly' states in Eastern Europe to act as a buffer against a future revived Germany, and on keeping Germany partitioned, was seen not as an indication of Russia's understandable sense of insecurity but of her unreasonable aggressiveness. They read and listened to Communist plans for world revolution and exaggerated the commitment of Stalin and his successors to such ideas. Without accurate knowledge of the extent of Russian economic exhaustion, they exaggerated both the Russian desire and ability to follow an aggressive policy. Moreover, they did not fully realize how the U.S. monopoly of nuclear weapons increased Russia's sense of insecurity. For their part, the Russians failed to realize how aggressive many of their policies seemed, particularly those affecting Eastern Europe. Their official newspapers and publications preached continually the inevitability of class war and international revolution. Their leaders exaggerated the aggressiveness of capitalism. They never properly understood what a threat their huge land army seemed to Western Europe, particularly when it was seen to be used for the suppression of democratic governments in Eastern Europe. For the next twenty years, each side tended to the mistaken view that the other was steadfastly and unceasingly aggressive, bent on increasing its power and influence whenever it had the chance.

By 1949 the shape of the new Europe was clear. Germany was divided. The French, British and American zones were amalgamated to form the Federal Republic of West Germany (usually known as the F.D.R.) with a democratic government and federal capital in Bonn. The Russian zone became the German Democratic Republic (known as the D.D.R.) with a communist government and capital in East Berlin. West Berlin, with its population of more than two million, became a West German island in a Communist sea. The western frontiers of Germany were hardly changed. The Saar with its rich coalfields was given temporarily to France but was returned after a plebiscite in 1955. In the east, in contrast, a huge area of Germany was lost to Poland. Russia was determined to keep the large section of Eastern Poland which she had gained in 1939 by the Nazi-Soviet Pact and aimed to compensate the Poles with parts of Germany. With the reluctant and provisional agreement of the Western allies the Polish-German border was moved therefore nearly 300 kilometres westwards to the line of the Oder-W. Neisse rivers. Most of the 9 million German inhabitants of this area either fled or were driven from their homes and replaced by 4,500,000 Poles. The West German government was prepared neither to accept the Oder-Neisse line nor the partition of Germany as permanent, nor to sign a peace treaty with Russia until 1970 (see pages 338–40).

Elsewhere in Europe there were a number of other important changes. Russia kept the three Baltic states of Lithuania, Latvia and Esthonia which she had conquered in 1940. She also took Karelia and Petsamo from Finland, the Subcarpathian region from Czechoslovakia, Bessarabia and parts of Bukovina from Roumania. Italy had to give the Dodecanese Islands to Greece and parts of Venezia-Giulia to Yugoslavia while Roumania yielded part of Southern Dobrudja to Bulgaria. Austria, like Germany, was divided into four occupation zones, but unlike Germany was allowed to elect a government of the whole country. In 1955 all the occupation troops were withdrawn on the understanding that Austria should join no military or political alliance.

The 'iron curtain' effectively divided the continent in two. Most states to the west either had or were soon to join N.A.T.O. and received economic assistance through the Marshall Plan. To the east, almost every state had a Communist government, was a member of Comecon and would join the military alliance, the Warsaw Pact, when it was formed in 1955. Exceptions were Yugoslavia and Albania which, while having fiercely communist governments, were independent of and sometimes hostile to Moscow.

This eventual post-war settlement reflected closely the positions of the Allied armies at the end of the war. In those areas conquered or liberated by British and American armies—e.g. West Germany, Italy, Greece—governments favourable to capitalism and the U.S.A. were established. Where the Red Army had fought its way by 1945—e.g. East Germany, Poland and Hungary—there Communist governments favourable to Russia were established. The Communist governments independent of Russia were in countries which liberated themselves by their own efforts.

The post-war settlement represented a Russian diplomatic and military triumph of the first order. The huge area of Eastern Europe which passed under their control was a gain of 220,000 acres for every day of 1945–48! At Yalta Stalin had skilfully played on differences between Roosevelt and Churchill. The latter was convinced that Stalin intended to expand Russian influence as far west as he could. He therefore pressed for British and American troops to get as far east across Europe as they could while the war lasted and hold their ground until a comprehensive European settlement had been made. Stalin, however, convinced Roosevelt that Churchill was unjustifiably distrustful of Russian intentions, and in the last weeks of the war the Americans stuck to the agreements made at Yalta. Consequently though Eisenhowever, the American Commander-in-Chief, could have reached Berlin, Vienna and Prague before the Russians, he left them to be liberated by the Russians and, when the war ended, withdrew to the agreed occupation zones. Then, in 1945 and 1946, while Britain and America rapidly demobilized, the Russians did not, so that by the end of 1946, less than a million troops to the west of the iron curtain faced more than 5 million to the east. The ability of the western allies to influence events in Eastern Europe was consequently limited.

The Russian Takeover of Eastern Europe 1945–48

The Russian takeover of Eastern Europe was ruthless and skilful. A similar pattern was followed in most of the countries affected. At first, a Communist party shared power in a coalition government; then with or without pressure from Moscow and the Red Army, communist ministers came to control the armed forces and the police; next, the most effective leaders of the opposition were discredited or imprisoned. Finally rigged elections returned to power one-party Communist governments whose leaders were to a man Moscow-trained and, with hardly an exception, loyal to Moscow policies. In Western eyes, the cynical perversion of democratic methods, especially of 'free elections', was particularly shocking. The word 'democracy' plainly meant one thing in the West, another in the East.

Events in Poland quickly demonstrated Russia's intentions and the strength of her position in Eastern Europe. Britain had gone to war to defend Poland from the Nazis and had recognized a Polish government in exile in London as the rightful government of post-war Poland. This government however was almost as anti-Soviet as it was anti-Nazi and Stalin, who simply regarded Poland as 'the corridor through which the enemy has passed to Russia', was not prepared to allow any Polish government to come into being other than a dependably pro-Russian one. As Poland began to be liberated by the Red Army in 1944, the mainly Communist 'Lublin Committee' was set up and in January 1945 declared itself, with Stalin's backing, to be the Provisional National Government of the Polish Republic. All Roosevelt and Churchill could do was to get Stalin to broaden this Provisional Government to include members of the London government-in-exile and to promise free elections. Mikolajczyk, the leader of the London government, and four of his colleagues became ministers, the other sixteen were Communists and, when 'free elections' were finally held in January 1947, Mikolajczyk and his anti-communist party were terrorized. In ten out of fifty-two constituencies his candidates were disqualified, and perhaps a sixth of his party workers were imprisoned. Hardly surprisingly, the result was a landslide victory for the Communists. The American ambassador resigned in protest, Mikolajczyk fled to the West. A new constitution creating a one-party communist state was adopted.

In Bulgaria, the Fatherland Front—a coalition of a number of political parties including the Communists—seized power as the Red Army drove out the Nazis. Thanks to Russian support and to differences among their opponents, the Communists increased their power. The Western allies did their best to see carried out the free elections agreed to by the Big Three at Yalta, but with little success. When the elections were finally held in October 1946, opposition leaders were in prison, voters got delayed at road blocks and voting-papers were mislaid. Nonetheless Petkov, the most influential non-Communist leader, claimed that his Agrarian party won at least 60% of the votes. Officially however, the Communist-controlled Fatherland Front won 364 seats against 101 to the Agrarians and took firm control of the government with Dimitrov as Prime Minister. In August 1947 Petkov was tried on trumped-up charges of treason, found guilty and executed. All opposition parties were banned.

The Roumanians fought with the Nazis until August 1944 when King Michael, realizing that the defeat of Hitler was certain, sacked his pro-Nazi government and formed a new coalition to make peace with the advancing Russians. At that time Anna Pauker, a leading Roumanian Communist, calculated that there were hardly a thousand Communists in the whole country. What they lacked in numbers they made up for in organization and resources. Forming themselves into the National Democratic Front, they organized, with the collaboration of the Red Army, strikes and demonstrations against the coalition. In February 1945 Russia's Deputy Foreign Minister, Vyshinsky, flew in from Moscow and, at the end of a series of interviews with King Michael, forced him to dismiss the coalition and appoint a new government headed by Groza, a Communist. Though Britain and America insisted that Groza's appointment was unacceptable to them, though other political leaders refused to join his government and though the king even demanded his resignation, Groza stayed in office. When elections were held in 1946, 200 opposition leaders were in prison, the Communist-controlled printers' union would only print pro-communist pamphlets, and Groza's party won 348 out of 414 seats. The king was forced to abdicate in December 1947, and a one-party state was established early in 1948.

Hungary had also sided with Germany and not until April 1945 was it completely conquered by the Red Army. Comparatively free elections were held in November 1945 in which the non-Communist Smallholders Party won about 60% of the seats, the Communists 17%. The Russians however insisted that the previous coalition government, which had included Communists, should continue and also that the key ministers of the Interior, Defence and Economics should be in Communist hands. In December 1946 the Minister of Defence and of the Interior claimed that there was a dangerous conspiracy against the Hungarian people. Without consulting their Prime Minister, Ferenc Nagy, leader of the Smallholders party, they arrested many leading politicians, one of whom, Kovacs, was handed over to the Russians and ended up in Siberia. Nagy's resignation was brought about by threats to his son, and Cardinal Mindszenty, outspoken leader of the Catholic party, was sentenced to life imprisonment. Another election gave pro-Communist parties a 90% majority and Rakosi, previously Minister of the Interior, became Prime Minister of Hungary.

In Czechoslovakia, a coalition government was set up at the end of the war which included Gottwald, a communist, as Minister of the Interior but had two non-Communists, well-known through Europe, Benes and Jan Masaryk, as President and Foreign Minister respectively. In contrast to other parts of East Europe, the Czechoslovak Communist Party was genuinely popular and, in a free election in May 1946, won the largest number of seats with 38% of the votes. Gottwald became Prime Minister of a coalition government. Between

1946 and 1948, however, the Communists lost support and Gottwald, with Russian backing, decided to stay in power by force rather than risk another free election. The police force was purged of non-Communists and huge street demonstrations were organized by Communist trades unions in support of Gottwald. Benes, fearing civil war, allowed Gottwald to form a totally Communist government without a general election. A fortnight later, the pyjama-clad body of Jan Masaryk, Czechoslovakia's leading non-Communist politician, was found on the pavement outside his apartment. Whether it was murder or suicide remains a mystery. Another election was held later in 1948. The Communists won 88% of the votes, Benes resigned and was replaced as president by Gottwald.

Also in 1948 and also in Eastern Europe occurred an event which was to have considerable significance both for Europe and for the world. The small and primitive country of Yugoslavia went Communist but somehow, despite all Stalin's considerable attempts to the contrary, managed to maintain her independence. Communism, the Yugoslavs demonstrated, did not necessarily mean slavish obedience to Russian orders.

The Nazis had conquered Yugoslavia in 1941 but the resistance movement, particularly the group of Communist partisans led by Josip Broz (alias Marshal Tito) was one of the most effective of Europe and managed to liberate most of the country, including the capital Belgrade, without the help of the Red Army. Consequently, Tito could and did act more independently of Moscow than most of his Eastern European counterparts. Tito, however, was a fanatical Communist. He had been in Russia in 1917 and had taken an active part in the Revolution. When, between the wars, the Yugoslav Communist party had been banned he had been hunted by the police, spent six years in prison and ended up in Moscow to work for Comintern. From 1945 to 1948 he seemed more Stalinist than Stalin. Without help from Moscow, he smashed the Yugoslav opposition with utter ruthlessness. King Peter was exiled, Mihailovic another resistance hero and rival executed and other politicians imprisoned. In a one party election in November 1945, Tito's Popular Front party won 90.5% of the votes. Moreover, his foreign policy was recklessly anti-American. In 1946 an American transport plane was shot down on his orders and active support given to the Greek Communists. He bitterly criticized the Truman Doctrine, joined Cominform as soon as it was formed and agreed to have its headquarters in Belgrade. In the autumn of 1947 the official Russian newspaper *Pravda* had nothing but praise for the Yugoslavs and their government. 'Events in Yugoslavia in the last four years', it said in its November 29 edition, 'speak eloquently of immense, truly gigantic strength of its people. The Yugoslav Communist

Party had followed the very road of which Lenin and Stalin have spoken.'

Within five months, however, Moscow had done a somersault. Bitter hostility became the official line towards Tito and his country. The reason for this was basically simple. Tito showed too much independence and too little respect. Stalin had disliked Tito and his attitudes for some time but what caused this policy somersault was a Yugoslav-Bulgarian scheme for a Balkan federation which did not include Russia. Stalin was easily able to dissuade the Bulgarians from participating and decided that Tito should be taught a lesson. In March 1948 Russian military advisers, whose behaviour had been roundly criticized by the Yugoslav leader, were suddenly withdrawn. In June, at a Cominform meeting in Bucharest which Tito was careful not to attend, he and his government were accused of a long list of misdeeds, including 'unfriendliness to the Soviet Union', 'betraying workers' international solidarity' and 'departure from the Marxist theory of class struggle'. 'We have evidence,' Zhdanov, the Russian delegate, told the meeting, 'that Tito is an imperialist spy', and the Yugoslav people were publicly invited to make their leaders 'admit their errors and correct them'. There can be no doubt that Stalin intended Tito's overthrow. 'I will shake my little finger,' he said, 'and there will be no more Tito.' When an internal plot failed, economic blockade was tried. Its effects were serious on the Yugoslav economy but, far from weakening Tito's support, further strengthened it. Finally, in the summer of 1949, Stalin gave serious consideration to the idea of an armed invasion. By that time, however, Tito not only had the overwhelming backing of his own countrymen but was likely to receive military support from the West. Stalin therefore made do with a war of words and Tito went his own way regardless.

The most immediate and serious consequences of Tito's defiance were felt in other parts of Eastern Europe. Stalin let loose a reign of terror to purge the newly established satellite governments of anyone who showed or might possibly show a trace of Tito-type independence. In scenes reminiscent of the Moscow show trials of the 1930's (see page 102) some of the leading Communists most loyal to Russia went to their deaths—Rajk in Hungary, Kostov in Bulgaria, Anna Pauker in Roumania, Slansky in Czechoslovakia. Some 530,000 party members were purged in Poland, another 300,000 in Czechoslovakia. Moreover, Stalin and his successors learnt a lesson from the Yugoslavian affair which they did not forget. Never again would they hesitate to use force when independence flickered into life in Eastern Europe. First in East Germany in 1953, then in Hungary in 1956 and again in Czechoslovakia in 1968, Russian tanks rumbled decisively into action.

Map 18 Cold-War Europe

Legend:

- - - - 1937 Frontiers

Allied Control Zones of Germany and Austria

Ceded to Russia by Britain and America

■ Cities divided into four Occupation Zones

Annexed by Russia in 1945

States which became Communist between 1945 and 1948

Yugoslav gains from Italy 1945

The 'Iron Curtain' from 1948

Germany since 1945

✈ The Berlin Airlift crisis 1947-48

★ Areas of unrest

Map labels:

FINLAND
Viborg
Leningrad
ESTONIA
Pskov
Riga
LATVIA
Memel
LITHUANIA
Konigsberg
Vilna
Minsk
BALTIC Sea
SWEDEN
DENMARK
1956
Danzig
EAST PRUSSIA (annexed by Poland)
RUSSIA
American
Bremen
Berlin
Szczecin (Stettin)
Poznan
Warsaw
Pinsk
HOLLAND
British
Russian
POLAND
BELGIUM
1953
annexed by Poland
Wroclaw (Breslau)
Lvov
LUXEMBURG
Erfurt
French
American
1968
Cracow
Nuremberg
Prague
CZECHOSLOVAKIA
Czernowitz
FRANCE
French
USA
Russian
Vienna
Uzhgorod
Kishinev
SWITZERLAND
French
AUSTRIA
British
1956
Budapest
HUNGARY
ROUMANIA
Trieste
YUGOSLAVIA
Belgrade
Bucharest
Black Sea
Pola
ITALY
Adriatic Sea
BULGARIA
Sofia
ALBANIA
Communist activity 1946
TURKEY
Mediterranean Sea
GREECE
Aegean

East Germany made a slow recovery from the War. The Russians insisted on reparations and refused to allow the acceptance of Marshall Aid. The Communist government of Ulbricht was rigidly pro-Russian. In the early 1950's 20,000 refugees on average each month were escaping to the obviously prospering West Germany. In May 1953, in a period of rising prices but stable wages, Ulbricht ordered a 10% increase in production without a corresponding wage rise. Massive unco-ordinated and sometimes violent demonstrations and strikes took place in Berlin and 300 other cities. The demand for free elections was heard, party and police headquarters attacked. Then Russian tanks were ordered in, the revolts put down. Ulbricht continued in power and became even more rigid and harsh in his policies.

In Poland, Khrushchev's de-Stalinization policies of 1956 (see page 303) caused great excitement since the Polish government between 1948 and 1956 had followed Stalinist policies, encouraging heavy industry at the expense of living standards. Strikes and anti-Soviet demonstrations in the industrial city of Poznan caused the government to fall and brought to power Gomulka, previously considered by Moscow to be an undesirable Tito sympathizer. As Russian tanks moved on Warsaw, Gomulka and Khrushchev met. The latter, judging that there was no real danger of Poland moving out of the Soviet bloc, allowed Gomulka to continue in power. It was a shrewd decision. Gomulka proved much less of a reformer than had been thought and in the 1960's his rule was as repressive as anywhere in Eastern Europe.

Hungary 1956

Events in Hungary were very different. The sinister Rakosi was feared and hated and duly kept himself in power by a policy of terror which Stalin himself would have found hard to outdo. The de-Stalinization policy inspired a massive movement within Hungary for greater freedom of speech and national independence. On October 23 a popular uprising took place in Budapest which brought to power Imre Nagy, committed to the granting of greater political freedom and to the withdrawal of Russian troops from Hungary. On November 4 Russian tanks entered Budapest and for the next ten days heroic but hopeless resistance continued in the capital. 20,000 Hungarians were killed, another 20,000 were arrested. In desperation, Budapest radio appealed to the West. 'We implore you in the name of justice, freedom and the binding moral principle of active solidarity to help us. Our ship is sinking, light is fading. The shadows grow darker every hour over the soil of Hungary. Extend to us your fraternal aid.' No aid came. The Western powers, deeply involved and divided by the Suez crisis (see page 283), were in no position to

Russian tanks in Budapest,
November 1956

act. By November 14 all resistance was crushed. Russian T.34 tanks lumbered along the main streets of Budapest dragging the bodies of freedom fighters behind them. Nagy, having been promised a safe-conduct, was arrested, taken secretly from Hungary to Russia where, two years later, he was shot with a number of his colleagues. A new government, led by Kadar, was put into power by the Russian army.

Between 1948 and 1968 the Czech governments, first of Gottwald and then of Novotny, were both rigidly Stalinist. By 1968 Novotny's ineffective economic policies and excessive censorship led to increasing discontent, and his party colleagues forced his resignation without Moscow raising any objection. The new government, however, led by Dubcek, began to experiment with a new brand of Communism. Greater freedom of speech, new economic policies with closer union with the west, less bureaucracy and more humanity, these were the elements of Dubcek's 'Socialism with a human face' and they had enthusiastic popular support. For Russia's leaders, Dubcek's success was intolerable. They saw a real possibility that Czechoslovakia would move first economically and then politically out of the Soviet bloc into Western Europe, and further that her example might prove infectious and thus ultimately destroy Russia's position in Eastern Europe. Despite Dubcek's promise that Czechoslovakia would remain a loyal member of the Warsaw Pact, armies of the Pact invaded Czechoslovakia on 21 August 1968, claiming

that the Americans and West Germans were plotting to seize power in Prague! The Czech leaders ordered that there should be no resistance. They were then summoned to Moscow and forced to drop most of their reforms. Dubcek had to resign (he eventually got the job of a mechanic in a provincial garage) and censorship was re-imposed. A new government, led by Dr. Husak, had the unenviable task of trying to revive a demoralized nation with the Russian army breathing down its neck.

Russian tanks again, this time in Czechoslovakia, August 1968

Russia since 1945

For all her tremendous military and diplomatic successes between 1942 and 1948, post-war Russia was not a happy place. Her wartime losses in lives and property had been the greatest of the warring nations, greater even than Germany's. Both agriculture and industry were dislocated by the war effort and especially in areas where the

LENIN (1870-1970) ON THE RUINS OF AN IDEAL...

How Lenin might have felt about the Russia of Stalin and his successors as sketched by a West German cartoonist

armies had fought, near-famine conditions continued until 1948. Furthermore, over this hard-pressed country brooded Stalin, more suspicious, repressive, ruthless and convinced of his own wisdom than ever. Despite the terrible hardships the Russians had suffered since the Revolution, he continued his pre-war policy of expanding heavy industry at the expense of living standards. Believing them to be disloyal to his government, he had perhaps as many as a million non-Russians like the Volga Germans and Crimean Tartars forcibly moved to Central Asia. More incredibly, he re-imprisoned nearly half of Russia's million surviving prisoners of war in labour camps on their return home to clear their heads of any dangerous Western ideas before they were allowed back into Russian society. Freedom of speech, indeed any sign of individuality, was stamped out, and Russian artists and writers found themselves controlled as never before. One of the few men close to Stalin, Zhdanov, was the judge of artistic activity. 'All forms of cultural activity,' he declared, 'must be placed in the service of the communist education of the masses.' Before long, he had put out of work two of Russia's finest writers—Zoshchenko and Akhmatova—and, arguably the greatest film director in the world, Eisenstein.

As Stalin grew older, he became even more isolated, and Beria, his chief of secret police, grew more powerful. In 1952 fear enveloped Moscow. One leading Party member remembered how 'you might go to Stalin on invitation as a friend. And when you sat with him, you did not know where you would be sent next, home or gaol'. In January 1953 a number of leading doctors were accused of plotting to assassinate party leaders. Stalin seemed on the point of launching yet another major purge. On March 4, however, aged 73, he had a stroke and died.

There was something larger than life about Stalin. His achievements and his vices were both colossal. By his single-minded insistence on 'socialism in one country', on rapid industrialization whatever the human cost, he more than anyone else transformed Russia into an industrial and military super-power. He led his country through 1941 and 1942, two of the darkest years in her history to 1945, her greatest victory. By his tough and resourceful diplomacy and his acute understanding of the realities of power politics, at Yalta and Potsdam, he won half of Europe and much needed security for Russia's western frontiers. The extension of national power has always been used as a measure of a ruler's ability. By it, Stalin must be rated among the greatest of Russia's leaders. Yet by all civilized standards of behaviour he was a monster. He caused the deaths of as many of his countrymen as Hitler, their most savage enemy. His collectivization policies both liquidated a peasant class and brought chaos to Russian agriculture for the rest of his lifetime. His lunatic suspiciousness made terror a constant element in Russian political life. It was unable to distinguish between friends and enemies and even poisoned his relations with his family. His wife committed suicide, his daughter fled the country and his son died an alcoholic at the age of 41.

Stalin's Successors

There was no obvious successor to Stalin and at first senior party members pursued a policy of 'collective leadership'. Their most significant achievement was to execute Beria and five other secret service chiefs. By 1955 Khrushchev, the First Secretary of the Party, was clearly the most powerful man in Russia but he never possessed nor appeared to seek the dictator-like authority of Stalin.

On the contrary, in 1956, he launched an astonishing attack on Stalin at the Twentieth Congress of the Communist Party which began a new if short-lived era in Russia and Eastern Europe. Stalin, he argued in a speech much discussed by party officials, was a tyrant who had perverted the ideals of Communism by his love of personal power. Overnight, Stalin was officially transformed from the hero of heroes to the most villainous of villains. His body was taken from its place beside Lenin in the Red Square mausoleum and Stalingrad was renamed Volgograd. The shock both to party officials and to the Russian people was considerable. For Russian writers, it was greeted as the thaw which precedes the spring and some superb novels, critical of Russia's recent past, were not only written but published with the government's blessing. The Politburo (the Party's highest committee) could not decide whether one of the most powerful of these, *One Day in the Life of Ivan Denisovitch*, by Alexander Solzhenitsyn, which was set in a labour camp of the Stalin era, should be published. Khrushchev insisted that it should, and it was an instant best-seller both in and outside Russia itself. De-Stalinization spread an infectious sense of freedom.

Khruschev also began other major reforms. With most of his senior colleagues he believed that priority must be given to raising the standard of living of ordinary Russians, of encouraging consumer as much as heavy industry. He also started ambitious new schemes to improve agriculture, tried to redirect the Moscow bureaucrats, whom he considered an obstacle to effective government, into the provinces and looked for new international approaches which might end the Cold War. Though he was a very tough politician, he could easily play the part of the warm and humorous peasant. The combination of his personality and policies made Russia seem a different place from the grey, forbidding fortress of Stalin's time.

His policies, however, had mixed fortunes. There was considerable economic progress and a boom in the consumer and housing industries. Russians grew more prosperous. Simultaneously, Russia's aerospace industry chalked up a number of spectacular triumphs over the Americans, being the first to send an unmanned satellite (Sputnik I) and then another manned by Yuri Gagarin into orbit round the earth. Other policies however were less successful. The Moscow bureaucrats liked living in the capital and proved hard to move. His great 'virgin lands' scheme to bring millions of acres under cultivation proved a fiasco, and his foreign policy (see pages 305–6) had as many downs as ups. His colleagues found him dangerously impulsive and unpredictable. They tried to get rid of him in 1957 but he outmanoeuvred them. They tried again in 1964 and this time, with his agricultural and foreign policies in tatters, they were successful. Condemned for 'adventurism', 'hare-brained schemes' and 'refusal to take advice', he was forced to retire and lived until his death in 1971 in a modest Moscow flat.

Another period of collective leadership followed from which Kosygin and Brezhnev emerged as the strong men. They projected a grey, solid image without Stalin's fearsomeness or Khrushchev's unpredictability. They continued to give priority to raising living standards, introduced more flexible economic policies and supervised a rapid expansion of higher education. In contrast to Khrushchev, however, they clamped down on their writers and re-imposed strict censorship. Publication of Solzhenitsyn's masterpiece *First Circle* was forbidden inside Russia and the author, an eloquent champion of artistic freedom, was expelled from the Writers' Union (membership of which was essential to him if he was to earn a living from his writings) in 1969 and from Russia itself in 1974. A still harsher warning to independently-minded writers was given in 1966 when Sinyavsky and Daniel were sentenced to seven and five years hard labour respectively for publishing books 'anti-Soviet in content ... which could be profitably used by enemies of communism'. Another tactic used increasingly was to imprison critics in psychiatric hospitals on the pretext that they were insane. This happened to Zhores Medvedev, a research scientist critical of many aspects of Soviet scientific policy and a friend of Solzhenitsyn, for a fortnight in 1970. Protests inside and outside Russia secured his release, but when in 1973 Medvedev left Russia on a visit to Britain he was forbidden to return. Another prisoner in a psychiatric hospital in 1973 was a critic of military policy, General Grigorenko.

The U.S.A. exploded her first atomic bomb in 1945, Russia hers in 1949. By 1953 both possessed hydrogen bombs, by the late 1950's intercontinental ballistic missiles, by the mid-1960's virtually undetectable missile-launching submarines and by the early 1970's multiple-targeted re-entry vehicles (M.I.R.V.s) which were computer-controlled missiles designed to dodge through any screen of defensive missiles. In 1963 Khrushchev calculated that the U.S.A. and Russia each possessed 40,000 hydrogen bombs and the capacity to destroy 800 million people. There was therefore good reason why both sides should get away from Cold War attitudes. Many years and some terrifying international crises however were to pass before there was significant easing of tension.

After the Berlin airlift of 1948, the focus of the Cold War shifted to the Far East. North Korea, supported directly by Communist China and distantly by Russia, went to war against South Korea, supported by the U.N. with the U.S.A. providing 90% of the U.N. forces. The war continued from 1950 to 1953, and as long as it lasted there could be no improvement in East–West relations. Stalin's successors made a positive effort to improve international relations. They helped to bring a temporary settlement in Indochina after the defeat of the French by the North Vietnamese (see pages 274–5),

Evidence of an arms race between the Superpowers of unprecedented scale and lunacy. May Day 1964: Russian rockets being paraded through Moscow

they agreed to end the occupation of Austria and withdrew their troops from Finland. 1955 was an encouraging year with both Eisenhower in the U.S.A. and Khrushchev in Russia stressing the need for peaceful coexistence and with a friendly if indecisive summit meeting, including Britain and France in Geneva. 1956, however, proved most unhappy. To the disgust of the West, the Russians ruthlessly crushed the Hungarian revolt and simultaneously threatened Britain and France with war if they continued their attack on Egypt (the Suez crisis, see page 283). In the following years, fences were mended, Khrushchev paying visits to both Britain and the U.S.A., the latter very successful, the former moderately so. Nothing came, however, of a summit conference arranged in Paris in 1960. 1961 and 1966 were two more dangerous years. In 1961, the Russians and East Germans once again threatened West Berlin. President Kennedy flew in to the city and made it clear to its citizens and the world that the U.S.A. would stand as firm as she had done in 1948. The East Germans then walled off East Berlin from the rest of the city to end the stream of refugees (40,000 in July 1961) escaping into West Berlin. The 'iron curtain' was thus

Above *The reality of divided Europe. East Berlin photographed from the Western sector*

Left *A cheerful President Nixon chats up first Mao Tse-Tung of Communist China and then Brezhnev of the U.S.S.R. The détenté between the U.S.A. and the Communist powers in the 1960's and 1970's was due largely to the growing distrust between Russia and China*

West Berliners clamber up ladders and lamp-posts to wave to their friends and relatives in the Communist parts of the city cut off from them by the wall built in 1961

vividly extended. In the year after the wall was built 50 people were killed trying to cross it and one student was left by the East German guards to bleed to death with West Berliners watching helplessly beyond the barrier. The nearest the two sides came to war was in the Cuba crisis of 1962. In 1959 the Caribbean island of Cuba had gone communist under the leadership of Castro. By 1962 the Russians were building missile-sites there aimed at the American mainland. President Kennedy decided that these sites must never be completed and the U.S. navy sailed into the Atlantic in October 1962 to prevent a Russian fleet with supplies for the missile sites from getting through to Cuba. With nuclear forces on full alert, the Russian fleet turned back in the nick of time. This crisis shocked Russia and the U.S.A. into positive measures to prevent such a terrifying situation arising again. A telephone link—'hot-line'—was made between Washington and Moscow. Moreover, negotiations led to the first restrictions on nuclear tests—the test-ban treaty signed by Russia, the U.S.A. and Britain. This was followed in 1969 by the Strategic Arms Limitation Talks (S.A.L.T.) which were not only lengthy but productive.

In the 1960's, the Russian/American domination of the world grew less and so did their rivalry. Communist China developed as a major power and the most direct threat to Russia's security. In the East, Japan and in the West the E.E.C. worked their way to greater independence from the U.S.A. as their economic power grew. Under de Gaulle's presidency France left N.A.T.O. and followed a more friendly policy towards Russia. After 1969, West Germany, with Brandt as Prime Minister, tried the 'Ostpolitik', a more friendly approach to Eastern Europe and Russia. With American backing, trade links between Eastern and Western Europe grew. By 1973 representatives of Comecon were ready to begin negotiations with the E.E.C. with the aim of co-ordinating their economic planning while the U.S.A. was seriously considering giving Russia 'most-favoured nation' status in matters of trade. By the early 1970's the Cold War seemed usually to be a thing of the past. Every now and then however, it reared its horrifyingly ugly head. In October 1973 President Nixon put American forces in Europe and elsewhere on full nuclear alert during the Israeli-Arab war. There were, he said, some dangerous-looking troop movements by the Russians. Nonetheless, the key word of the early seventies was 'détente', the relaxation of tension. In early 1974, President Nixon visited Brezhnev in Moscow where both leaders stressed their determination to continue the détente policy and when President Ford succeeded to the American leadership after Nixon's resignation in the August of that year, he was quick to commit himself along the same lines.

Chapter 17
Economic and Social Developments 1945-1973

Postwar Chaos

Europe in 1945 was enough to fill with despair the stoutest of hearts, especially the economic experts whose responsibility it was to plan for the future. They could not but remember the destruction caused by World War I and realize that it was small compared with that caused by World War II. Nor could they forget that the European economy never recovered in the inter-war years and that poverty and unemployment had been the fate of millions of Europeans for years on end. The omens for the future, seemed bleak and the chances of a genuine economic recovery within a generation slim.

The first three post-war years, 1945-48, were exceptionally difficult. They were also similar in many ways to the years immediately after World War I, and the worst fears of the pessimists seemed justified. As in 1919, things briefly looked bright. The immense immediate problems of refugees, disease and hunger were energetically taken in hand by the United Nations Relief and Rehabiliation Administration (U.N.R.R.A.) which had been set up for this purpose in 1943. At the height of its activities it employed 25,000 people and distributed 3,700 million dollars worth of relief, most of which was supplied by the U.S.A., Canada, and the U.K. and was given to Central and Eastern Europe and to China. Individual nations also began to tackle their own particular economic problems with energy. Britain and France nationalized important sectors of their economies and launched ambitious schemes of social welfare. Austria and Italy moved in the same direction if more cautiously.

In 1946, however, this limited recovery faltered, for reasons similar to those of 1919 and 1920. Again, the U.S.A. seemed to be moving towards isolation. Five days after the war with Japan ended, Truman's government cancelled Lend-lease, the economic aid which had been the financial life-blood of Britain's and to some extent Russia's war effort. As far as Britain was concerned, the cost of World War II had been so heavy that some form of continuing U.S. aid was essential even though the war was over. Churchill had tried to persuade Roosevelt to continue Lend-lease for at least a year but Roosevelt had died undecided and his successor Truman was not to be persuaded. He and the American public knew that Lend-lease had cost the U.S.A. 46 billion dollars. Now the enemy was defeated, it was time to call a halt. Britain had no choice therefore but to negotiate a loan from the U.S.A. on terms she could ill afford. 'Now we faced not war any more,' said Dalton, the Chancellor of the Exchequer, 'only total economic ruin.'

As in 1919, the recovery of Germany was hindered by the question of reparations and by divisions among the victorious allies. Before the war ended, there had been much talk about 'making Germany pay'. 'If I had my way,' said Roosevelt, 'I would keep Germany on the breadline for the next twenty-five years,' and Stalin demanded reparations to the tune of 20,000 million dollars. When the war ended, disagreements between Russia on the one hand and Britain, France and the U.S.A. on the other prevented Germany being treated as an economic unit. While in Western Germany the allies soon gave up any attempt to extract reparations from their zones of occupation, in the East, Stalin stuck to his demand and systematically plundered East Germany and other parts of Eastern Europe.

Another hindrance to economic recovery in Central and Eastern Europe—again reminiscent of 1919-23—was an acute currency problem. Inflation was so bad in Germany that people preferred to barter goods rather than trust coins or notes. Cigarettes, followed by coffee and chocolate, were the most popular exchange units. In Italy prices rose to 35 times their pre-war level, while in Hungary the old currency had to be abolished when the exchange rate reached 11,000 trillion pengoes to the dollar. Then came an over-hot summer in 1946 which damaged the crops, the worst winter in living memory in 1946-47 which paralysed first coal supplies and then heavy industry and then another poor harvest in the summer of 1947. Two years after the war ended, the European economy was on the point of collapse.

The Marshall Plan

What saved it—and in fact redirected it into the most rapid and lasting expansion ever—was the Marshall Plan. G. C. Marshall was the American Secretary of State (i.e. the minister for foreign affairs) and on 5 June 1947 he made a speech at Harvard University outlining the new policy of the U.S.A. towards Europe. Instead of withdrawing, the U.S.A. was ready to make financial assistance available to Europe on an unprecedented scale. This change of policy was caused to a large degree by anxiety about Russian expansion in Europe. In March 1947 the 'Truman Doctrine' (see page 287) had committed the U.S.A. to the active political and military backing of anti-communist governments in Europe. Part of the thinking behind the Marshall Plan was that a poverty-stricken Europe was more likely to fall to communism than was a prosperous one. Thus helping Europe back to prosperity was very much in America's interest. The Marshall Plan however was more than just an exercise in power politics. It was an imaginative scheme of great generosity. 'Our policy,' said Marshall at Harvard, 'is directed not against any country or doctrine but against hunger, poverty, desperation and chaos . . . Any government that is willing to assist the task of recovery will find co-operation on the part of the United States government.' This economic aid was offered without strings attached. 'The initiative,' he said, 'must come from Europe.' Marshall Aid was offered to every country of Europe including Russia and the U.S.A. insisted that European nations organized themselves to decide how the aid should be distributed. That it eventually went only to the nations of Western Europe was mainly the responsibility of the Russians. Britain, France and Russia met to consider Marshall's proposals but after some days of fruitless discussions, Molotov, the Russian Foreign Minister, walked out and dismissed the Plan as 'an American design to enslave Europe'. When a further conference was organized in Paris, only Czechoslovakia of the East European countries attended and her representatives were withdrawn after their Prime Minister had paid a visit to Moscow. The communists insisted on treating the Marshall Plan as nothing more than the Truman Doctrine in economic terms.

In contrast, sixteen countries of Western Europe responded enthusiastically to Marshall's ideas. In June 1947 they set up a Committee of European Economic Co-operation which, in less than ten weeks, drew up a four-year recovery programme. The Organisation for European Economic Co-operation (O.E.E.C.) was formed in the summer of 1948 to distribute Marshall Aid. The U.S.A. did not become a member but made 9,500 million dollars available to the O.E.E.C. between the end of 1947 and 1950. Of this Britain with more than 3,000 millions and France with 2,700 millions received the lion's share. Italy, West Germany and Holland all received more than 1,000 millions each. This was the medicine that the frail European economy needed. Recovery began almost at once. By 1950, industrial production was on average 30-35% above 1938 levels. In major industries progress was even better. Steel production was 70% higher, cement 80% and motor vehicles 150%. In the next twenty years there was no looking back.

From 1948 to 1973 the growth of the European economy was spectacular. Prosperity increased with barely a break. A generation of Europeans grew up who did not remember the war or had any experience of a serious economic depression. They expected that they both would and should grow richer each year. Such lasting growth and such expectations have no parallel in European history. For example, in the seventeen years before World War I, when

Germany was enjoying the greatest economic boom she had then experienced, she achieved a 5% annual growth rate just once. In the seventeen years after World War II a 5% growth rate was achieved nine times and a 10% one three times. Even Britain, which performed much worse than most of her European competitors, managed to achieve a 2.5% annual growth rate between 1948 and 1962, which was better than any other period in the last 100 years (1.7% from 1870 to 1913 and 2.2% from 1924 to 1937).

Between 1948 and 1963 the average rate of growth of the eleven major nations of Western Europe was 4.4%. Between 1965 and 1970 it was 4.3%. In the 1950's, the leader was Germany, the sluggards Britain and Belgium. In the 1960's Italy and France moved among the leaders with Britain still at the back. Rates of expansion varied not only from country to country but from year to year. In 1952 a serious depression seemed to threaten as a result of the Korean war but it turned out to be mild and short-lived. In 1957-8 a currency crisis slowed down the rate of growth. The 1960's turned out to be more bumpy. Between 1963 and 1965 the economies of France and Italy faltered and 1967 was a year of general crisis, particularly affecting Britain, West Germany and Austria. But even in the bad years Europe continued to prosper, if only growing at one or two per cent rather than the usual four or five. Compared with the depressions of the inter-war years, and nineteenth century Europe, the 'bad' years of the 1950's and 1960's were extremely good.

Eastern Europe

The economy of Eastern Europe also enjoyed an unprecedented rate of growth. In 1949 Russia established its alternative to the Marshall Plan and the O.E.E.C. in the shape of Comecon, the Council of Mutual Economic Assistance which co-ordinated the economic activities of communist Europe and also of non-European communist states like China. At first this meant the organization of the East European economy for the benefit of Russia with Russia selling at high prices to and buying at low prices from her satellites. Nonetheless, there was real economic progress throughout Eastern Europe and, especially after Stalin's death, exploitation by the U.S.S.R. lessened. The 1950's were a period of particularly rapid growth, the early 60's saw a slowdown and the late 60's another acceleration. Between 1948 and 1955 industrial production in East Germany and Czechoslovakia doubled. The most rapid increase, however, occurred in countries like Roumania and Bulgaria, which before the war had been almost entirely agricultural. In 1964 Bulgaria's industrial production was six times higher and Roumania's five times higher than in 1938. While Stalin lived the emphasis both in the U.S.S.R. and her satellites was on the expansion of heavy industry, or armaments and on quantity rather than quality.

Though there was a swift increase in national wealth, the standard of living of ordinary people, unlike in Western Europe, rose only slowly. After Stalin's death there was a change of emphasis. Farming, housing and light industry (e.g. the manufacture of household goods) were expanded and there was a real improvement in living standards. In the 1960's industries like chemicals and electronics grew fast, and in the late 1960's the benefits of twenty years of economic growth began to mean significant improvements for the man in the street. Over the years, Comecon changed. As Russia and China began to quarrel more openly, China left Comecon and other members won greater freedom for themselves. Hungary and Czechoslovakia were able to introduce some limited economic reforms and, with Poland, increased the porportion of their trade with the non-communist world to between 40% and 45%.

There has been a tendency in the U.S.A. and Western Europe to underestimate the communist economic achievement since the war. Cold war suspicions have encouraged this attitude as have exaggerated communist propaganda, the misery and waste caused by Stalin's economic policies and the apparent drabness of much of Eastern Europe when compared with the West. Nonetheless in its own way the economic progress of Eastern Europe between 1948 and 1973 was spectacular, particularly in comparison with the economic progress of the preceding fifty years in the same area.

A housing development in the U.S.S.R. These 'neighbourhood units' housing thousands of families were built in the early sixties in Vladivostok

Index of Industrial Production, 1958—100

	1938	1948	1958	1967
W. Germany	53	27	100	158
France	52	55	100	155
Italy	43	44	100	212
Britain	67	74	100	133
Sweden	52	74	100	176
U.S.A.	33	73	100	168
Japan	58	22	100	347

Average Growth Rates of the Gross National Product* of Seven West European Nations between 1950 and 1970.

	1950-60	1960-64	1965-70
France	4·5	5·6	5·4
W. Germany	7·8	4·9	4·3
Italy	5·4	5·5	6·3
Sweden	3·3	5·5	3·5
U.K.	2·7	3·6	2·4
Netherlands	4·8	4·9	4·7
Switzerland	4·4	5·5	3·5
U.S.A.	3·3	4·5	3·7

Gross National Product (G.N.P.) is the money value of all the goods and services produced in the year by a particular nation.

Five Countries

France A succession of government plans were a central feature of the expansion of the French economy in the post-war years. In January 1946 an overall planning department was set up with Jean Monnet as its first director. In comparison with the Russian Five Year plans, the French plans were more flexible. Nor had French planners any power to compel private industry to follow their recommendations. What they did was to suggest the most desirable directions in which the economy might move and where modernisation would be most effective. Then the government, which took a thoroughly Keynesian view of its duties (see page 77), used its control of the nationalized industries and its powers of encouraging investment and of adjusting taxes to push the economy in the direction advised by the planners. Industry immediately prospered. Between 1946 and 1954 the steel industry was completely modernized and the output of electrical industry doubled. In the aerospace, railway and automobile industries moreover, the French displayed a technological and managerial flair which took them to the forefront of

European development in these fields. Agriculture in contrast stagnated, the peasant farmer proving remarkably resistant to the modernization schemes of the planners. From 1958, however, the Common Market (E.E.C.) gave a new boost to the economy, including agriculture. France in fact benefited from the Market more than any of its members. Between 1958 and 1968 the proportion of its foreign sales going to E.E.C. countries rose from 22% to 41%, of its agricultural products from 18% to 50%. After a troubled period in the mid-sixties with a rapid increase in the cost of living and major strikes culminating in the political crisis of 1968, the economy forged ahead once more. The managerial talent of France was the envy of Europe, and in 1973 the Hudson Institute, an American organization which specialized in predicting future trends, prophesied that by 1980 France would be the most prosperous nation in Europe.

West Germany In January 1945 an English prisoner of war found himself part of a long column of people being marched westwards from Eastern Germany away from the advancing Russians. Thousands of German refugees moved with the prisoners. 'They were a pitiable sight', he remembered, 'frozen, hungry, their shoes and clothes falling apart, dragging themselves along to an unknown destination, hoping only that it might be beyond the reach of the Russian army. It was so cold that . . . at night men and women could only keep alive by huddling together in a wagon . . . Those who fell asleep in the snow were dead within a few minutes.'

No-one knows how many Germans died in the economic and social chaos between 1945 and 1947; probably as many as had died in the previous six years of war. Between 10 and 15 million refugees moved into Western Germany, 9 million from lands lost to Poland, 3 million from the Sudetenland and another 3 million from East Germany. Most of them came with no belongings other than what they could carry and they arrived in cities devastated by Allied bombing where the necessities of life were already in desperate demand. During these years managing to survive was the aim of many Germans. 'You had to be clever', said one who lived through it, 'I spent two years doing nothing but scrounge food for my family.' Since the German currency was virtually worthless, barter became more usual than payment in cash. A black market flourished out of which some Allied soldiers made their fortunes. A non-smoking corporal of the British Women's Army Corps was reported to have made 500 dollars a month out of her cigarette ration. In the bitterly cold winter of 1946-47 cases of acute malnutrition rose sharply and the future of Germany looked hopeless.

In 1948, however, there was a sudden and permanent change for the better brought about by a dramatic currency reform and the

Scenes in Germany immediately after the war
A queue for water in Magdeburg

Left *An elderly looter in the ruins of Cologne*
Right *A few of the millions of refugees sleep in a former air-raid shelter at Bielefeld. In the foreground is a former German officer with his wife asleep on his shoulder and his child sleeping in a blanket beneath his father's cap*

Marshall Plan (see page 309). The economy began to expand and Dr. Erhard, the Economics Minister of the new Federal government, took immediate and intelligent steps to keep this expansion going. Arguing that the quickest way to get West Germany back on her feet again was to give her businessmen and industrialists the greatest possible freedom he persuaded the Allies to allow him to abolish rationing and to lower taxes. Then followed the outstanding achievement of the European economy in the 1950's, West Germany's 'economic miracle' *(Wirtschaftswunder)*. It began in 1949 when foreign trade doubled in a single year and increased again by 75% the next. Between 1948 and 1964 industrial production increased sixfold and unemployment fell from 9% to less than 1%. There was less planning than in France. Erhard, Economics Minister in Adenauer's government from 1949 to 1963, continued to encourage the free play of market forces. His policy was 'competition as far as possible, planning as far as necessary'. Agriculture was also thoroughly modernized. In 1965 there were 1,164,000 tractors where in 1950 there had been only 138,000. Though the numbers employed in farming fell from five to three millions, productivity rose 250% and in contrast to many parts of Europe, the rural villages were nearly as prosperous as the urban centres. In the mid-sixties the rate of growth slowed and 1965–66 was a period of crisis. The growth-rate dropped to 1% and Volkswagen, the automobile firm whose export achievements in many ways had typified the miracle, had to put its employees on short-time. Though an improvement soon followed, the growth-rate steadied down and many experts believed that the spectacular expansion of the 'miracle' years would not be repeated.

Italy Italy's 'economic miracle' attracted less attention than West Germany's but it was equally remarkable. The decade after the war was uncertain, with rapidly rising prices and a backward agriculture, but from 1956 expansion was rapid. Certain industries were outstanding—for example, automobiles (in 1973 Fiat was the largest motor manufacturer in Europe), office machinery (Olivetti) and electrical and household goods (one in three refrigerators produced in Europe in 1967 were Italian). Tourism was another valuable source of income and, by the mid-sixties, the standard of living of northern Italians, particularly in major industrial cities like Milan and Turin, was as high as anywhere in Europe. Two things flawed this achievement. First, unemployment remained high even in the years of most rapid growth. Secondly, the gulf between the prosperous north and the poverty-stricken south remained wide, despite government attempts to encourage industrial investment south of Rome. As in France, the government played an active part in the direction of the economy, and by the mid-sixties five of the nine

The banner headline of Die Neue Zeitung *announces the currency reform in the American, British and French zones which began both West Germany's 'economic miracle' and the political crisis of the Berlin airlift*

largest companies and nearly a third of all Italian industry was under state control. In contrast to her neighbours, Italy was barely affected by the slowdowns of the 1960's but ran into difficulties in the early 1970's.

The United Kingdom Britain's economic performance in the twenty-five years after World War II was the worst of the major industrial nations of Europe. Not only was she only able to achieve an annual growth-rate of between 2%-3% while her neighbours averaged 4%-6% but her economic expansion was disrupted by constant balance of payments crises—in 1948, 1951, 1955, 1961, 1964 and 1967-8. These were the result of a constant inability to export as much as she imported and were only temporarily solved by government measures which cut back expansion. 'Stop-go' was the phrase coined to describe this hesitant development. The pound sterling was devalued against the dollar in 1948, again in 1967 and, in the early seventies, dropped further in value against some European currencies, notably the West German mark. In the immediate post-war years, the standard of living of the average Briton was higher than that of most Europeans; in the mid-fifties he was overtaken by the French and Germans, and by the mid-sixties by many other Western European nations too.

Many factors combined together to cause this poor performance. British investment rates (i.e. the proportion of profits which companies used to modernize and expand their factories) were singularly low, 7.7% in 1962 as against 13.1% in West Germany and 25.2% in Japan. Continuous low investment meant out-of-date machinery and industrial plant. Industrial relations were bad and grew worse rather than better with the passing of time. Trade-union pressure led to wage rises in excess of productivity and increased costs which reduced the competitiveness of British goods in international markets. Strikes occurred so often over such trivial-seeming issues that strikes came to be referred to as 'the English sickness' on the continent. Furthermore, management was too conservative and too amateur. A revealing study showed that British firms under

One view of Britain's industrial relations problem, from the Daily Sketch

317

American management were 50% more profitable than similar firms under British management. Nor was government policy helpful. Both the Conservative and Labour parties maintained for too long an exaggerated notion of Britain's importance in the world and pursued imperial, defence and financial policies which were too heavy an economic burden. Both parties refused to join the Common Market in the 1950's and spent the 1960's trying to reverse this decision (see Chapter 18). The early seventies saw signs of expansion at a new more typically European rate of 4%-5%, but industrial relations continued to worsen and the balance of payments remained obstinately unhealthy.

Russia In the last years of his life, from 1945 to 1953, Stalin relentlessly imposed economic policies similar to those of the inter-war years, a rigidly-controlled expansion of heavy industry at the expense of lighter industry, agriculture and the standard of living of ordinary Russians. The production of steel was trebled, of coal doubled and of oil increased sixfold. By the time of his death Russia was the world's second industrial power. His successors changed the emphasis while maintaining strict government control. Malenkov, for example, promised 'an abundance of food for the people and of raw materials for consumer goods industries in the next two or three years.' This promise was largely fulfilled. Industrial wages were increased by 38% between 1952 and 1958 (a rate not much slower than West Germany's) and there was a dramatic increase in the sale of consumer goods (360,000 refrigerators for example were produced in 1958 as against 49,000 in 1953). Though there was some improvement in agriculture—a 56% increase in production in the same period—the huge and vital agricultural sector never made the same progress as heavy industry. Neither Stalin's collectivization nor Krushchev's later ambitious 'virgin lands' schemes nor more recent plans have had the success hoped-for and in the early 1970's, after a series of disastrous harvests, basic foodstuffs were rationed and huge supplies imported from North America and Western Europe. In the early sixties the whole economy slowed down to a 3% growth rate and did not pick up again until 1966-67. In some ways, the Russian planners were too ambitious. The rapid growth-rate of the fifties had convinced some of them that the U.S.A. might be caught and overtaken before long. Krushchev in fact predicted that by 1970 the U.S.S.R. would be the first industrial nation in the world. Moreover not only did they try to outproduce the U.S.A. industrially but spent heavily on space exploration, defence spending and overseas aid. Krushchev's prediction never looked like coming true. Though in the 1950's Russian industrial production had risen from 32% of the U.S.A.'s to 47% in 1967 it was still only 48% and the gap narrowed slowly if at all in the years that followed. And the standard of living of the average Russian remained markedly lower than that of much of Europe as well as of the U.S.A.

Social Trends

In the quarter of a century after World War II, there was a quiet social revolution. Before the war the limited social benefits provided by the state—and some had been provided in Germany since the 1880's—were usually available only to certain social groups and based on the principle of giving the bare essentials to the most poor. By 1970, however, most Europeans could expect at least some degree of state assistance towards insurance against unemployment, accidents and ill-health, medical expenses, family allowances, old-age pensions, education and housing. And this assistance was generally regarded no longer as charity but the duty of every civilized state towards its citizens. In the late 50's Western Europe was spending four times as much on social security than in the thirties. The Swedes had increased their spending six times, the French seven and the Italians fourteen. Within this general trend, there were many variations. On the whole, northern countries provided fuller services than the southern with Sweden and Britain the most comprehensive. For example, while virtually all Swedes and Britons could claim state medical aid in the fifties, only 80%-90% of Germans, Frenchmen and Italians could do so, and the rural peasantry of the Soviet Union were not included until 1965. In Britain and Russia medical services were free while in France, Belgium, and Switzerland, up to a quarter of the costs had to be paid for. West Germany and Belgium provided the best old-age pensions, France the most generous family allowances. In Britain, most welfare services were run by state departments like the Ministry of Social Security or the National Health Service. In Scandinavia special independent agencies, in Germany, France and Italy, co-operative insurance companies, and in Eastern Europe the trade unions played an active part in the administration of the social services.

Education also became increasingly available, either virtually free as in Britain and France or at a very cheap rate. In Scandinavia and Holland higher education was either free or financed by a state loan to be paid back on easy terms over a ten-year period. Education, therefore, absorbed a growing proportion of the national budget, from an average of about 2%-3% in 1938 to 4%-5% in the late fifties, to about 6% in 1970. The total amount spent on social services varied considerably in different countries. Between the wars, Britain and Germany spent much the most. After the war the picture changed as the following table on the left shows.

Social Security Spending per head of Population in

	1930	1957
Belgium	12	148
France	17	136
Germany	54	132
Italy	5	54
Sweden	22	135
U.K.	59	93

A major problem in 1945 was housing. The destruction of World War II which varied from 25% of the houses of Germany and Greece to 6%-7% in Britain and France had transformed the housing shortage of the inter-war years from a serious problem to an acute one. Moreover the existing stock of houses left much to be desired. Only one in two houses in France and one in three in Italy, Spain and Austria had running water. Useful progress was made in the fifties, the number of houses built roughly doubling. In a good year, the West Germans built half a million houses, Britain, France and Spain between 250,000 and 400,000. Most of this housing was for renting. Private housing did not really boom until the sixties when growing prosperity allowed an increasing number of young people to take out a mortgage. The Eastern European countries gave housing lower priority. Russian standards were particularly low with the average Soviet family occupying a single room and often having to share a kitchen with their neighbours. Better standards were achieved in the sixties when two or three-roomed flats became more usual in the cities. In other parts of Eastern Europe, housing was better than in Russia though worse than in the West. Quite a useful measure of housing standards is the average number of rooms per person. In 1960, Britain scored 1.5, West Germany 1.1, France 1.0, Hungary 0.7, and Bulgaria 0.5. Another approach to lessen the dangers of overcrowding was to develop completely new towns well away from the old urban centres. London's new towns, e.g. Harlow and Stevenage, won a high international reputation as did those of the Paris, Berlin and Stockholm areas.

Another trend threw the housing situation in Europe into sharp relief. Low unemployment and high wages attracted foreign workers to Western Europe in their millions. In 1973 there were 3.7 millions in France of whom the majority were Algerians, 2.4 millions in West Germany (mainly Turks and Yugoslavs) and 1.5 millions in Britain (mainly citizens of the old British Empire, West Indians, Indians and Pakistanis). They tended to move into slum areas, or in France, into shanty towns, forming distinct ghettoes which sometimes became the focus of serious racial conflict.

Increased incomes led to changing spending patterns. Proportionately less was spent on food. In 1865 it has been calculated that the average German spent 65% of his income on food, a century later he paid 36%. Yet though they spent proportionately less, Europeans nonetheless actually spent more and ate better. Consumption of cereals and potatoes dropped, of meat, fish, sugar and eggs rose. The danger was no longer of having too little but too much to eat. Proportionately less was spent on clothing too, 13% to 15% in the inter-war years to 11% or so in the mid-sixties. Much more was spent on so-called consumer durables like automobiles,

An immigrant shanty-town in Nanterre, a suburb of Paris

refrigerators and washing-machines and on leisure activities like
hobbies, entertainment and travel. Tourism became a major industry. By 1965 more than 6% of Austria's G.N.P., 5% of Spain's
and 4% of Portugal's came from tourism. Italy in 1966 achieved a
2 billion dollar surplus, and in that year Britain, France and
Germany spent 3.3 billion dollars abroad on their holidays. By the
late sixties, tourism was breaking through the iron curtain. Over
four million visitors were reaching Czechoslovakia and Hungary
annually and Roumanian and Bulgarian resorts on the Black Sea
began to attract package tours from the West.

In the nineteenth and early twentieth centuries, European trade
unions had had to struggle to be recognized by the governments of
the time and they had been closely connected with socialist, communist and anarchist political parties which had tried, above all in
the years immediately before and after World War I, to change

*Tossa de Mar on the Spanish
Costa Brava, a piece of the
Mediterranean particularly
popular with British tourists*

society by violent means. After 1945, however, they were generally recognized to be important and valuable institutions, and tended to prefer respectability, to work for reform rather than revolution. Though their sympathies still lay with left-wing political parties like the socialists and communists, their chief concern was the improvement of their members' living standards within the existing social framework. The largest and politically the most powerful trade union movement was the British Trade Union Congress which numbered eight million in the sixties and was closely linked to the Labour Party. British trade unions enjoyed a comparatively strong financial position which enabled them to maintain strikes lasting weeks or months rather than days. There were also hundreds of unions and demarcation disputes over which union should do which job was also a source of much industrial strife. The British economy was more seriously affected than any in Europe by strike action and a strong body of opinion developed which held that excessive trade union bargaining power coupled with irresponsible trade union leadership was a major cause of Britain's economic weakness.

The German movement was, at 6.5 million, only slightly smaller than the British. It was however made up of only 16 unions, with usually one union for one industry. The trade-union leaders drove hard bargains with their employers and secured above European average living standards for their members. Strikes however were virtually unknown in the fifties and sixties in marked contrast to the situation in Britain, France and Italy. The West German trade union movement showed itself more ready to come to terms with the problems of modernization and believed that in the long run their members would benefit from co-operation rather than conflict with their employers.

Both the French and Italian movements were dominated by communist unions, the C.G.T. (1.5 million members) in France and the C.G.I.L. (3.5 million members) in Italy. There were also small non-communist unions. The C.G.T. and the C.G.I.L. were linked with the communist trade unions of Eastern Europe and were more obviously concerned with politics than other movements. However their determination actually to bring about revolution rather than just talk about it could be questioned. In the potentially revolutionary weeks of May and June 1968 when de Gaulle's government seemed about to be toppled by the student riots, the C.G.T. took the opportunity to negotiate better wages for its members, but otherwise kept its head down. Strikes were frequent both in France and Italy. They tended to be shorter and less economically damaging than the British ones.

There was a slow but steady movement towards greater social equality between 1945 and 1973. The development of social services for all citizens, movements in wages and salaries, the use of graded taxes which took more from the rich than from the poor, have all contributed to greater equality. But there continued to be marked differences in wealth and opportunity, with income from the ownership of property still being a major source of wealth. However as a proportion of the national income, income from property fell between 1956 and 1963, in Britain from 33% to 26%, in France from 47% to 33%, and in Germany from 39% to 30%. The greatest social equality was achieved by the Scandinavian countries, the least by southern states like Spain, Portugal and Greece. In Eastern Europe and in Russia particularly, there were marked inequalities of wealth. In Stalin's time, government officials and military leaders did particularly well. These grew less in later years but were by no means eliminated.

Chapter 18
Western Europe and the EEC

The Four Major Nations
Great Britain In 1945 Britain's international reputation and national morale were as high as at any time in her proud history. Britain alone had fought Germany from the beginning to the end of the war and, with Germany and Japan now totally defeated, she ranked unmistakably as one of the 'Big Three' powers of the world, less strong than the U.S.A. and Russia but, backed by her world-wide empire, towering far above other nations. However, this appearance of international might hid from many, Britons included, her desperate economic position. The war, though won, had proved terribly costly. £1,000 million of foreign investments had had to be sold; £1,000 millions worth of bomb damage had to be repaired; £3,000 millions of foreign debt had still to be repaid. In order to pay her way economically, Britain needed to improve her export performance radically, which in turn required radically new approaches to management, industrial relations and education as well as a cautious approach to her responsibilities as a world power. Too many years were to pass, before British politicians and the British people came to terms with this post-war reality.

The years immediately after the war saw great difficulties and achievements. To almost everyone's surprise, Churchill lost the 1945 election, the Labour Party, led by Attlee, winning by an overwhelming majority. There were a number of reasons for this result. While there was almost unanimous gratitude for Churchill's record as a war leader, the electors remembered the Conservatives' poor record in home affairs between the wars. Unlike in 1918, there was no general desire to get back to the good old days of before the war. For millions in 1945 'before the war' had been bad old days and there was a general determination to create a better and fairer society. In 1942 Sir William Beveridge had outlined in what became known as the Beveridge Report how the five 'giant' enemies of human happiness—'want, disease, squalor, ignorance and idleness' —could be overcome by what he termed the 'welfare state'. The Labour Party looked the best-equipped party to give reality to these ideas.

The new Prime Minister, Clement Attlee, was, Churchill sneered, 'a modest little man with a great deal to be modest about'. In fact, despite his insignificant appearance and boring public speeches, Attlee is likely to be remembered as one of Britain's more effective peace-time prime ministers. An admirable chairman, he held together a government of gifted and temperamental ministers for six testing years and supervised the carrying out of several major social reforms which amounted to a social revolution.

In 1946 the National Insurance Act was passed, in 1948 the Industrial Injuries Act, the National Assistance Act and the National Health Service Act. The total effect of these was to give effective insurance compensation, medical services and a minimum living standard to all citizens whatever their private means. The cost was borne by contributions from employers and the employed and from rates and taxes. The most controversial of these Acts was the National Health Service Act. The opposition of the medical profession to the idea of a national service was at first strong and only after many compromises and skilful negotiations by Aneurin Bevan, the Minister of Health, was this opposition overcome. Thus in these years a major step was taken in the creation of a 'welfare state' in Britain. The state took on vastly increased responsibilities for the care of all its citizens. The huge cost was borne by greatly increased direct progressive taxation—i.e. the richer you were the more taxes you paid. Labour also nationalized the Bank of England, the mines, the railways, civil aviation, the docks, the road transport industry and electricity and gas supplies. Of these, only road transport was later denationalized.

Otherwise, it was a bleak age. To push up exports, the government had to demand austerity at home. Wartime rationing of food, fuel and clothing was continued, travellers could only take a limited amount of currency abroad. An abnormally cold winter in 1946-47 caused an acute fuel shortage, so badly disrupting industry that over a million men were temporarily out of work.

In foreign affairs, Labour pursued a policy of withdrawal, sometimes voluntarily, as in India and Pakistan, sometimes because the situation was getting beyond them as in Greece and Palestine. Yet its military spending remained high, about 7% of the national income each year, in keeping with its 'Big Power' attitude. An atomic bomb was secretly built and conscription reintroduced in 1948.

The continuation of rationing, failures in housing and in an ambitious but ill-informed scheme to grow groundnuts in East Africa lessened Labour's popularity. Attlee was only able to win the 1950 election with a majority of six and in another election the following year the Conservatives under Churchill got back to power.

They remained in power for the next thirteen years. Though they appeared a mainly middle and upper class party with close connections with big business while Labour seemed a mainly working-class party with close connections with the trade unions, there was a broad measure of agreement on major issues between both parties. The Conservatives made no attempt to dismantle the welfare state, indeed government spending on the social services rose rapidly while they were in power. They accepted, at least after Churchill's resignation as Prime Minister in 1955, that the former imperial possessions must be allowed their independence and they continued spending much the same proportion of the national income on defence. Like Labour, they went for sustained export-led growth and, like Labour, achieved it only in fits and starts. The chief characteristic of Conservative Britain in the 1950's was steadily increasing prosperity. Britain shared the generally improving living standards of the Western world even if her rate of growth was slower than that of her neighbours. Another major achievement was in housing where there was a rapid expansion of both private and council building. The Conservatives won the 1955 election with an increased majority. Eden succeeded Churchill as Prime Minister

Churchill, on the right, with Eden in 1954 while he was still Prime Minister and Eden Foreign Secretary

but, after the fiasco of the Suez affair (see page 283) retired on grounds of ill-health. He was succeeded by Macmillan, who displayed an image of unruffled confidence to the world, and fighting the 1959 election on the slogan 'You've never had it so good' won a majority of 107 seats, over a Labour party bitterly split on foreign policy. Then the Conservatives faltered. Split themselves over the future of the African colonies, unable to master the balance of payments problem (see page 317), refused entry to the Common Market and shaken by the Profumo scandal of 1963 (when the Minister for War lied to Parliament about his association with a prostitute who was also associated with a Russian security officer) their popularity ebbed away. Macmillan retired in 1963, to be succeeded, after a most bitter and public struggle for the leadership, by Sir Alec Douglas-Home, who lost the election of 1964 to Labour.

The new Prime Minister was Harold Wilson, a clever economist who had already had ministerial experience in Attlee's government. A skilful speaker with a gift for the telling phrase, he seemed to understand and to offer realistic solutions to the country's economic maladies and to have a genuine vision of modern technology harnessed in the service of a fairer society.

Having inherited a serious balance of payments crisis, however, the Wilson government never got on top of its economic problems. Growth continued but in the usual stop-go rhythm. The Labour application to join the Common Market was also turned down. The white settlers of Rhodesia, whose unilateral declaration of independence was described by Wilson in 1965 as a rebellion which would be over in a matter of days rather than weeks, were still successfully defying him when he lost office five years later. Attempts by his senior ministers to regulate Britain's increasingly chaotic industrial relations were stifled at birth by the trade union leadership whose influence within the party grew. In the 1970 election, the Conservatives, now led by Edward Heath, regained power.

In the meantime, tragedy had come to Northern Ireland (Ulster) which was still part of the United Kingdom. The majority of the population of Ulster was originally of Scots stock and fanatically Protestant in contrast to that of the rest of Ireland (Eire) which, overwhelmingly Catholic, had fought for and gained its independence from Britain earlier in the century (see page 109). Within the boundaries of Northern Ireland however was a sizable Catholic minority. The two religious groups had always regarded each other with fear and suspicion, and in 1968 violence erupted between them. Catholics demanded greater civil rights, Protestants assumed that what they were really after was the union of Ulster with the rest of Catholic Ireland. The province moved towards civil war. Demonstrations were followed by bombings, torture, mutilations and murder. The British government imprisoned many suspects without trial and sent the British army in to keep the warring groups apart. The violence continued. In 1972 the separate Ulster parliament was suspended and the British government took over the 'direct rule' of the province. By the end of 1973, over 200 British soldiers and nearly 700 civilians had been killed in Ulster since the 1968 troubles began. A tiny glimmer of hope appeared in an otherwise hopeless situation when some Protestants and Catholics agreed to co-operate in a new Northern Irish Assembly. Early in 1974, however, a Protestant strike caused this Assembly to be dissolved. Violence spread to England and a solution seemed as far away as ever.

Ulster barbarism, November 1971 : a teenage Catholic girl, tied to a lamp-post, her head shaven, tarred and feathered by the woman's section of the I.R.A. Her crime : to go out with one of the British soldiers serving in Ulster to prevent civil war

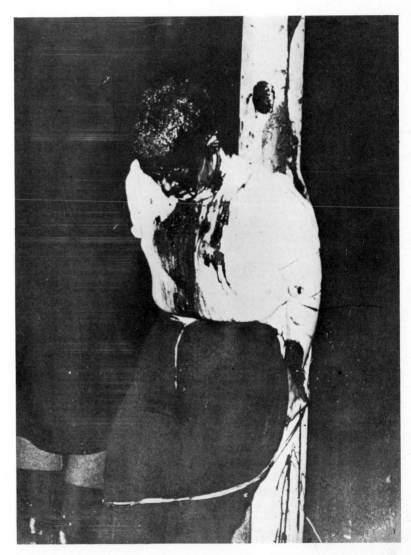

Sunday Express

DECEMBER 9

"Careful, Mr. Acheson! <u>We</u> turned up for that funeral, but our own came first."

London Express Service.

In 1962 Dean Acheson, a distinguished American expert on foreign affairs, remarked that 'Great Britain had lost an Empire but had not yet found a rôle'. The remark caused 'sharp official resentment in London', *The Times* noted, mainly because it was true and reflected what many Britons knew but did not like to admit. By the early seventies there was considerable confusion and lack of confidence in British political life. Politicians were less respected than at any time in modern political history. While foreign commentators and even British government advisers were predicting that at the end of the century the British because of their economic difficulties were likely to have become 'the peasants' of Western Europe, neither of the major parties was able to pursue consistent policies let alone put forward a political philosophy relevant to the needs of the time. Heath took Britain belatedly into the Common Market but there was little popular enthusiasm and some noisy opposition. 1972 and 1973 were characterized by bitter industrial strife and the appearance of a considerable protest vote with the Welsh and Scots turning towards local 'nationalist' parties and the English turning to the Liberal party which had been virtually dead for twenty years. In the election of February 1974, the nationalists and Liberals won

Cummings of the Sunday Express *was not impressed by Acheson's remarks*

enough seats to hold the balance of power between the two major parties whose share of the vote dropped sharply. Writing in 1970, W. Laqueur, a distinguished contemporary historian of the University of Tel Aviv in Israel, concluded: 'In its long history the British people had emerged from worse calamities with flags flying, revealing sterling qualities of calmness and determination. But then it is always far easier to mobilize a nation to cope with a sudden emergency than with a long-drawn out creeping crisis such as it faced after World War II.'

France When the Nazi attack overwhelmed the Third Republic in 1940, General de Gaulle had fled to England and established himself as leader of the 'Free French'. As France was liberated in 1944, he won the support of the Resistance leaders and set up a provisional government of National Union. Leaders of the Vichy government and other collaborators were put on trial. About 800 were executed (de Gaulle changed Pétain's death sentence to life imprisonment) and another 40,000 imprisoned for pro-Nazi activities. There were many acts of private vengeance and in some areas lynch law operated.

As in Europe generally, the French economic situation was terrible. The war had devastated a considerable area, work had to be found for two million prisoners of war and forced labour workers returning from Germany, there was an acute shortage of vital supplies and soaring inflation. In the face of these challenges, French political leaders were deeply divided. A new constitution, similar to

Lynch law or justice? A French collaborator shot near Rennes soon after the liberation, August 1944

that of the Third Republic, brought the Fourth Republic into existence in 1946. Though a majority of Frenchmen had voted in favour of it, two important minority groups strongly opposed it. De Gaulle believed that the French Head of State must have greater powers, more like the President of the U.S.A. and refused to have anything to do with the new parliament. He resigned as Acting President and created a new party, the Rassemblement du People Français (the Union of the French People) or R.P.F. The Communists also had no intention of making the new constitution work. Since they were well-organized, helped from Moscow, and could rely on the support of one Frenchman in five, they were a force to be reckoned with. In between were fourteen separate parties hoping to make the system work. Short-lived coalitions were inevitable and between 1944 and 1958 there were 27 separate governments. In 1947 and 1948 there was massive industrial strife, inspired by the Communist-dominated C.G.T., which culminated in October 1948 in a six-week strike of 300,000 miners. Democracy in France seemed close to overthrow either by the Communists or by the anti-Communist R.P.F. Marshall Aid, however, brought some economic recovery and the Fourth Republic survived though it failed to gain strength. First the war in Indo-China and then the Algerian revolt further divided the various parties. By 1958 effective parliamentary rule was becoming impossible and popular disillusion with the Fourth Republic was almost total. On 13 May 1958, army officers and white settlers seized power in Algeria and a few days later occupied Corsica without opposition. Since the government of Pflimlin could not be sure that the army and police in France would obey its orders, a civil war or a military dictatorship seemed near. On 15 May de Gaulle, who had been living in retirement since 1955, made his move. 'France has for the last twelve years followed a disastrous road,' he wrote in a statement to the Press. 'In the past, the country from its very depths entrusted me with the task of leading it to salvation. Today, with new ordeals facing it, let the country know that I am ready to assume the powers of the Republic.' The country rallied to him. On 1 June 1958 he was voted the authority to rule by decree for six months and the Fourth Republic effectively came to an end.

The dramatic failures of the Fourth Republic can easily conceal its achievements which were less dramatic but nonetheless real and lasting. Thanks to men like Monnet and a well-trained and forward-looking civil service, French economic planning was skilful and effective. The foundations of the sustained growth of the French economy in the sixties and early seventies were laid during the Fourth Republic. There was also achievement in foreign policy.

The Communist newspaper L'Humanité protests against de Gaulle being asked to form a government by President Coty. It would, it argued, mean fascism for France

`l'Humanité`

ORGANE CENTRAL DU PARTI COMMUNISTE FRANÇAIS

FONDATEUR : JEAN JAURÈS	DIRECTEUR : ÉTIENNE FAJON	6, boulevard Poissonnière - PARIS-9^e
RÉDACTEUR EN CHEF (1926-37) VAILLANT-COUTURIER		Tél. : PRO 13-05 Nlle série - N° 4273
DIRECTEUR (1918-1958) MARCEL CACHIN	**VENDREDI 30 MAI 1958**	et la suite 150° jour de l'année

PRIX : 20 FR.

DEFI A LA REPUBLIQUE

COTY PRESSENT DE GAULLE POUR FORMER LE GOUVERNEMENT

Pour barrer la route au fascisme, pour un gouvernement de défense républicaine, rassemblez-vous et unissez-vous

MULTIPLIEZ LES ARRETS DE TRAVAIL

MANIFESTEZ CONTRE LA DICTATURE

Exigez des députés, par d'innombrables messages, qu'ils disent NON A DE GAULLE

Appel du Comité Central du Parti Communiste Français

Travailleurs et républicains,
Français et Françaises,
L'Assemblée nationale vient d'être mise en demeure par le président de la République d'avoir à désigner le général de Gaulle comme chef du gouvernement. Cette sommation intolérable intervient au moment où la majorité républicaine a manifesté par des votes massifs et répétés son refus de précipiter le pays dans les aventures et la guerre civile et où, à travers toute la France, s'affirme avec une puissance imposante la volonté de défense républicaine.

Au même instant, à Alger, le général factieux Massu, au nom des rebelles que de Gaulle n'a cessé de couvrir et d'encourager, se déclare prêt à porter celui-ci au pouvoir par la force armée sur une décision de sa part.

Ni le Parlement ni le pays n'acceptent ce double défi.

Le pays veut que soient respectées la loi et la Constitution et que soit formé sans délai un gouvernement s'appuyant sur la majorité républicaine de l'Assemblée.

Le Comité Central du Parti Communiste Français appelle solennellement tous les travailleurs, tous les démocrates, tous les patriotes à se tenir en permanence en état d'alerte, à riposter énergiquement à toute tentative fasciste, à multiplier les comités de défense républicaine, à manifester sous toutes les formes leur résolution d'épargner à la France les hontes et les malheurs d'une dictature militaire et fasciste.

Le Comité Central lance un pressant appel à la jeunesse, aux jeunes travailleurs ouvriers et paysans ainsi qu'aux étudiants, il s'adresse aux soldats, aviateurs et marins pour qu'ils accomplissent fidèlement leur devoir civique et qu'ils agissent partout aux côtés du peuple contre les hommes de la guerre civile et du fascisme, pour la défense de la République.

Par son unité, par sa détermination, le peuple de France brisera le complot des généraux factieux et des hommes de la guerre civile.

Vive la République !
Vive la France !

Le Comité Central du Parti Communiste Français.
Le 29 mai 1958.

Grèves et manifestations
dans les grandes villes de France

- ◆ 30.000 à Lyon
- ◆ 15.000 à Marseille
- ◆ 10.000 à Grenoble
- ◆ 10.000 à Nice
- ◆ 8.000 à Boulogne
- ◆ 5.000 à Lorient
- ◆ 4.000 à Rouen
- ◆ 3.000 au Havre

PAGE 9
Marcel PRENANT
professeur à la Sorbonne :
« En signe de dégoût pour la dictature du sabre »
☆
André WURMSER :
« Contre quoi luttent les instituteurs »
Nos inf. p. 3, 4, 5, 8 et 9

25.000 métallos de la Sambre arrêtent le travail cet après-midi

---

— MOI aussi, je suis pour la République

GRÈVE DES ENSEIGNANTS
aujourd'hui dans toute la France

- ◆ Les parents d'élèves sont invités à ne pas envoyer leurs enfants à l'école.
- ◆ L'Union Nationale des Étudiants de France s'associe aux manifestations prévues.
- ◆ Dans les lycées, les comités de vigilance appellent les élèves à ne pas se présenter en classe.
- ◆ La C.G.T., l'Union des syndicats parisiens et l'U.G.F.F. apportent leur appui total au mouvement.
- ◆ Meeting cet après-midi à la Bourse du Travail (à l'appel de la F.E.N., du S.G.E.N. et de l'U.N.E.F.).

(Voir en page 8.)

UNE DÉCLARATION DU GROUPE PARLEMENTAIRE COMMUNISTE

ALORS qu'avant-hier les partisans du général de Gaulle ne groupaient que 166 voix à l'Assemblée contre 408, le président de la République, dans un message qui confirme une véritable sommation à la représentation nationale, prétend imposer au Parlement le général de Gaulle comme chef du gouvernement.

Cette mise en demeure se produit au moment où, à Alger, le général Massu et les hommes de la rébellion et de la guerre civile proposent de porter au pouvoir, par un coup de force militaire, le général de Gaulle sur une simple décision de celui-ci, ce qui plongerait inévitablement le pays dans la guerre civile.

Il n'y a pas d'autres moyens pour écarter la guerre civile, que d'imposer le respect de la loi et de la Constitution.

Le groupe communiste condamne toute tentative d'association des hommes de la rébellion avec quelque forme que ce soit, le général de Gaulle.

Ce qu'il faut au pays, c'est un gouvernement de défense républicaine créé selon les principes de la Constitution et s'appuyant sur la majorité républicaine de l'Assemblée et le pays républicain.

LE GROUPE PARLEMENTAIRE DU PARTI COMMUNISTE FRANÇAIS.

Without French vision between 1948 and 1957 it is hard to see how the Common Market could have come into existence.

De Gaulle was one of the more extraordinary figures of post-war Europe. Born in 1890, he followed a military career, first making his name in 1934 with a book on military strategy which predicted that future wars would be won by mobile tank divisions supported by aircraft and infantry. His superiors in the French army ignored him but when the Germans proved him right in 1940, he was made, in the last days of the defeated Third Republic, Under Secretary for War. When, having escaped to Britain, he created the Free French forces and was condemned to death in his absence by the Vichy government, he came to represent for millions of Frenchmen the honour of his country. A proud and prickly patriot whom Roosevelt and Churchill found very troublesome to deal with, he insisted that he spoke for France and should be treated with the consideration due to a major head of state. Then came his triumphant return to liberated Paris in 1944. He was a national hero and though busily involved in politics between 1944 and 1955, he managed to seem somehow above normal politics which gave him great popular appeal in 1958.

His political outlook belonged more to the eighteenth than to the twentieth century. 'France is nothing without greatness,' he began his memoirs and his major concern was to revive France's greatness and his countrymen's confidence in it. With a superb command of the French language, in both writing and speech, with a formidable appearance and personality, a majestic calm and almost arrogant detachment from the daily concerns of ordinary people, his style was kingly in a way no other politicians could rival. His use of mass media was masterly. Seldom were there airport press conferences or hurried kerb-side interviews, instead carefully stage-managed press conferences or television broadcasts at a time and place chosen by de Gaulle himself. He was, however, no dictator. Though he insisted on much greater power for himself as President in the new consitution of 1958, he also believed that ultimate power rested with the French people and he gave them frequent chances to express their will both in elections and in referenda on major issues. As soon as a majority was clearly against him, as the referendum on local government reforms showed in 1969, he resigned, apparently without regret.

From 1958 to 1966 he was most successful. His chief interest was foreign policy. He solved the Algerian problem in 1962 by a solution which few expected of him and no-one else had dared try— the granting of complete independence to Algeria. Some diehard white settlers tried to assassinate him for such 'treachery' but the overwhelming majority of Frenchmen supported him. He also

De Gaulle's second veto on Britain's E.E.C. application. Mr Wilson, Britain's Prime Minister from 1964 to 1970, is coming down the ladder

DAILY EXPRESS   NOVEMBER 28, 1967

granted the rest of France's African colonies their independence. He was extremely suspicious of the U.S.A., whose military and economic influence he considered a serious obstacle to the real independence of France and of Western Europe. He pulled French forces out of NATO and forced the transfer of its headquarters from Paris to Brussels in 1966. He created a special relationship with West Germany and tried, with partial success, to do the same with Russia, China and Eastern Europe. High priority was given to strengthening the French armed forces, especially in nuclear weapons.

He had originally opposed the Common Market. The hopes of men like Monnet and Spaak for a Europe united both economically and politically conflicted with his pride in France's national character and his confidence in her future greatness. He was interested in the E.E.C. only in so far as it directly helped France. For these reasons he opposed Britain's entry, which he believed would have the effect of increasing America's and lessening France's influence in Europe. Moreover in 1965 he completely disrupted the E.E.C. by boycotting its proceedings after disagreements over the Common Agricultural Policy. Though differences between the six were patched up and, after his retirement, Britain, Denmark and Ireland were able to join the E.E.C., it did not function with its former sense of common purpose.

At home, the first eight years of Gaullism saw rapidly increasing prosperity, a new spirit of dynamism in French industry and a stability which most Frenchmen seemed to appreciate. However, a considerable minority had always opposed Gaullism and when after economic stagnation in 1964 and 1965 the government was neither able to reduce unemployment nor hold down prices, discontent grew.

In 1967, sociology students and teachers went on strike at the new university of Nanterre in the suburbs of Paris. This strike was in protest partly against the state of education in France but also against the capitalist system in general and the American involvement in the Vietnam War in particular. In February 1968, students of the University of Paris also went on strike and violent police action was met by counter-violence. A week of street-fighting in Paris temporarily united the trade unions, socialist and Communist parties behind the students, and on 17 May ten million workers went on strike against the government. De Gaulle was disgusted. 'Reform yes—bed-messing no,' he commented. After some hesitation, he made a secret visit to the French army in Germany to make sure that he could rely on its support. He then called a general election. By this time, the unity between students and workers had begun to weaken, the Communist-dominated unions,

May 1968. French students and riot police face to face in the Boulevard St Michel in Paris' Latin Quarter

having won considerable wage rises, were not interested in total revolution. At the election, the Gaullists won a considerable majority which they were able to preserve, despite the General's retirement in 1969. He was succeeded as President by Georges Pompidou, under whose less majestic and less controversial guidance rapid economic growth and stability were maintained.

West Germany In 1945 the Germans exchanged one dictator, Hitler, for four, the commanders of the four Allied occupation zones. 'We come as conquerors but not as oppressors', the Allied posters read in 1945, and the German people were left in no doubt that conquered they were. The Allies immediate aim was to thoroughly de-nazify German society. While the top surviving Nazis were brought to trial at Nuremberg (see page 284), investigation and trials of lower ranks were carried out in all the zones. The Americans were much the most thorough. They issued 12 million questionnaires and found 930,000 persons guilty. For more serious crimes, they charged 170,000, against 22,000 by the British, 18,000 by the Russians and 17,000 by the French. While awaiting trial, prominent Nazis, especially ex-members of the S.S., were thrown into internment camps, conditions in some of which were little better than in their own infamous concentration camps. The de-nazification programme was strongly criticized by many Germans including respected anti-Nazis. It was, they said, inefficient. Many Nazis were going free while many innocent were punished. Moreover it was preventing the vital task of reconstruction from beginning.

As the Cold War developed, the Western Allies decided to treat West Germany as a partner in the struggle to prevent communism spreading further across Europe and to re-establish parliamentary democracy. By the Basic Law of 1949, the Federal Republic of (West) Germany came into existence. It was a federation of ten states, or *Länder*, each with its own state parliament. The central, federal parliament was in Bonn. The Basic Law included measures which made difficult the existence of the small parties which had so weakened the Weimar Republic. Only three were to have any real power in the next twenty-five years; the C.D.U. (Christian Democrats), mainly Catholic and moderately conservative, the S.P.D. (Socialists), and the F.D.P. (Free Democrats), a rather curious combination of North German conservatives and South German liberals. The C.D.U. and S.P.D. were much larger than the F.D.P. but were so evenly balanced that the F.D.P. often held the parliamentary balance and took part in coalitions.

After the first, 1949, election to the federal parliament, the C.D.U. and F.D.P. combined to form a coalition, with Adenauer, the C.D.U. leader, as the first Chancellor. Adenauer was already 73. He was born in Cologne, the son of a minor and poorly paid civil servant, and spent almost all his life in the Rhineland. Having married into an influential family, he moved into local government work. In 1909 he became Chief Clerk of Cologne and, continuing there during the war, became Lord Mayor of Cologne in 1917. He stayed in this office until 1933, a hard-working, competent and respected civic leader. However, he made no secret of his disapproval

of the Nazis, and, while they were in power, lived a kind of 'twilight existence' in retirement outside Cologne. For a short time in 1944 he was in a concentration camp. Since he was one of the few Christian Democrats of the Weimar Republic to survive Nazism without collaborating with it, he quickly became the leader of the C.D.U. and, from 1949, of West Germany.

He was to remain in power for another fourteen years, until he was 87. His aim was to give his shattered country much needed stability and in this he was thoroughly successful. Leaving economic affairs in the more than competent hands of Erhard, his chief interest was foreign affairs. Towards Russia and East Germany he followed a policy of continual hostility, and took West Germany into NATO in 1955. Towards Western Europe he worked for reconciliation and played a major part in the creation of the Common Market. He became a close friend of de Gaulle and between 1958 and 1962 Germany and France co-operated more closely than ever in their history. During the 1950's he had considerable popular backing. In both 1953 and 1957 the C.D.U. won a clear majority over the S.D.P. and F.D.P. combined. His popularity then began to de-

Adenauer, on the left, with de Gaulle and Dr Lübke. The photo was taken in 1962 when Adenauer was Chancellor and Lübke President of the F.D.R.

cline.Adenauer had grown up in imperial Germany and he had held considerable authority for much of his long life. He was not at heart a convinced democrat and he tended to treat his colleagues as a father might treat his children. He seemed to be growing too old for the job yet was unready to retire gracefully. In home affairs, moreover, his government was increasingly criticized for lack of achievement, most of all in education.

Various pressures brought about his resignation in 1963 but the fortunes of the C.D.U. did not improve. After a confused period of manoeuvring between the parties, the S.P.D. was able to secure the support of the F.D.P. in 1969 and formed a government with Brandt as Chancellor.

Brandt began a new era in the history of West Germany. He was born Herbert Frahm in Lübeck in 1913. His mother was a nineteen year old shop assistant. He never knew his father and was brought up by his grandfather, a lorry driver and convinced socialist. An active Social Democrat in 1933, he joined the anti-Nazi underground movement in Berlin, his code-name being 'Willy Brandt'. As the Nazi purge of socialists and communists grew fiercer, he fled to Norway, married a Norwegian girl, and joined the Norwegian resistance movement against the occupying German forces before fleeing again, this time to Sweden. He had intended to remain in Norway after the war and became a Norwegian citizen. However, working as a journalist, he was sent to cover the Nuremberg trials. He then decided to return to his old country and came to live in

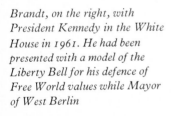

Brandt, on the right, with President Kennedy in the White House in 1961. He had been presented with a model of the Liberty Bell for his defence of Free World values while Mayor of West Berlin

Berlin. Ten years later, in 1957, he became Mayor of West Berlin, and, in the front line of the Cold War, was always in the headlines. His courageous opposition to Russian and East German threats made him an international figure and a hero to many West Germans. By 1961, he had become the leader of the S.P.D.

Once Chancellor, Brandt intended to reverse the West German foreign policy of the previous twenty years and to make his country a 'bridge' between East and West, building up confidence and co-operation rather than suspicion and division. He negotiated treaties with Poland, East Germany, Czechoslovakia and Russia, the main results of which were to recognize the boundaries of Germany and Eastern Europe as they had existed since World War II, to begin normal diplomatic contact with Russia and East Germany and to develop trade links between West Germany and communist Europe. This 'Ostpolitik' or Eastern policy seemed to be an important step towards ending the Cold War in Europe, and in 1971 he was awarded the Nobel Prize for his services to the cause of peace. The following year, in the 1972 elections, his countrymen gave him their vote of confidence, returning him to power with an improved majority.

Italy The years immediately after World War II were surprisingly lacking in bitterness. Much of the credit for this must go to De Gasperi, leader of the mainly Catholic Christian Democratic party, who formed a coalition government of the three main parties— Christian Democrats, Socialists and Communists—and to Togliatti, the Communist leader who was Minister of Justice. There were no war-crimes trials and little attempt to punish the Fascists. A new democratic constitution was drawn up, women were given the vote and, following a special referendum in 1946, the monarchy was abolished. The Communist Party had in 1945 the support of one in every five Italians, and while the Communists were successfully taking over Eastern Europe, a Communist attempt to seize power in Italy seemed likely. De Gasperi, however, was a shrewd politician. He ended his coalition with the Communists in 1947 and playing on Cold War fears won a large majority in the election of 1948. For the next fifteen years the Christian Democrats were the largest party in parliament. As in France, however, small parties and therefore coalitions continued in Italian politics.

Despite rapid economic growth and Italy's membership of the E.E.C., the record of the Christian Democrat dominated governments was not impressive, especially in social matters. Italy's educational and welfare services were the most backward in west Europe, her legal and civil services the most inefficient. Left-wing parties like the Socialists and Communists won greater popular support and in 1963 a Left-Centre coalition of Socialists and Christian Democrats took office. It began well. Funds were made available for education on an unprecedented scale and a five-year plan produced to get to grips with Italy's regional problem. In the bandit-infested islands of Sardinia and Sicily, the police were told to take sterner measures. By the time of the 1968 election however too many of its promises seemed unfulfilled and while the Communists increased their vote, the Christian Democrats were split by the issue of whether divorce should be made legal. Governments again rapidly came and went. Though two major reforms were passed in 1970—the creation of new regional governments and the legalization of divorce—the political situation continued to deteriorate. With no stable government, a civil service of quite extraordinary incompetence, a legal system on the verge of breakdown and inadequate educational and welfare services, extremist groups began to emerge. In the early seventies, a neo-Fascist party, the M.S.I., grew in popularity and acts of sabotage and murder were committed by political extremists.

The Common Market

Since the Middle Ages some men had dreamed of a united Europe, but until 1945, such dreams seemed hopeless fantasies, so great were the differences between the various European nations. In the years after 1945, however, unity seemed to many European statesmen no longer an idle fantasy. Without unity, they argued, Europe had no future. There was plenty of evidence to support their view. A disunited Europe, a Europe of quarrelling nation-states, had brought upon herself the catastrophe of World War I. If that was not enough, she had repeated the performance, with worse results, in World War II. Consequently where as in 1900 Europe had been the dominant continent of the world, the centre of civilization, of artistic and scientific achievement, in 1945 she was dwarfed by Russia to the east and by America to the west. No one European nation could compete with these super-powers. Real independence, real revival and long-term security for Europe seemed possible only by reaching a degree of unity never before attempted.

A major step towards this unity and the most important development in Western Europe since World War II was the foundation of the European Economic Community (E.E.C. or Common Market) which came into existence on 1 January 1958. At first it consisted of six nations (Belgium, the Netherlands, Luxembourg, France, W. Germany and Italy). Fifteen years later, on 1 January 1973, it was joined by three more, Britain, Denmark and Ireland. The formation of the E.E.C. brought a large measure of economic unification to Western Europe and the intention of its founders was that at

least some social and political unification should follow economic unity in the not-too-distant future.

The beginning of the E.E.C. The signing of the Treaty of Rome, March 25th, 1957

The architect of the E.E.C. was Jean Monnet, a Frenchman with unusual international experience and outlook. He was born in Cognac in 1888, the son of a brandy seller. At the age of 18 he was selling brandy in the Canadian backwoods; at the age of 28, unable because of kidney trouble to join the French army, he was organizing a joint supply system for the French and British forces on the Western Front. So impressive was his performance that when the war ended the two governments asked him to become Deputy Secretary of the League of Nations. Resigning from this post in 1923, he concentrated on building up the family firm for a couple of years and then he began an international business career which took him to the U.S.A., Poland, Sweden and China, and included an elopement with the wife of an Italian diplomat whom he married after a Moscow divorce. When World War II broke out, he was acting on behalf of France in the U.S.A. and, as a result of his World War I experience, was contacted by the British. After the Fall of France, he worked in America as a British civil servant but kept in close contact with de Gaulle. He returned to France at the end of the war full of ideas—most of which were put into effect—first for getting France moving (see page 313) and secondly for getting Europe (including Britain) more united. Monnet was a man who knew the right questions to ask. He also asked them of the right people and went to endless trouble to be sure he had got the right answers. Moreover, once he had made up his mind he got things done.

Where European unity was concerned, the vital first move was made in 1950. Monnet hoped that France and Germany might be persuaded to combine their coal and steel production along with other European nations. If agreement could be reached over such important industries as coal and steel then full economic co-operation might follow. He wrote a memo to this effect and sent it to the French Prime Minister, who, it seems, put it away in a drawer and forgot about it. Monnet, however, got a friend to give a copy to Schuman, the French Foreign Minister. Schuman took it away with him to his country home one weekend. He telephoned Monnet on his return to Paris. 'I agree, for me it is decided,' he declared. He then persuaded the French government to back the scheme which was presented to the world at a press conference on 9 May 1950. What Schuman suggested (and became known as the Schuman Plan) was the European Coal and Steel Community (ECSC).

The idea of the ECSC was that coal and steel should be regarded as a single industry and be run by a single international authority. The Belgians, Dutch, Luxemburgers, Italians and Germans leapt at the idea, the British, though Monnet did his best to win them over, did not. 'Our people,' said Harold Macmillan, 'will not hand on to any supranational authority the right to close our pits and our steel works,' and both the Labour and Conservative parties believed that Britain should stay outside. The Headquarters of E.C.S.C. was established at Luxemburg with Monnet as the first president. It became effective in 1953 and was soon very successful. Between 1953 and 1958 steel products in the six participating countries rose by 42%. Equally important, European economic co-operation was seen to work for the benefit of all concerned.

Monnet's next scheme was a common defence organization, a European Defence Community, which he hoped would lead eventually from a unified defence policy to complete political unity. During the Korean War, when Western Europe felt dangerously dependent on the U.S.A. for defence against Russia, the scheme came close to success. However Britain again stood aloof. France also decided against it, so nothing came of the scheme.

The E.C.S.C. however was going from strength to strength and in 1955 the Foreign Ministers of its member-states met at Messina in Sicily to consider further economic co-operation. Here the decision was taken to set up the E.E.C. (the European Economic Community or Common Market) and a European nuclear research institute, Euratom. The Belgian Foreign Minister, Paul-Henri Spaak, was appointed to work out the details of the new organisations. Meanwhile Monnet, having resigned from the E.C.S.C., formed 'the Action Committee for the United States of Europe,' the aim of which was to persuade European governments and public opinion that European unity was now both possible and desirable. Spaak's committee planning the organization of the E.E.C. worked fast and well. Spaak, a convinced European, was a brilliant chairman who used his lively temper to keep things moving. One day when progress was slow because the experts could not agree on tariff duties for bananas, he walked out saying 'I give you two hours. If it's not settled by then, I shall call the press in and announce that Europe won't be built after all, because we can't agree about bananas.' On his return, banana tariffs were no longer a problem. After only ten months, the details of both the Common Market and Euratom were complete. On 25 March 1957 the fundamental agreement of the Common Market, the Treaty of Rome, was signed by the Six to take effect from January 1958.

The founder-members are, so the Treaty begins, 'determined to establish the foundation of an even closer union among the European people and are decided to ensure the economic and social progress of their countries by common action in eliminating the barriers which divide Europe.' This was to be done by 'establishing a Common Market and progressively approximating the economic policies of member states to promote throughout the Community a harmonious development of economic activities . . . and closer relations between member states.' Any other European nation could join. Colonies or former colonies of member-states would become associates. All customs barriers between the member-states were to be abolished by 1967, by which time also the organization of the Common Market, E.C.S.C. and Euratom was to be merged. The day-to-day running of the E.E.C. was to be the responsibility of a nine-man commission, the first President of which was Hallstein, former head of the German foreign office. Major decisions were to be taken by the Council of Ministers (i.e. the relevant ministers, usually Foreign, Finance or Agriculture of the member-states). There was also to be a European parliament, the members of which were to be nominated by the parliaments of the member-states. The parliament was to be a centre for discussion, not decision. Finally there was to be a Court of Justice to make sure that the terms of the Treaty were fairly carried out.

At Messina, the Foreign Ministers of the Six specifically asked Britain to join the Common Market. As with the E.C.S.C. Britain refused. The Conservative government argued that to join the E.E.C. would weaken the Commonwealth and that the Treaty of Rome meant too great a surrender of Britain's economic independence. It also suspected though it did not say so publicly that the E.E.C. like the E.D.C. would fail.

It did not. In its first phase it was a triumphant success. Ten years after the signing of the Treaty of Rome the volume of trade between the Six had risen fourfold. Between 1955 and 1964 their combined G.N.P. rose by 63% and by 1964 the E.E.C. was the world's greatest trading power, the biggest exporter and buyer of raw materials. Moreover harmony among the members was generally preserved. A common agricultural policy (C.A.P.) was agreed which protected European farming (though setting high prices and causing near-scandalous surpluses in some products such as butter), and customs union was established eighteen months ahead of schedule. Britain then tried to establish a rival trading area, the aim of which was to encourage trade among the member-states without subordinating the various governments to a central policy-making body as in the E.E.C. This European Free Trade Association (E.F.T.A.) consisted of Austria, Britain, Denmark, Norway, Portugal, Sweden and Switzerland and was established by the Stockholm Convention in 1959. It became effective on 1 January 1960. Though trade within the Association rapidly expanded and the members prospered (though less fast than the Six) E.F.T.A. never had the dynamism nor the geographical compactness of the E.E.C. For Britain it was a salvage operation and not a particularly successful one. Hardly had the Stockholm Convention been signed than Britain let it be known that she was reconsidering the whole of her policy towards the E.E.C. 'I believe we made a mistake,' admitted Selwyn Lloyd the British Foreign Secretary in January 1960, 'in not taking part in the negotiations which led to the formation of the E.C.S.C.,' and in the summer of 1961 Macmillan's Conservative government formally applied for membership.

There were a number of reasons for Britain's change of mind. For one thing, Britain was becoming more aware of her economic weakness in comparison with the Six and her military weakness in comparison with the U.S.A. and U.S.S.R. The Suez adventure of 1956 had made clear that she could no longer play at being a big power without American support and that to count on Britain keeping a 'special relationship' with the U.S.A. for any length of time would be rash. Moreover the Commonwealth, instead of becoming more united and a powerful support for Britain throughout the world, showed signs of disintegration.

Intensive negotiations then took place in Brussels between the British delegates led by Heath and the Six. After fourteen months hard work all the major difficulties including agriculture and Commonwealth trade had been settled. Then de Gaulle, who had been President of France since 1958, stepped in. Without consulting any of the other members, he decided to veto Britain's application. Britain, he argued, was not yet ready to join. 'England', he told a press conference in January 1963, 'is insular, maritime, bound by her trade, her markets, her supplies to countries that are very diverse and very far away . . . How can England as she lives, as she produces, as she trades, be incorporated in the Common Market.' His real objections he left unsaid. His main one was Britain's close links with America especially in nuclear matters. He believed that through Britain America might extend her influence over the E.E.C. and he was determined that a French-dominated E.E.C. should act as independently of the U.S.A. as of the U.S.S.R. Moreover, he feared that Britain once inside the Common Market would be strong enough to prevent France from dominating the E.E.C. in her national self-interest.

Since the constitution of the E.E.C. gave each member powers of veto and since the Common Market without France was unthinkable, the other five members could do little to counter his actions. Consequently between 1963 and 1967 de Gaulle came close to wrecking the E.E.C. His first (1963) veto of Britain caused the first serious crisis. A second, more serious still, occurred in 1965 over agriculture. The French walked out and virtually boycotted the E.E.C. for seven months. Then in 1967 Britain, this time through Wilson's Labour government, re-applied for membership. The E.E.C. welcomed the application and suggested that negotiations should begin. De Gaulle soon held another press conference. To bring Britain into the Common Market as it then stood was, he said, 'an impossibility'. To his 1963 objection he added another. Britain's economy must recover before membership could be considered. None of the other members agreed with him but there was nothing they could do. As long as de Gaulle remained in power in France, the E.E.C. stagnated. Not until after his resignation in 1969 could the question of British entry along with many other urgent problems be given active consideration.

As soon as Pompidou, de Gaulle's successor, made it clear that he did not object in principle to Britain's membership, the Six invited Britain to re-apply. Negotiations began in 1970 with first Barber and then Rippon as the chief negotiators for Heath's Con-

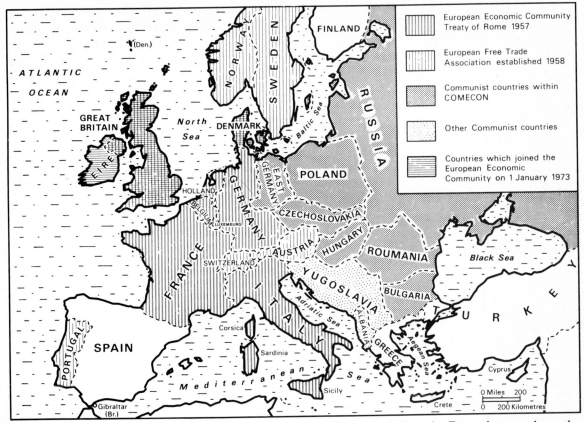

	European Economic Community Treaty of Rome 1957
	European Free Trade Association established 1958
	Communist countries within COMECON
	Other Communist countries
	Countries which joined the European Economic Community on 1 January 1973

Map 19 The economic groupings of postwar Europe

servative government. The negotiations in Brussels were brought to a successful end in June 1971. The British government then had to persuade Parliament to accept the negotiated terms. It had a hard fight on its hands. Public opinion in Britain was deeply divided. At no time during the negotiations nor during the debates in parliament did public opinion polls indicate a clear majority in favour of joining. Opponents to the E.E.C., who were convinced that they represented the majority of the British people, demanded a referendum. This demand the government refused. Opposition to entry was led by about two-thirds of the Labour Party (despite Wilson's 1967 application for membership), the trade union movement and a right-wing group of Conservatives. Their chief concerns were independence and prices. Like de Gaulle, they disliked the supranational powers of the Common Market Commission. To join the Market would mean a major sacrifice of Britain's independence. They also disliked the Common Agricultural Policy of the E.E.C. To join would mean higher prices immediately while economic benefits would appear, if at all, only after some years. The government, which was also backed by the small Liberal Party and about one-third of the Labour Party, argued that economically Britain had no choice but to join. In comparison with the E.E.C. her economic performance since 1958 had been poor and, as long as she continued outside this dynamic trading bloc, it was unlikely to improve. Furthermore, her position as a world power which had declined rapidly since World War II would decline faster still when the E.E.C. achieved greater political as well as economic unity. Inside the E.E.C., the economic opportunities were immense and Britain, with her long experience of successful democracy and world-wide diplomacy, would have a major role to play in the further political and social development of Europe. In an early debate on the general principle of joining, the government, with the help of a group of Labour members, gained a comfortable majority of 112. From then on, the pro-Marketeers of the Labour Party came under heavy pressure to toe the official party line which was that the terms negotiated by the Conservatives were not good enough and that there must be re-negotiation before Britain could join. As the debate on details of the negotiated terms continued, the pro-market majority was whittled away. On one critical night, it fell to single figures. Finally, however, Heath got the bill through, with the result that on 1 January 1973 Britain became a member of the E.E.C. Three other countries also negotiated successfully to join—Denmark, Ireland and Norway. Unlike Britain, however, these three held referenda before actually joining. The Danish and Irish people voted in favour, the Norwegians against. From 1 January 1973 therefore, the Six became Nine.

'What design should we seek for the new Europe?' said Heath in a speech made at the Egmont Palace in Brussels in January 1972 after the ceremonial signing of the Treaty of Accession to the Common Market by Britain, Denmark and Ireland. 'It must be a Europe which is strong and confident within itself. A Europe in which we shall be working for the progressive relaxation and elimination of East-West tensions. A Europe conscious of the interests of its friends and partners. A Europe alive to its great responsibilities in the common struggle of humanity for a better life. Thus this ceremony marks an end and a beginning. An end to the divisions which have stricken Europe for centuries. A beginning of another stage in the construction of a new and greater united Europe. This is the task for our generation in Europe.'

It is a noble vision but one on which both Britain and her Common Market partners proceeded to turn their backs.

Epilogue
Whither Europe?

by the General Editors

1973 marks the end of the long stretch of history covered by the present series of Portraits of Europe. The first of these began in 300 A.D. when the concept of Europe as we know it did not exist. The Roman empire was the dominant entity and that was approaching collapse. It was a time when innumerable people lived close to fear, hunger and death and looked out upon a black and threatening world; a time when barbarism was rapidly taking the place of Roman order. This last portrait ends when signs of a new barbarism are abundantly evident. The news media resound with grim reports of bombing, assassination, riots and open war, in Europe no less than elsewhere. At the far extremities of what was once the Roman empire Northern Ireland and Israel are torn with bloody conflicts, conflicts of rights which have their roots far back in the past and admit of no quick or easy solutions. Nor is this all. Throughout Europe, except where communist dictatorships exist, we see governments which, in varying degrees, are either unstable or operating on such slender majorities that they are partially impotent. We feel the threat of economic collapse in the wake of financial inflation and industrial unrest. These agonies and confusions may not be new but they are on a scale greater than at any time before and they tend to develop with a speed which defies many of the most manful attempts to deal with them. As we look round it is easy to become utterly pessimistic and defeatist about our future and indeed some people question whether Europe can recover from the present turmoils and troubles. There is even a fear that some Western European democracies may yield to dictatorships either of the right or left.

All through her history Europe has been plagued by waves of political instability, by economic disasters, by outrageous outbreaks of violence, and in every generation thoughtful men have wondered if there existed any hope for the future. There have been many times when existing institutions have been deeply shaken, some even destroyed, others changed almost beyond recognition; when the outlook was black and threatening. The present time is one of these. Now political and economic instability of very grave dimensions exist, presenting some dangers and problems which are new, some which are made worse by the characteristics of the modern world. For instance, the speed of communication means that news of calamities reaches us everywhere with a swiftness and exactness which is both terrifying and numbing; in a flash the present moment becomes of world-wide interest. The use for destructive purposes of applied science can put weapons of death within the grasp of any terrorist. The scale is thereby altered: a man could always commit a murder with a sword and dagger, but he can do much worse with a time-bomb.

Yet in spite of all the conflicts and problems which have torn Europe over the past one thousand four hundred years she has produced a civilisation which scientifically has brought us into the nuclear age, and in creative achievement in art, music, literature and government has reached heights perhaps unparalleled in the experience of man. Nevertheless it must be remembered that these high achievements were seldom born of tranquillity but over and over again out of turmoil and tension. The restless energy of Europeans has ceaselessly forged a way into the future while always drawing heavily on the past. In the Middle Ages, for instance, Europe reached back eagerly to the lost world of Greece and Rome.

The scientists of the nuclear age possess their own startling originality and make great strides forward into new achievement, but behind them lies a long inheritance of experiment and inventive genius going back to Newton, Galileo and Copernicus on which in each succeeding generation many of them have built.

In 1973 Britain took her first steps towards sharing in a new-found European unity. One of the greatest architects of the European Community, M. Monnet, welcomed Britain's entry because he believed that she could bring to Europe a well-tried skill in creating and developing institutions. These institutions, parliament, the law courts, the universities among them, have their origins far back in the Middle Ages, but they have been continually modified by generations of men who have contrived to carry the past into the future.

The challenge which lies ahead of us now as we face the uncertainties of the future is whether we can yet again delve into the immensely rich and varied inheritance which we possess for the inspiration to grapple with the daunting problems of our present situation. The challenge is to the Europeans of this generation. To despair is usually all too easy; the truly great in European history have been those who refused to give up hope.

Bibliography

There are two disadvantages which writers of twentieth century history face compared with historians concerned with more remote centuries. For one thing the written, visual and aural material at his disposal is both vast and incomplete. A vital document may not yet be published or if published may not have become noticed amongst the tons of published material which Europe produces year by year. Secondly, the events which he describes are too close for him to be detached and balanced about them. No European can yet do justice to the catastrophe of World War II since his outlook will be coloured either directly by its effects on his own life or indirectly by its effects on his relatives and friends. Only when time enough has passed to allow for the collection and the sifting of all the relevant records and for the cooling of present passions and prejudices will fair accounts and balanced explanations of recent events become possible.

The books listed here have been chosen for those whose appetite for recent European history has been whetted by their reading of this volume.

I. GENERAL

(a) *Europe*

Wide-ranging, thought-provoking and difficult is J. M. Roberts, *Europe 1880–1945*. More straightforward are C. E. Black and E. C. Helmreich, *Twentieth Century Europe* and Gordon A. Craig, *Europe since 1914*.

A luxurious combination of superb illustrations and interesting essays in the Thames and Hudson *Twentieth Century* ed. Alan Bullock. Also lavishly illustrated and containing a text of a generally high standard contributed by historians from all over the world is Purnell's *History of the Twentieth Century* ed. A. J. P. Taylor and J. M. Roberts (6 volumes, first published in magazine form).

In a war-torn century, Field Marshal Montgomery's *History of Warfare* is a good descriptive survey and A. Buchan's *War in Society* an excellent analysis.

(b) *Individual countries*

Austria 1918–72: E. Barker

Belgium: Vernon Mallinson

Britain—A. Marwick: *Britain in the Century of Total War* is good on the social and economic aspects while A. J. P. Taylor: *English History 1914–1945* is more directly political.

Central and Eastern Europe—A. W. Palmer: *The Lands Between*

Czechoslovakia: A Short History: J. F. N. Bradley

A History of Modern France Vol. 3: A. Cobban

Bibliography

France: Douglas Johnson
Germany since 1789: Golo Mann
Modern German History: R. Flenley
The Story of Modern Greece: C. M. Woodhouse
Hungary: C. A. Macartney
A Short History of Ireland: J. C. Beckett
Italy: D. Mack Smith
A Short History of Modern Italy: ed. Hearder and Waley
The Netherlands: Max Schuchart
A History of Modern Poland: H. Roos
Portugal: J. B. Trend
A History of Russia: N. Riasonovsky
The Scandinavians: D. Connery
Modern Spain: A. Hennessy (History Association Pamphlet G59)
Yugoslavia: S. Pavlowitch

II. INDIVIDUAL CHAPTERS

1. *Europe in 1900*

A comprehensive and scholarly American survey is C. J. H. Hayes, *A Generation of Materialism 1871–1900*. For racialism and nationalism, see L. C. Snyder, *Race: A History of Ethnic Theories* and E. Kedourie, *Nationalism*. For particular countries, see above.

2. *World War I*

Of many accounts of the war, C. Falls, *World War I* and B. H. Liddell-Hart, *A History of World War I* can be particularly recommended and two interesting studies of the military leadership are C. Barnett, *The Swordbearers* and A. Clark, *The Donkeys*. Good examples of the superb literature inspired by the war are Robert Graves: *Goodbye to All That* and Edmund Blunden, *The Undertones of War*.

In *The Break-Up of the Habsburg Empire*, Z. A. B. Zeman deals with an important but often neglected aspect of the war.

3. *The Versailles Settlement and the League of Nations*

The standard work is in 6 volumes; H. W. V. Temperley, *A History of the Peace Conference in Paris*. H. Nicolson's *Peacemaking 1919* is a vivid and beautifully written account by a member of the British delegation. J. M. Keynes expressed his own critical viewpoint in *The Economic Consequences of the Peace* and was himself criticized by E. Mantoux in *The Carthaginian Peace or the Economic Consequences of Mr. Keynes*. International diplomacy and the League of Nations are well covered by G. M. Gathorne-Hardy, *A Short History of International Affairs 1920–39* and F. P. Walker, *A History of the League of Nations*.

4. *Russia 1900–41*

One of the major works of twentieth century history is E. H. Carr's *History of Soviet Russia* (7 vols. so far and still continuing). See also G. Vernadsky, *The Russian Revolution 1917–21*; A. B. Ulam, *Lenin and the Bolsheviks*; I. Deutscher, *Stalin* and a study of Stalin's purges, *The Great Terror* by R. Conquest. A brilliant imaginative reconstruction of the 'purges' is A. Koestler's *Darkness at Noon* and an eye-witness account of the whole period can be found in the memoirs of Ilya Ehrenburg, *Men, Years, Life.*

5. *The European Economy to 1939*

Two good introductions are W. A. Lewis, *Economic Survey 1919–39* and W. Ashworth, *A Short History of the International Economy.* *The Great Crash* by J. K. Galbraith is an intelligent and witty analysis of the Wall Street Crash of 1929. On Keynes, see R. Harrod, *The Life of J. M. Keynes.* Note also the works by Keynes and Mantoux listed under Chapter 3.

6. *Liberal Democratic Europe*

C. L. Mowat, *Britain between the Wars* and A. Werth, *The Twilight of France 1933–40* are solidly researched studies; R. Graves and A. Hodge, *The Long Weekend: A Social History of Great Britain 1918–39* and M. Muggeridge, *The Thirties* are more impressionistic.

7. *Dictators*

The best introduction is E. Wiskemann, *Europe of the Dictators.* For Mussolini, see L. Fermi, *Mussolini* and for Hitler, *Hitler, A Study in Tyranny* by Alan Bullock. Essential reading is Hitler's *Mein Kampf* which is available in translation. An excellent work by an outstanding journalist is *The Rise and Fall of the Third Reich* by W. Shirer. Compulsive reading is provided by the memoirs of one of Hitler's most able and intelligent ministers, A. Speer, *Inside the Third Reich.* For the Spanish Civil War, see Hugh Thomas, *The Spanish Civil War.* G. Brenan, *The Spanish Labyrinth* is also of great merit.

8. *Hitler's Triumphs*

On the origins of the war, see M. Gilbert, *The Roots of Appeasement.* See also the stimulating but controversial *The Origins of World War II* by A. J. P. Taylor. The most comprehensive coverage of the war years is provided by G. Wright, *The Ordeal of Total War* and the purely military aspects of these years are well covered by B. H. Liddell-Hart, *A History of World War II*, and by Brigadier Peter Young, *A Short History of World War II.* Of the innumerable books dealing with particular events of the years 1933–42, W. Shirer's description of the Fall of France in *The Collapse of the Third Republic* and P. Fleming's *Operation Sea-Lion* which deals with the Battle of Britain can be particularly recommended.

9. *The Defeat of Nazism*

See the works of Liddell-Hart, Young and Wright listed in Chapter 8. See also F. W. Deakin, *The Brutal Friendship, Mussolini, Hitler and the Fall of Italian Fascism,* and A. Clark, *The Russian-German Conflict 1941–45.* The most serious plot against Hitler is the subject of *The Conspiracy against Hitler in the Twilight War* by H. Deutsch and the final struggle for Berlin in 1945 of *The Last Battle* by C. Ryan.

10. *Science and Technology*

The five volume *History of Technology* ed. C. Singer is a useful work of reference as is J. D. Bernal's *Science in History* vols. 3 and 4 despite the author's rather obtrusive left-wing outlook. *Science and Society* by H. and S. Rose is a useful analysis of the increasing governmental involvement in the development of science and technology. More exciting and better written than most books on modern science is *The Double Helix,* Watson and Crick's account of the epoch-making discovery of DNA.

Even better written is *Silent Spring* by Rachel Carson which more than any one book began the environmental movement in the USA. Of the many books on the threat which population and industrial growth pose to our environment, the best is P. and A. Ehrlich's *Population, Resources, Environment.*

11. *Urbanization*

Wide-ranging, profound and full of relevant illustrations is L. Mumford, *The City in History.* See also P. Hall, *World Cities.* Though written for geographers, the Oxford *Problem Regions of Europe* series contains some useful information on twentieth century urban growth; particularly relevant are G. R. P. Lawrence, *Randstad Holland* and I. B. Thompson, *The Paris Basin.* Dramatically illustrated and despairing in tone is *Venice: How Long?* by G. Obici.

12. *Religion and Psychology*

The best introduction of modern Christian history is A. Vidler, *The Church in the Age of Revolution.* A more detailed survey is the five volume *Christianity in a Revolutionary Age* by K. Latourette. For the Church in Communist Russia, see M. Bourdeaux, *The Opium of the People.* A clear introduction to some of the major figures in modern psychology, including Freud and Pavlov, is George A. Miller, *Psychology, The Science of Mental Life.* See also J. A. C. Brown, *Freud and the Post-Freudians.* Of Freud's own books, *The Psychopathology of Everyday Life* is probably the best starting point.

13. *The Arts*

H. Read, *A Concise History of Modern Painting* and *A Concise History of Modern Sculpture*, N. Pevsner, *Pioneers of Modern Design* and J. M. Richards, *An Introduction to Modern Architecture.* Ed. P. Cowie, *A Concise History of the Cinema.* Useful works of reference are the Penguin *Dictionaries of Architecture* and of *Art and Artists* and the Oxford *Companions to Music* and *to the Theatre.*

14. *Women and Society*

Good examples of the feminist viewpoint are S. de Beauvoir, *The Second Sex*, G. Greer, *The Female Eunuch* and K. Millett, *Sexual Politics.* See also C. Thomas, *Women in Nazi Germany* and D. Brown (ed.), *Women in the Soviet Union.*

15. *Europe and the World*

K. Irvine, *The Rise of the Coloured Races* is a thorough survey and R. Oliver and J. Fage, *A Short History of Africa* and Ian Thomson, *The Rise of Modern Asia* can also be recommended. A lively account of the decline of the largest of the empires is C. Cross, *The Fall of the British Empire.* On the vexed subject of South Africa, the two volume Oxford *History of South Africa* ed. Wilson and Thomson is excellent; and on the equally vexed subject of Vietnam, R. B. Smith achieves an admirable impartiality in his *Vietnam and the West.* Interesting biographies are B. R. Nanda, *Mahatma Gandhi*, M. Edwardes, *Nehru*, and J. Lacouture, *Ho Chi Minh.*
Postwar Europe

Chapters 16, 17 and 18 are all concerned in different ways with Europe since 1945 and the following books are relevant to them all:

W. Laqueur, *Europe Since Hitler*

A. J. Mays, *Europe since 1939*

The Times *History of Our Times* is a well-illustrated 'coffee-table' production with some interesting individual essays.

16. *The Cold War*

There are two good analyses; H. Seton-Watson, *Neither War nor Peace* and P. Seabury, *The Rise and Decline of the Cold War*. Events in Russia are covered in R. W. Pethybridge, *A History of Postwar Russia* and E. Crankshaw, *Khrushchev; A Career*. A superb if awful imaginative portrait of the last years of Stalin's rule is to be found in *The First Circle* by the novelist A. Solzhenitsyn. See also P. Auty, *Tito* and *The Czech Crisis* 1968 ed. R. R. James.

17. *European Society since the War*

A comprehensive if rather dull survey is C. Waterlow and A. Evans, *Europe 1945–70*. More lively and more impressionistic views of particular countries are J. Ardagh, *The New French Revolution*, P. Nichols, *Italia, Italia* and B. Levin's savage yet witty commentary of Britain in the 1960's, *The Pendulum Years*.

18. *Western Europe and the EEC*

W. O. Henderson, *The Genesis of the Common Market*

R. Mayne, *The Recovery of Europe*

F. Roy Willis, *France Germany and the New Europe*

For individual countries, see the titles listed at the beginning of this bibliography. See also B. Crozier, *De Gaulle* (2 vols.), A Crawley, *The Rise of Western Germany 1945–72*, S. Hughes, *The Rise and Fall of Modern Italy*, T. O. Lloyd, *Empire to Welfare State: English History 1906–67*.

Historical Atlas: The best is M. Gilbert, *Recent History Atlas 1860–1960*.

Further Reading: A more detailed bibliography is provided by the Historical Association pamphlet, *Modern European History: A Select Bibliography, 1789–1945*.

Index

Abercrombie Plan, 206–9
Abortion, 251, 253, 254, 255, 257, 258
Abyssinia, 20, 61, 148, 263–4
Adenauer, German Chancellor, 337–9
Adler, Afred, 228, 229
African independence, 265–7, 275–7
African National Congress (A.N.C.), 278–80
Afrikaners, 263, 267, 277–82
Agriculture, 65–7, 93–4, 99–101, 309, 311, 314, 316, 318
Albania, 34, 153, 166, 292
Alexandra, Empress of Russia, 79–81
Algeria, 24, 265, 267, 332, 334
Anti-Comintern Pact, 150
Anti-semitism, 106, 131, 133, 142–4, 228; see also Jews
ANZACS, Australian and New Zealand troops, 33, 34
Apartheid, 279–81
Arab-Israeli war, 283, 307
Architecture, 237–9
Asquith, Herbert, British Prime Minister, 45, 107
Atomic bomb, 186, 193–5, 304; see also Nuclear weapons
Attlee, Clement, British Prime Minister, 325–6
Auchinleck, General, 172
Austria, 60, 72, 150–1, 291, 305, 311, 345; see also Austrian Empire; Habsburg Empire
Austrian Empire (Austria-Hungary), 21, 24–6, 30–1, 33, 47, 52, 53; see also Habsburg Empire
Automobile industry, 65, 190, 199, 212, 310, 316

Badoglio, Marshall, 177
Balance of payments, 317, 328
Baldwin, Stanley, 64, 110, 111–13, 14
Balkans, 21–2, 24–5, 33–4, 166–9
Ballet, Russian, 241
Baltic-White Sea canal, 99
Barth, Swiss theologian, 218
Basques, 144, 145
Battle of Britain, 161, 164–6
Battle of the Atlantic, 171
Battle of the Bulge, 185
B.B.C. (British Broadcasting Company), 113, 196

Beckett, Samuel, English writer, 242, 243
Becquerel, French scientist, 188
B.E.F., British Expeditionary Force, 28, 29, 159
Behaviourism, 225–7
Behrens, Peter, German architect, 237, 239
Bela Kun, Hungarian leader, 62, 105
Belgian Congo, 265, 275–7
Belgium, Belgians, 18–19, 26, 28–9, 53, 63, 66, 71, 105, 159, 161–2, 311, 319 341
Benes, Czech President, 153, 294–5
Beria, Russian Chief of Secret Police, 302
Berlin, 127, 133, 217, 288, 289, 291, 305–7; in World War II, 164, 186, 187, 292
Biological sciences, 189, 197–8
Bismarck, German Chancellor, 19, 24
Blériot, 189
'blitzkrieg', 159
Blockade, in World War I, 41–3, 48
Blum, Leon, French Socialist, 118
Boccioni, Umberto, sculptor, 235
Bohr, Niels, Danish scientist, 188, 194
Bolsheviks, 85–98, 221–2, 226, 255, 290
Bonhoeffer, German pastor, 218
Boris III, King of Bulgaria, 105
Brancusi, Roumanian sculptor, 235, 237
Brandt, West German Chancellor, 307, 339–40
Braque, French artist, 230, 231, 233
Brecht, Bertolt, German writer, 243–4
Brest-Litovsk, Treaty of, 44, 90, 97
Brezhnev, Russian leader, 304, 307
Briand, French Prime Minister, 116, 117
Britain, 11, 17–18, 68, 202, 217, 266–72, 277; in World War I, 24–9, 32–4, 38–45, 47; between the Wars, 50–2, 55–6, 58, 60–1, 67, 69, 71–2, 73–4, 103, 105–14, 148, 150–6, 196; in World War II, 156–65, 171, 172–3, 176–80, 185, 186, 190, 192–3, 264–8; after World War II, 239, 249, 251, 259, 278, 283, 286–7, 288–93, 308–11, 313, 317–8, 319, 320, 324, 325–31, 341, 343–8

Brusilov, Russian general, 36–7, 38–9, 79
Brussels, 346, 347
Bulgaria, 27, 47, 52, 60, 284, 291, 293, 296, 311, 320

Camus, Albert, French writer, 242–3
Capitalism, 78, 84–5, 262, 290
Carol II, King of Roumania, 105
Catholic Church, 16, 20, 125, 139, 144–6, 218–21, 224
Chamberlain, Neville, British Prime Minister, 114, 152–3, 155–6, 158
CHEKA, 92
Chicherin, Russian Minister for Foreign Affairs, 63
China, 13, 58, 61, 307, 311, 312
Churchill, Winston, 27, 33–4, 72, 112, 158–66, 169, 177, 194, 284, 285, 287, 290, 292–3, 308, 325, 326, 327, 334
Cinema, 245–7, 248
City of London, 67–8
Civil War in Russia, 90–3
Clemenceau, French Prime Minister, 45–6, 50–1, 58, 115–6
Collectivization, 98–101, 302
Cold War, 277, Chapter 16, 337, 340
Collins, Michael, 110
Comecon, 288, 291, 307, 311–2
Cominform, 288, 295, 296
Common Market (E.E.C.), 209, 307, 314, 318, 328, 330, 333, 335, 338, 340, 341–8
Communism, Communists, principles of, 84–5, 104–5, 224, 262; Russian, 62, 78, 94, 98, 101–3, 222, 226, 254–8; European, 105, 106, 114, 118–9, 122, 127, 130, 133–4, 287, 291–8, 323, 332, 340–1; Cold War and, 284–6, 290
Communist International, 62
Computer industry, 190, 197
Congo, Belgian, see Belgian Congo
Connolly, Sinn Fein leader, 108
Concentration camps, 182–3
Concrete, 199
Conservative Party, British, 107, 110, 113–4, 325–31
Contraception, 220, 221, 253, 254, 255
Convoy-system, 42, 176–7
Crick, British scientist, 199, 200–1

Cuba crisis, 307
Cubist style of painting, 230–2, 233–4
Currency difficulties, 71–2, 117, 309,
  314–5; *see also* Sterling
Czechoslovakia, 52, 53, 55, 63, 103, 106,
  151–3, 291, 294–5, 296, 298–300,
  310, 311–2

Daladier, French Prime Minister, 152–3
D'Annunzio, Italian poet, 120
Danzig, 53, 59, 60, 153
Dawes Plan, 56–7
D-Day, 179–80
De Gasperi, Italian leader, 340
De Gaulle, French leader, 166, 307,
  323, 331–6, *338*, 346
Denmark, 53, 65–6, 335, 341, 345, 348
Depression, Great, 57, 71, *74*, 113–4,
  129–30, 132; *see also* Slump, Great
D'Espérey, Franchet, French
  commander, 47
De Valera, President of the Irish Free
  State, 109–10
Dictators, 64, 105–6, Chapter 7
Dirac, scientist, 189
D.N.A. (deoxyribonucleic acid), 199,
  200–1
Doenitz, Admiral, 133, 176–7, 186
Dogger Bank, Battle of the, 41
Dreadnoughts, 25, 39
Drummond, Sir Eric, Secretary-
  General of the League of Nations,
  59
Dual Alliance, 21, 24
Dual Monarchy, 21
Dubcek, Czech Prime Minister,
  299–300
Dublin, 108, 109
Dunkirk, 161–3
Dzerzhinsky, chief of CHEKA, 92

East Germany (German Democratic
  Republic), 291, 296, 297–8, 309, 311,
  314
Ebert, German Socialist, 127–8
Economy, European, 49, 64–5, 67–75,
  93–4, 99, 308–19, 324, 327–8,
  341–3, 345
E.C.S.C. (European Coal and Steel
  Community), 343–4
Ecumenical Movement, 223–4
Eden, Anthony, British Prime Minister,
  283, 327–8
E.E.C., *see* Common Market
E.F.T.A. (European Free Trade
  Association), 345
Egypt, 166, 171, 263, 265, 283
Einstein, scientist, 188, 189, 193
Eisenhower, General, 177, 292, 305
Eisenstein, Russian film producer, 245,
  247, 302

El Alamein, battle of, 172–4
Electrons and electronics, 189, 190,
  196–7
Eliot, T. S., poet, 242
Empire, British, 11, 15, 260–3, 264–72,
  277–82
Empires, European, 11, 14, 15, 19, 20,
  21, 22, 52, 55, 58, 131, 261, 263,
  265–8, 272–5, 275–7
Engels, Friedrich, 254–5, 257, 258
Erhard, Dr., 316, 338
Erzberger, German politician, 129
Europe, Eastern, 284–307, 311–2, 320,
  324
Europoort, 209–10
Existentialism, 242–3

Facta, Italian Prime Minister, 123
Falkenhayn, German Commander, 36
Fascists, Fascism, 104–5, 106, 121–7,
  138–40, 219, 224, 237
F.D.R. (Federal Republic of West
  Germany), *see* West Germany
Feminism, 250–1, 252, 257, 258–9
Ferdinand, Archduke, 25
Finland, 53, 60, 239, 284, 291, 305
Five-Year Plans, Russian, 74, 99
Florey, British scientist, 193
Foch, French general, 46–7
Fourteen Points, Wilson's, 50, 59, 261
France, 15, 16–17, 202; in World War
  I, 24, 26–9, 32, 36–9, 44–7, 49, 50–3,
  58; between the Wars, 63, 71, 73–4,
  103, 105, 114–9, 152–6; World War
  II, 156–64, 186, 217; after World
  War II, 212–4, 250, 259, 283, 291,
  309, 310, 311, 311–6, 319, 320, 331–6,
  341
Franco, General, Spanish dictator, 105,
  144–6, 150, 234
Franz Josef, Habsburg Emperor, 21
French, Sir John, 28, 29, 38
Freud, German psychologist, 225,
  227–9, 232

Gagarin, Yuri, Russian astronaut, 303
Gallipoli, 33–4, 158
Gandhi, Mohandas, 269–71
Geddes, Patrick, 202–4, 205
Geddes Committee, 107
General Strike of 1926, 72, 106, 112–13
Genotdel, 255, 257
German Democratic Republic
  (D.D.R.), *see* East Germany
Germany, 18, 19, 24–6, 65–7, 69, 72,
  73, 116–7, 202, 209, 250, 252, 260,
  284, 288–90, 311, 313; World War I,
  27–49; Versailles Treaty and, 52–7,
  60, 63, 155; Nazi, 74, 102–3, 118,
  127–42, 147–54, 252–4; World War
  II, 156–71, Chapter 9

G.N.P. (Gross National Product), 192
  310–11, 313, 345
Goebbels, Nazi Minister of Propaganda,
  138–9, 142–3, 186
Goering, leading Nazi, 135, 162, 186,
  284
Gold Standard, 72, 76
Gomulka, Polish communist, 298
Goode, Sir William, 70, 71
Gottwald, Czech communist, 294–5,
  298
Gravier, French geographer, 213
Greece, 27, 60, 166, 286–7, 291, 292
Green Belt, 206, 211
Gropius, German architect, 237, 239
Groza, Roumanian communist, 294
Guderian, German commander, 159,
  161

Haber, German scientist, 69
Habsburg Empire, 15, 21, 47, 52, 70;
  see also Austrian Empire
Haig, British commander, 38, 45, 46
Haile Selassie, Emperor of Abyssinia,
  61, 62
Haliliva, Zarial, Russian feminist, 257
Heath, Edward, British Prime Minister,
  328, 346, 348
Heisenberg, scientist, 189
Henlein, leader of Sudeten German
  Party, 151, 152
Hepworth, Barbara, English sculptor,
  235
Himmler, Head of the S.S., 142–3,
  181, 186
Hindenburg, German Commander and
  President, 30, 134, 136, 137
Hiroshima, 194–5
Hitler, 56, 60, 61, 103, 105, 119, 120,
  127–42, 145, 146, Chapter 8, 175,
  183–4, 185–7, 219, 252–3, 267
Hoare-Laval Pact, 61
Ho Chi Minh, Vietnamese Communist,
  273–5
Holland, 18–19, 66, 105, 159, 162, 196,
  209–10, 283, 310, 313, 319, 341
Hopkins, F. G., British biochemist, 198
Housing, 204, 205–6, 211, 212, 320, 327
Hungary, 52, 53, 62, 72, 105, 224, 284,
  292, 294, 296, 298, 309, 312, 320;
  see also Austrian Empire; Habsburg
  Empire

India, 263, 264, 268–72
Indochina, 262, 265, 267, 272–5, 304,
  332
Indonesia, 263, 264, 265
Industry, 13, 18, 22, 65–7, 73–4, 189–90,
  199, 202, 209, 213, 215, 313, 316,
  317
Inflation, 72, *128*, 129, 309, 331

International Labour Organization
(I.L.O.), 59
I.R.A. (Irish Republican Army), 109–10
Ireland, 17, 108–10, 218, 341, 348
Iron Curtain, 224, 291–2, 322
Isolationism, 57–8, 71, 308
Israel, 282–3, 307
Italy, 20, 27, 52, 53, 58, 63, 73, 214–6,
239, 250, 251, 291, 292, 309, 310, 311,
313, 316–7, 340–1; under
Mussolini, 60–1, 120–7, 146, 148;
in World War II, 166, 177, 186,
284

Japan, 11–12, 27, 58, 60, 61, 102–3,
150, 169, 172, 180, 186, 194, 262,
264, 273–4, 307, 313, 317
Jellicoe, British naval commander, 41
Jews, 182–3, 186, 219, 282–3; *see also*
Anti-semitism
Joffre, French Commander-in-chief, 29,
36, 44
John XXIII, Pope, 219–20, 224
Joliot, French scientist, 193
Jung, psychologist, 228, 229
Jutland, battle of, 41

Kafka, Czech writer, 242
Kaiser, German, *see* William II
Kapp, German Nationalist, 128
Kemal, Mustapha, later Ataturk,
Turkish commander, 34, 62, 263
Kennedy, John, American President,
305, 307, *339*
Kerensky, Russian socialist leader, 44,
82, 86, 87–8
Keynes, J. M., British economist,
50, 56–7, 68, 76–7, 107, 114
Kirov, Russian Party member, 102
Kluck, von, German commander in
World War I, 29
Kolchak, Admiral, White Russian, 90–2
Kollontai, Alexandra, 255
Korea, North and South, 304, 343
Kornilov, Russian general, 86
Kosygin, Russian leader, 304
Kronstadt rebellion, 1921, 92–3, 94, 97
Krushchev, Russian First Secretary,
298, 302–3, 304, 305, 318
Kuznetsstroi, new city in W. Siberia,
99

Labour Party, British, 107, 110–11,
113–4, 323, 324ff, 347–8
Lateran Treaties, 1929, 125–6
Lausanne, Treaty of, 52, 62
Laval, French Foreign Secretary, 61
Lawrence, T. E., 27
Le Corbusier, French architect, 237,
239
League of Nations, 53, 55, 57, 58, 59–61,
62, 64, 72, 103, 148, 261, 264

lebensraum, 132–3, 147–8
Lend-lease, 308
Lenin (Vladimir Ulyanov), 44, 52, 63,
78, 82–90, 93–6, 98, 104, 221, 222,
226, 262
Leningrad, 169, 211; *see also* Petrograd
Liberal Party, British, 107–10, 330–1,
347
Liebknecht, German Communist, 127
Literature, 242–3
Lithuania, 53, 60, 153, 154
Lloyd George, David, British Prime
Minister, 42, *44*, 45, 50–2, 76, 107–10,
158
Locarno Pact, 63–4, 129
London, 164–6, 183, 187, 205–9, 211,
232
Ludendorff, German commander, 30,
45, 46–7, 48, 131
Luftwaffe (German Air Force), 157,
162, 164
Lusitania, Cunard liner, 41, 43
Luxemburg, 105, 341, 343
Luxemburg, Rosa, German
communist, 127

MacDonald, Ramsay, British Prime
Minister, 11, 113–4
Madrid, 144, 145, 146
Maginot line, 156, 159, *160*
Magnitostroi, Russian city in Urals,
99, *100*
Malta, 172
Manchuria, 60
Mandates, 55, 261–2
Manstein, German Commander, 159,
161
'March on Rome', 123
Marconi, 196
Marne, battle of the, 28–9, 48
Marshall Plan, 288, 291, 309–11, 332
Marx, Karl, 78, 84–5, 104–5, 255
Masaryk, Jan, Czech Foreign Minister,
294–5
Mass media, 248, 251
Mass production, 190, 199, 203
Matteotti, Italian Socialist, 123–4, 126
May Committee, 113–4
Medvedev, Zhores, Russian scientist,
304
Mein Kampf, 132–3, 147
Metaxas, Greek dictator, 105
Michael, King of Roumania, 293–4
Midway, battle of, 172
Mikolajczyk, Polish anti-communist,
293
Molotov, Russian Foreign Minister, 310
Moltke, General von, 26, 28–9, 36, 48
Mondrian, abstract artist, 232
Monnet, Jean, 313, 332, 335, 342–4
Montgomery, General, 172–4

Moore, Henry, English sculptor, 235,
236–7
Morocco, 24, 144, 263, 265
Moscow, 87, 88, 99, 168, 210–12
Mosley, Sir Oswald, 106
Mountbatten, Earl, 271, *272*
Muller, Ludwig, Bishop of Third Reich,
140
Mumford, Lewis, 204–5, 239
Munich, 131, 152–3
Music, 239–41, 249
Mussolini, Italian dictator, 60–1, 104,
105, 120–7, 145, 146, 147, 148, 156,
162, 177, 186, 219, 264

Nagy, Ferenc, Hungarian Prime
Minister, 294
Nagy, Imre, 298
Nasser, Egyptian President, 283
Nationalism, 15, 19, 21, 23, 27, 124–5,
260–8, 272–3
Nation-states, 15, 58, 341
N.A.T.O., (North Atlantic Treaty
Organization), 288, 291, 307, 335,
338
Naval powers, 18, 25, 39–43, 48
Nazis, 104, 129–44, Chapter 8,
Chapter 9, 180–4, 217, 228, 239,
252–4, 284, 290, 337–8
Nazi-Soviet Pact, 103, 154, 290, 291
Nehru, Motilal and Jawaharal, 268–9,
271
N.E.P. (New Economic Policy), 94–5
Netherlands, *see* Holland
Neuilly, Treaty of, 52
New Towns, 206–9, 320
Nicholas II, Russian Tsar, 22, 26, 44,
79–82
Niemöller, Pastor, 140
Nivelle, French commander, 44
Nixon, American President, *306*, 307
Norman, Montague, Governor of the
Bank of England, 107
North Africa, in World War II, 166,
171, 172–4
Norway, 19, 157–8, 159, 250, 259, 345,
348; *see also* Scandinavia
Novotny, Czech Prime Minister, 298–9
Nuclear energy, 190, 193–5, 344;
weapons, 201, 290, 304, 307;
disarmament, 201, 307
Nuremberg Laws, 142
Nuremberg Trials, 284, 337, 339

O.E.E.C. (Organisation for European
Economic Co-operation), 310
Office, modern, 203, 206
'Operation Barbarossa', 166
Operation Overlord, 177–80
Oppenheimer, American scientist,
193–5

'Ostpolitik', 340

Painting, 230–5
Pakistan, 271
Palestine, 282–3
Paris, 230, 232, 233, 241, 273; in World War I, 28–9; in World War II, 161, 162, 179–80, 334; after World War II, 205, 209, 211, 212–4
Parliamentary democracy, 104, 105, 106, 114, 127
Partition of Ireland, 109–10
Party purges, in U.S.S.R., 101–2
Patton, American General, 179
Pauker, Anna, Roumanian Communist, 294, 296
Paul VI, Pope, 219–21
Paulus, von, German commander, 174–5, *176*
Pavlov, Russian psychologist, 225–7
Pearl Harbor, 169, *170*
Penicillin, 190, 192–3
Pétain, French general, 36, 44, 162, 331
Petkov, Bulgarian anti-communist, 293
Peter, King of the Serbs, 31, 34
Peter, King of Yugoslavia, 295
Petrograd (St. Petersburg), 82, 85–8, 96, 97; *see also* Leningrad
'Phoney War', 156–9
Picasso, Pablo, artist, 230, 232–5
Pius XI, Pope, 125–6, 219
Plekhanov, Russian Marxist, 83, 85
Poincaré, President of France, 115–7
Poland, 52–3, 55, 60, 62, 63, 70, 92, 103, 153–4, 156–7, 159, 180, 186, 284, 286, 291, 292, 293, 296, 298, 312, 314
Politburo, 96, 98, 303
Pollution, 201
Polymers and Plastics, 197
Pompidou, Georges, French President, 336, 346
Popular Front, in France, 118–9
Population, 12, 17, 18, 65–6, 202, 209–13
Portugal, 15–16, 267, 345
Pschyology, 225–9
Putnik, Serbian general, 34–6

Racialism, 13–14, 132–3, 263, 267, 278–81
Radar, 164, 190
Radio, 189, 190, 196, 248
R.A.F., 162, 164–5
Rakosi, Hungarian Prime Minister, 294, 298
Randstad, Holland, 209–10
Rapallo, agreement between Germany and Russia, 63
Rasputin, 81

Rathenau, Walter, German industrialist, 69, 129
Reichstag, German Parliament, 134, 135, 136, 252
Religion, 217–24; *see also* Catholic Church
Reparations, after World War I, 55–7, 71, 116; after World War II, 309
Revolution, in Russia, 44, 62, 78ff; of 1917, 82–90; in Germany, 47–8
Rhineland, 148–50
Rhodesia, 13, 328
R.I.C. (Royal Irish Constabulary), 109
Riefenstahl, Leni, German film producer, 245
Riga, Treaty of, 53, 62, 92
Rivera, General Primo de, 144
Rodin, Auguste, French sculptor, 235
Rohe, Mies van der, German architect, 237, 239
Röhm, Captain, leader of S.A., 131, 136–7
Rome, Treaty of, 259, 344–5
Rommel, German general, 159, 171, 172–4, 184, 187
Röntgen, German scientist, 188
Roosevelt, American President, 77, 177, 193, 284, 285, 290, 292–3, 308, 334
Roumania, 27, 52, 53, 63, 180, 284, 291, 293–4, 311
Ruhr, 56, 63, 72, 116–7, 155, 202, 209
Russia, 13, 22, 24–7, 58, 65, 196, 202, 221, 250; World War I, 29–30, 36–9, 44, 52–3, 69–70; Communist, 60, 61, 62–3, 71, 78–103, 146, 199, 217, 221–3, 226, 242, 248, 254–8, 262; World War II, 154, 166–9, 171, 174–6, 177, 180, 181, 182–7, 194; after World War II, 285–307, 309–10, 311–2, 318–9
Rutherford, British scientist, 188, 193

Saar coalfields, 53, 59, 63
St-Germain, Treaty of, 52
Salonika, 47
S.A.L.T. (Strategic Arms Limitation Talks), 307
Samsonov, Russian commander, 30
Sanctions, against Italy, 61–2
Sanders, Liman von, German commander, 34
Sarajevo, 25
Sartre, Jean-Paul, French writer, 242, 243
Scandinavia, 19–20, 105, 251; *see also* Norway; Sweden
Scapa Flow, 40, 55
Schacht, Dr., German economist, *140*, 141–2
Scheer, Admiral of German High Seas Fleet, 40–1
Schlieffen Plan, 26, 27–9
Schönberg, musician, 241
Schrödinger, scientist, 189
Schuman, French Foreign Minister, 343
Schuschnigg, Austrian Chancellor, 151
Science and Society, 201
Science and Technology, 69, Chapter 10
Sculpture, 234, 235–7
Second Vatican Ecumenical Council, 219–21, 224
Self-determination, principle of, 53
Serbia, *23*, 24–5, 31, 34–5, 47, 52
Sèvres, Treaty of, 52, 62
Sidi Barrani, battle of, 166
Singapore, 169
Sinn Fein, 108–10
Slump, Great, 72–6, 125; *see also* Depression, Great
Smithson, Alison and Peter, British architects, 239
Snowden, Philip, 111, 113
Social Security, 319–24, 325–7
Solid-state Physics, 189, 196–7
Solzhenitsyn, Alexander, 303–4, *304*
South Africa, 263, 267, 269, 277–82
Spaak, Paul-Henri, Belgian Foreign Minister, 335, 344
Space exploration, 303, 318
Spain, 15–16, 144–6, 218, 234, 251
Spanish Civil War, 118, 144–6, 150, 234, 242
Spee, von, German admiral, 39
Speer, German Minister for Armaments, 185, 186
Sport, organized, 249
Stalin, 63, 78, 88, 97–8, 120, 154, 290; domestic policy, 98–103, 247, 257–8, 311–2, 318; in World War II, 169, 177, 223, 284; after World War II, 286, 288, 292, 293, 296, 301–2
Stalingrad, 171, 173–6, 183, 187, 303
Stauffenberg, von, 184
Stavisky, French swindler, 117–8
Sterling, 67–8, 71, 72, 317
Stravinsky, Russian musician, 240, 241
Stresemann, German Chancellor, *64*, 129
Sudetenland, 151–3, 314
Suez Canal crisis, 283, 298, 345
Sweden, 19–20, 60, 65, 202, 223, 250, 313, 319, 345; *see also* Scandinavia
Switzerland, 20, 105, 223, 319, 345

Taberin, Mme., French journalist, 63–4
Technology, 13, Chapter 10, 214
Television, 196, 248, 249
Theatre, 243–4
Thomson, J. J., British scientist, 188,

Index

189, 196
Tikhon, Russian Patriarch, 221–2
Tirpitz, von, German admiral, 19, 25
Tito, Marshall (Josip Broz), 295–6
Tourism, 316, 322
Towns and Cities, 202–5; see also
  Housing; New Towns
Trade, international, 13, 67, 70–2, 312,
  314, 316
Trade Unions, 142, 317, 322–3, 328
Transistor, 196–7
Transport, urban, 202–4, 205–9,
  211–12
Treaty of Rome (E.E.C.), 259, 344–5
Treaties, at end of World War I, 52
Trenches, in World War I, 31–2
Trianon, Treaty of, 52
Triple Alliance, 20, 24
Triple Entente, 24
Trotsky, 62, 87–9, 90, 92–3, 96–8, 255
Truman, American President, 194,
  286–8, 308–9
Truman Doctrine, 287, 288, 309, 310
T.U.C. (Trades Union Congress), 112
Tukhachevsky, Red General, 93, 102
Turkey, 21, 27, 33–4, 52, 55, 62, 110,
  260, 262

U-boat, German submarine, 39–42, 43,
  48, 69, 171, 176–7
Ulbricht, East German leader, 297–8
Ulster, 108–10, 328–9
Unemployment, 73–5, 77, 108, 113–4,
117, 125, 130, 141
United Kingdom, see Britain
United Nations, 265, 277, 284
United States of America, 11–12, 18, 67,
  190, 239; in World War I, 27, 42,
  43–4, 48, 261–2; between the wars,
  57–8, 63, 70–3, 95, 196; in World
  War II, 169, 172, 177–80, 185, 186,
  193–5; after World War II, 197, 201,
  275, 286–8, 304–7, 308–11, 312–3
U.N.R.R.A. (United Nations Relief and
  Rehabilitation Administration), 308
U.S.S.R., see Russia

Van der Lubbe, 135
Vatican City, 126
Vatican Council, 219–21, 224
Venice, 214–6
Verdun, 36–7
Versailles Settlement, 50–9, 60, 63, 76,
  128, 147–8, 150, 261–2
Vichy, 162, 331, 334
Vienna, 130–1, 151, 227–8, 292
Vietnam, 272–5, 304
Vitamins, 197–8
Von Papen, German Chancellor, 134–5

Wall Street Crash, 57, 64, 72–3, 113,
  117
'War Communism', 93–4
Warsaw Pact, 291–2, 299
Watson, British scientist, 199, 200–1
Weil, Simone, French writer, 218

Weimar Republic, 127–30, 133–5, 138,
  141, 252, 253, 338
West Germany, 219, 288, 292, 297, 310,
  311, 313, 314–6, 317, 319, 320, 335,
  337–40, 341; Federal Republic of,
  291, 307, 337; see also Germany
Whites, anti-communist Russians, 90–3
William II, German Kaiser, 19, 25–7,
  42, 48, 127
Wilson, Harold, British Prime Minister,
  328, 355, 346
Wilson, Woodrow, American President,
  43, 50–1, 53, 57, 60, 61, 105, 261–2
Winter Palace, Petrograd, 87–8
Women, votes for, 17, 250, 252–9
Women's Liberation Movement, 258–9
World Council of Churches, 224
World War I, Chapter 2, 116, 131, 190,
  314; in 1914, 27ff; in 1915, 46ff;
  economic effects of, 68–76; in
  Russia, 79–89; naval warfare in,
  39–43, 48
World War II, 49, 76–7, 154–71, 190,
  264, 284, 308, 325
Wright Brothers, 189, 191

Yalta, 284, 286–7, 292, 302
Young Plan, 57
Yugoslavia, 52, 166, 186, 291, 292, 295

Zhukov, Russian commander, 169, 175
Zimmermann, German Foreign
  Secretary, 43
Zionism, 282–3